Advances in Experimental Medicine and Biology

Volume 1142

More information about this series at http://www.springer.com/series/5584

Qing Yang · Tamo Fukamizo
Editors

Targeting Chitin-containing Organisms

Springer

Editors
Qing Yang
State Laboratory of Biology for Plant Diseases and Insect Pests
Institute of Plant Protection at Chinese Academy of Agricultural Sciences
Beijing, China

School of Bioengineering
Dalian University of Technology
Dalian, China

Tamo Fukamizo
Biochemistry–Electrochemistry Research Unit, Institute of Science
Suranaree University of Technology
Nakhon Ratchasima, Thailand

ISSN 0065-2598 ISSN 2214-8019 (electronic)
Advances in Experimental Medicine and Biology
ISBN 978-981-13-7317-6 ISBN 978-981-13-7318-3 (eBook)
https://doi.org/10.1007/978-981-13-7318-3

This Springer imprint is published by the registered company Springer Nature Singapore Pte Ltd.
The registered company address is: 152 Beach Road, #21-01/04 Gateway East, Singapore 189721, Singapore

Contents

Contributors

Finn L. Aachmann Department of Biotechnology and Food Science, NOBIPOL, NTNU Norwegian University of Science and Technology, Trondheim, Norway

Yasuyuki Arakane Department of Applied Biology, Chonnam National University, Gwangju, South Korea

Han Bao School of Bioengineering, Dalian University of Technology, Dalian, China

Qi Chen Institute of Plant Protection, Chinese Academy of Agricultural Sciences, Beijing, China

Gaston Courtade Department of Biotechnology and Food Science, NOBIPOL, NTNU Norwegian University of Science and Technology, Trondheim, Norway

Tamo Fukamizo Biochemistry-Electrochemistry Research Unit, School of Chemistry, Suranaree University of Technology, Nakhon Ratchasima, Thailand

Takafumi Itoh Faculty of Bioscience and Biotechnology, Fukui Prefectural University, Eiheiji-cho, Yoshida-gun, Fukui, Japan

Xi Jiang School of Bioengineering, Dalian University of Technology, Dalian, China

Hisashi Kimoto Faculty of Bioscience and Biotechnology, Fukui Prefectural University, Eiheiji-cho, Yoshida-gun, Fukui, Japan

Ashutosh Kumar Laboratory for Structural Bioinformatics, Center for Biosystems Dynamics Research, RIKEN, Tsurumi, Yokohama, Kanagawa, Japan

Xiaojian Liu Research Institute of Applied Biology, Shanxi University, Taiyuan, Shanxi, China

Hans Merzendorfer Department of Chemistry and Biology – Molecular Biology, University of Siegen, Siegen, Germany;
Institute of Biology, University of Siegen, Siegen, Germany

Bernard Moussian Université Côte d'Azur, CNRS, Inserm, Institute of Biology Valrose, Parc Valrose, Nice, France

Subbaratnam Muthukrishnan Department of Biochemistry and Molecular Biophysics, Kansas State University, Manhattan, KS, USA

Deliang Peng Institute of Plant Protection, Chinese Academy of Agricultural Sciences, Beijing, China

Janin Rösner Department of Chemistry and Biology – Molecular Biology, University of Siegen, Siegen, Germany

S. Shinya Laboratory of Molecular Biophysics, Institute of Protein Research, Osaka University, Osaka, Japan

Lea Steinfeld Department of Chemistry and Biology – Molecular Biology, University of Siegen, Siegen, Germany

Guillaume Tetreau University of Grenoble Alpes, CNRS, CEA, IBS, Grenoble, France

Ali Vafaei Department of Chemistry and Biology – Molecular Biology, University of Siegen, Siegen, Germany

Ping Wang Department of Entomology, Cornell University, Geneva, NY, USA

Jinkui Yang State Key Laboratory for Conservation and Utilization of Bio-Resources in Yunnan, and Key Laboratory for Microbial Resources of the Ministry of Education, Yunnan University, Kunming, Yunnan, China

Qing Yang State Laboratory of Biology for Plant Diseases and Insect Pests, Institute of Plant Protection at Chinese Academy of Agricultural Sciences, Beijing, China;
School of Bioengineering, Dalian University of Technology, Dalian, China

Jianzhen Zhang Research Institute of Applied Biology, Shanxi University, Taiyuan, Shanxi, China

Kam Y. J. Zhang Laboratory for Structural Bioinformatics, Center for Biosystems Dynamics Research, RIKEN, Tsurumi, Yokohama, Kanagawa, Japan

Ke-Qin Zhang State Key Laboratory for Conservation and Utilization of Bio-Resources in Yunnan, and Key Laboratory for Microbial Resources of the Ministry of Education, Yunnan University, Kunming, Yunnan, China

Kun Yan Zhu Department of Entomology, Kansas State University, Manhattan, KS, USA

Chapter 1
An Introduction to the Book

Qing Yang and Tamo Fukamizo

Abstract Chitin is a linear biopolymer composed of β-1,4-linked *N*-acetylglucosamine (GlcNAc), and an essential component in exoskeleton of insects and crustaceans, the egg shells of parasitic nematodes, and the fungal cell wall. Since the chitin-containing organisms often threaten human health, food safety, and agricultural production, it has been highly desirable to control the hazardous chitin-containing organisms. This book will provide researchers and students with information on the recent research progress about the biology of chitin-containing organisms and their cross-talk with other organisms. This book also contains essential knowledge of drug design for controlling chitin-containing organisms. The authors deeply hope this book brings more attention to the fascinating yet unexploited world of chitin. We would like to thank all contributors for their expertise and generous support.

Keywords Chitin · Structure remodeling · Biodegradation · Cross-talk · Drug design

This book contains comprehensive contributions to chitin biology and chitin-metabolism-related enzymes. Chitin, the second most abundant biopolymer next to cellulose in nature, is a linear biopolymer composed of β-1,4-linked N-acetylglucosamine (GlcNAc) (Chap. 2), and an essential component in

---

Q. Yang (✉)
State Laboratory of Biology for Plant Diseases and Insect Pests,
Institute of Plant Protection at Chinese Academy of Agricultural Sciences,
No. 2 Yuanmingyuan West Road, Beijing 100193, China
e-mail: qingyang@dlut.edu.cn

School of Bioengineering, Dalian University of Technology, No. 2 Linggong Road,
Dalian 116023, China

T. Fukamizo
Biochemistry-Electrochemistry Research Unit, School of Chemistry,
Suranaree University of Technology, 111, University Avenue, Muang District,
Nakhon Ratchasima 30000, Thailand
e-mail: tamo0111fuka@gmail.com

Q. Yang and T. Fukamizo (eds.), *Targeting Chitin-containing Organisms*, Advances in Experimental Medicine and Biology 1142,
https://doi.org/10.1007/978-981-13-7318-3_1

exoskeleton of insects (Chap. 5), mites, ticks, and crustaceans, the eggshells of parasitic nematodes, and the fungal cell wall (Chap. 3). Some of these chitin-containing organisms threaten human health, food safety, and agricultural production. On the other hand, non-chitin-containing organisms like human, mammals, and plants do have an innate immune response to the hazardous chitin-containing organisms (Chap. 4). It is short of a book focusing on chitin biology, though more and more fascinating research progress on chitin-containing organisms has been achieved. The aim of this book is to provide researchers and students with information on the recent research progress about the biology of chitin-containing organisms as well as cross-talks between chitin-containing and non-chitin-containing organisms. By paying great attention to chitin remodeling enzymes and inhibitors (Chaps. 6–12), this book will also provide drug designers with essential knowledge of designing specific molecules for the control of hazardous chitin-containing organisms (Chap. 13).

The identification of chitin has gone through a long process. The history of chitin identification is detailed in Chap. 2. In 1901, Fraenkel and Kelly proposed for the first time that chitin is a polysaccharide. Until 1963, people found chitin fibers interacting with each other to adopt one of three possible crystalline organizations, i.e., α-, β-, or γ-chitin. Chapter 2 also introduces chitin bundles which interact with various chitin-binding proteins to form higher order structures, such as tissues and organs, with different mechanical properties.

How can chitin be synthesized in simple creatures? Chapter 3 introduces the prevalence and synthesis of chitin in bacteria, fungi, and protists. The bacterium itself does not contain chitin, but Rhizobium can synthesize chitin oligosaccharide which is a backbone of signaling molecules called nodulation factors (Nod factors) through the enzyme NodC. In fungi, the expression of many chitin synthase genes helps the formation of cell walls and other structural matrices of mycelia, stalks, and spores. Chitin in protists plays a role in protection against the mechanical and chemical stress.

After long-term coevolution, non-chitin-containing organisms have evolved into a relatively complete signal pathway responsive to the invasion of chitin-containing organisms. Chapter 4 introduces how the hosts (mammals and plants) recognize chitin-containing pathogens through specific pathogenesis-related proteins (PRPs) followed by the activation of intracellular signaling cascade.

Chitin is the extracellular matrix polysaccharide of insects and arthropods. After synthesis by membrane-bound chitin synthases, it is extensively remodeled before incorporation into divergent matrices with wide-ranging physical and biological properties. Chapter 5 discusses the properties of a variety of insect enzymes and proteins involved in this process, and proposes a model of chitin synthesis, assembly, and degradation.

Researchers found that the activity of glycoside hydrolases mixture is significantly lower on crystalline substrates than on soluble carbohydrate chains. This is because water molecules are precluded from penetrating the tightly packed and hydrogen-bonded network in crystalline chitin fibrils. Lytic polysaccharide monooxygenases (LPMOs), discovered in 2010, circumvent the challenges faced by GHs and have brought about a paradigm shift regarding the degradation of crystalline polysaccharides. Chapter 6 introduces the structural features and catalytic mechanism of chitin-active LPMOs, and some multi-modular LPMOs with carbohydrate binding modules. The chemoenzymatic modification of LPMO products, which represents a simple and environmentally friendly method for the greener production of functionalized chitin-based biomaterials, is also discussed in Chap. 6.

As chitin is the second most abundant biomass next to cellulose, the homeostasis of chitin in nature is important and maintained by marine and soil-dwelling bacteria. Chapter 7 introduces bacterial chitinolytic enzymes. These enzymes include processive-chitinases, endo-acting non-processive-chitinases, lytic polysaccharide monooxygenases, and *N*-acetyl-hexosaminidases. Their modes of action and structural features are summarized. The application of these enzymes to produce biocontrol agents against pathogenic fungi and insects is discussed.

Chitin is an important virulence factor for pathogenic fungi. Chapter 8 analyzes the amino acid sequences of fungi chitin synthases and chitinases from several typical species. The functions of chitin synthases and chitinases in typical fungi are summarized, and the crystal structures of chitinases are also discussed.

Arthropod growth and morphogenesis are dependent on the capability of remodeling chitin-containing structures. Chapter 9 discusses the biochemical processes of chitin biosynthesis, modification, and degradation, as well as various enzymes involved in these processes. The cuticular proteins and peritrophic membrane (PM) proteins, which largely determine the physicochemical properties of the cuticle and PM, are also discussed.

Human chitinases, chitotriosidase 1 (CHIT1), and acid mammalian chitinase (AMCase) play a protective role against chitin-containing pathogens through their capability to degrade chitin present in the cell wall of pathogens. The roles of both CHIT1 and AMCase in the development of various diseases have been revealed and several classes of inhibitors have been developed. Chapter 11 reviews the structural features and the progress in understanding the role of human chitinases in the development of various diseases. This chapter also summarizes the inhibitor discovery efforts targeting both CHIT1 and AMCase.

Plant chitinases are major enzymes acting in plant–microbe interactions, and are involved in self-defense against fungal pathogens. Chitosanases from soil bacteria are also involved in plant defense by hydrolyzing chitosan components of the fungal cell wall. In Chap. 12, the crystal structures of these enzymes are elucidated, and the physiological roles are speculated based on the structures. The enzymes are found to form an appropriate modular organization to fulfill their roles in plant–microbe interactions.

Chitinous structures in insects are complex and their formation and maintenance are dynamically regulated with the growth and development of insects. Chapter 13 introduces the physiological functions, compositions, structural formation, and regulation of the chitinous structures. The mechanisms to disrupt chitinous structures in insects and strategies for potential development of new tools of insect control by targeting chitinous structures are discussed.

The authors and coauthors are gracefully acknowledged for their joint effort in contributing state-of-the-art chapters. We would like to thank all contributors for their expertise and generous support. Finally, we deeply hope this book brings more attention to the fascinating yet unexploited world of chitin.

Chapter 2
Chitin: Structure, Chemistry and Biology

Bernard Moussian

Abstract Chitin is a linear polysaccharide of the amino sugar N-acetyl glucosamine. It is present in the extracellular matrix of a variety of invertebrates including sponges, molluscs, nematodes and arthropods and fungi. Generally, it is an important component of protective or supportive extracellular matrices that cover the tissue that produces it or the whole body of the organism. Chitin fibres associate with each other adopting one of three possible crystalline organisations, i.e. α-, β- or γ-chitin. Usually, chitin fibre bundles interact with chitin-binding proteins forming higher order structures. Chitin laminae, which are two-dimensional sheets of α-chitin crystals with antiparallel running chitin fibres in association with β-folded proteins, are primary constituents of the arthropod cuticle and the fibrous extracellular matrix in sponges. A tri-dimensional composite material of proteins coacervates and β-chitin constitute hard biomaterials such as the squid beak. The molecular composition of γ-chitin-based structures that contribute to the physical barrier found in insect cocoons is less well studied. In principle, chitin is a versatile extracellular polysaccharide that in association with proteins defines the mechanical properties of tissues and organisms.

Keywords Extracellular matrix · Cuticle · Body shape · Evolution · Barrier

2.1 Introduction

In his main work about animal biology *Historia animalium* (Greek: Τῶν περὶ τὰ ζῷα ἱστοριῶν, Ton peri ta zoia historion), Aristotle (384–322 BC) named the casted case of the larvae of a moth he found in clothes, probably the common clothes moth, *Tineola bisselliella*, (Hummel 1823), *χιτών* (kithon), which means 'sheath' in Greek (Fig. 2.1a). In his description of the integument of other insect species, he used the terms 'skin' and 'clothing'. Thus, the term *χιτών* does not seem to specifically assign

B. Moussian (✉)
Université Côte d'Azur, CNRS, Inserm, Institute of Biology Valrose,
Parc Valrose, Nice CEDEX 2 06108, France
e-mail: bernard.moussian@unice.fr

Q. Yang and T. Fukamizo (eds.), *Targeting Chitin-containing Organisms*, Advances in Experimental Medicine and Biology 1142,
https://doi.org/10.1007/978-981-13-7318-3_2

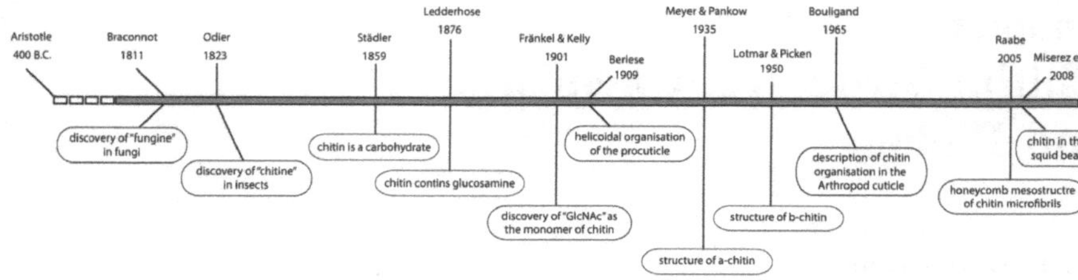

Fig. 2.1 *History of chitin*. The term chitin was coined by Aristotle more than three centuries BC. It took 18 centuries until the first scientific experiments were done to start unravelling chitin chemistry that was achieved with Fränkel and Kelly's work in 1901. Thereafter, the second era of chitin research focussed on chitin structure. The third era of chitin research, starting with Berlese in 1909, dealt with chitin organisation at the ultrastructural level especially in Arthropods

the insect cuticle in Aristotle's work but is rather loosely defining any structure that surrounds an insect. In 1822, the French naturalist Auguste Odier nevertheless used this term to denote a substance he extracted from the elytral cuticle of the scarab beetle, probably *Melolontha melolontha*: *Chitine* (Odier 1823). Initially, incubation of the elytra in water or warm alcohol did not change the appearance of the tissue. In the next step, Odier extracted chitin by cooking the elytra in potash. He found that the obtained transparent substance represented 25% of the initial weight of the elytra. He mentioned that chitin contains carbon but only very little nitrogen as compared to other animal substances like hairs and horn.

Odier was probably unaware of the work by another French scientist namely Henri Braconnot, who more than a decade before Odier's discovery, extracted a substance, probably chitin, from the cell wall of fungi that he had termed *fungine* (Braconnot 1811). In this case, fungi including *Agaricus* species were boiled in alkali to obtain fungine (detailed protocols of this extraction and of the following experiments were not reported in the respective article). Fungine was subsequently treated with potash, ammoniac and ether to show that fungine is composed of carbon, nitrogen, hydrogen and acetate. A few decades later, Georg Städler proposed in 1859 that chitin from crabs is a carbohydrate (Städeler 1859). In the last quarter of the nineteenth century, Strasbourg, a city now in France, at that time in Germany, became a centre of chitin chemistry research. There, in 1876, the German scientist Georg Ledderhose identified the monomers of chitin to be a sweet sugar that can be consumed by yeast as an energy source (Ledderhose 1876). Subsequently, in the laboratory of Felix Hoppe-Seyler, chitin was defined as poly- or oligosaccharide containing nitrogen and acetic acid. In the late nineteenth century, the polysaccharide in fungi (*Boletus edulis*, *Agaricus campsetris*, *Morchella esculenta*) was named *Pilzcellulose* (Engl. fungal cellulose) (Winterstein 1893–95). In Strasbourg, it was soon discovered that Pilzcellulose and the arthropod chitin are the same molecule (Hoppe-Seyler 1894). In this work, Hoppe-Seyler also identified a deacetylated form of chitin he named chitosan. At the beginning of the next century, in 1901, Fraenkel and Kelly published two important findings on chitin. First, after extraction with sulphuric acid and acetone followed by neutralisation with barite, alkalisation and precipitation with alcohol, they determined the composition of chitin demonstrating that it consists of an acetylated chitosamine (glucosamine) with the acetyl group bond to a nitrogen yielding N-acetyl-chitosamine (Fränkel and Kelly 1901). Second, they proposed for the first time that chitin is a polysaccharide. However, for more than a decade, the constitution of chitin, especially its nitrogen content was still under debate (Morgulis 1916). In 1926, finally, the structure proposed by Gonell (1926), confirmed in 1935 by Meyer and Pankow (1935) was widely accepted. Meyer and Pankow used X-ray diffraction to analyse chitin organisation in the spiny lobster *Palinurus vulgaris* (or *elephas*), an organisation that later was named α-chitin with antiparallel chitin fibres (Fig. 2.2a).

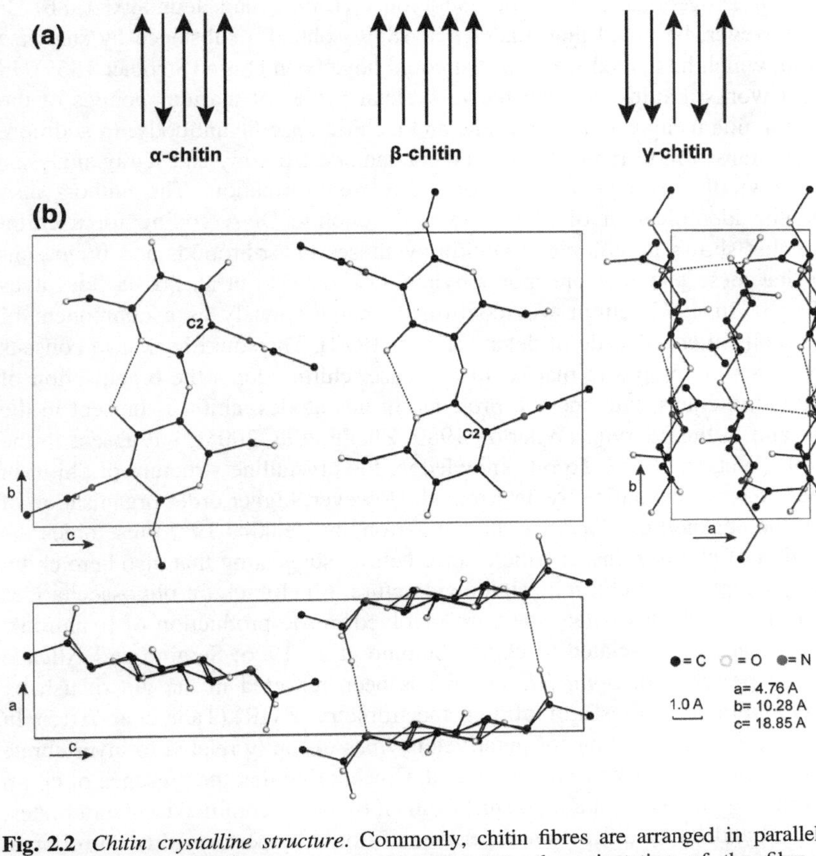

Fig. 2.2 *Chitin crystalline structure*. Commonly, chitin fibres are arranged in parallel. Three different possible organisation types with respect to the orientation of the fibre can be distinguished, α, β or γ (**a**). Here, the chitin crystallite according to Carlstrom (1957) in the α-form is shown (**b**). This model is detailed in the CIF file (from The Cambridge Crystallographic Data Centre (CCDC), Deposition Number: 1264751). We used the PyMOL software to draw the three projections ab, bc and ac of the crystallite. Hydrogen bonds are represented by dotted lines. The carbon number 2 (C2) is shown in the bc projection. Please note that the molecular and cellular mechanisms of chitin crystallite production are unknown

Besides α-chitin, two other types of chitin organisation were described, namely β-chitin with parallel running chitin fibres and γ-chitin with alternating parallel and antiparallel aligned chitin fibres (Fig. 2.2a). The organisation of β-chitin was unravelled in 1950 by Lotmar and Picken analysing chitin crystals by X-ray diffraction in bristles of the polychaete *Aphrodite aculeate* (Lotmar and Picken 1950), while γ-chitin was first described in 1962 or 1963 in Loligo, a squid genus (Rudall 1963).

Discovery of chitin in different phyla occurred at different dates and was often under debate. It was, for instance, ultimately in 2007, when chitin was discovered as a component of the skeletal fibres in marine sponges (Ehrlich et al. 2007a, b; Brunner et al. 2009) after some confusion as to whether chitin in sponges may derive from exogenous sources of endobionts (Dauby and Jeuniaux 1986). It should, however, be noted that Städeler in 1859 isolated a substance by stepwise extraction, which he named *spongin* that could have been chitin (Städeler 1859). In the recent works, chitin was detected in skeletal pieces of marine sponges of the order Verongida using Calcofluor White and a chitin-specific antibody. In addition, by Fourier-transform infrared (FTIR) and Raman spectroscopy and X-ray analyses, it was shown that sponge chitin adopts the α-conformation. The authors also reported the identification of transcripts and genomic DNA coding for a chitin synthase that shows homologies to chitin synthases in Arthropods and fungi suggesting that these enzymes are monophyletic (Zakrzewski et al. 2014; Goncalves et al. 2016). In 1972, chitin was confirmed unambiguously as a component of mollusc shells after a decade of debate (Peters 1972). The squid beak also consists of a chitin-based composite matrix. In this case, chitin adopts the organisation of β-chitin and interacts with specific proteins. In nematodes, chitin is present in the eggshell and in the pharynx (Wharton 1980; Zhang et al. 2005), but absent in the body wall (Watson 1965). To our knowledge, the crystalline structure of chitin in nematodes has not been directly determined. However, higher order organisation of chitin in the eggshell of *Trichuris suis* (Wharton and Jenkins 1978) resembles the organisation of chitin in insect cuticles (see below) suggesting that also here chitin preferably adopts the α-chitin crystalline structure. Finally, chitin oligosaccharides are found in vertebrates where they are involved in the production of hyaluronic acid, a polysaccharide related to chitin (Semino et al. 1996; Semino and Allende 2000). The presence of *bona fide* chitin has been reported in the gut of fish by histology and Fourier-transform infrared spectrometry (FTIR) (Tang et al. 2015). In the same work, genes coding for putative enzymes distantly related to invertebrate and fungal chitin synthases were mentioned. Considering that the presence of chitin in vertebrates would be a paradigm shift from chitin being confined to invertebrates, biochemical and X-ray diffraction experiments should be performed to corroborate these findings.

Commonly, chitin is found exclusively in the extracellular space as a protective and supporting component of complex extracellular matrices like the cell wall of fungi, the shell of molluscs or the cuticle of arthropods. Correlating with its requirement in different contexts, chitin organisation varies. In this chapter, we summarise the chemistry and structural organisation of chitin that together accounts for its versatile use in the majority of living organisms. Besides in nature, chitin and its derivative chitosan are used in biomedicine and biotechnology (Crini et al. 2007). This chapter, however, will not address this issue.

2.2 Chemistry

Chitin is a linear polymer of the amino sugar *N*-acetyl-D-glucosamine (GlcNAc). The β-glycosidic bonding between GlcNAc residues entails a repetition of di-saccharides with respect to the position of the N-acetyl group (Fig. 2.2). Chitin $(C_8H_{13}O_5N)^n$, hence, can be considered as polymer of two GlcNAcs, i.e. chitobiose. Despite its charges especially at the acetyl group, chitin is insoluble in aqueous and non-polar solutions.

A number of attempts were undertaken to assess the molecular organisation of chitin fibre. Three types of crystalline structure of chitin, namely α-, β- and γ-chitin (Fig. 2.2), have been observed in especially protostome animals and fungi (Wester 1909; Jeuniaux 1982). In 1957, by X-ray diffraction Diego Carlström analysed in a very comprehensive way the structure of α-chitin from the lobster (*Homarus americanus*) cuticle (Carlstrom 1957). According to his data, a chitin crystallite unit cell is composed of two antiparallel oriented cellubioses. The axes are a = 4.76 Å, b = 10.28 Å and c = 18.85 Å. This orthorhombic conformation is stabilised by hydrostatic bonds in the a and b direction, whereas binding forces in the c direction are rather weak. In principle, this structure has been confirmed repeatedly with only minor modifications. Minke and Blackwell by analysing X-ray diffractions determined the dimensions of the α-chitin unit cell as a = 4.74 Å, b = 18.86 Å (corresponding to c according to Carlström) and c = 10.32 Å (corresponding to b according to Carlström) (Minke and Blackwell 1978). In 2009, Sikorski et al. described the axes of the α-chitin unit cell as a = 4.72Å, b = 18.89 Å (corresponding to c according to Carlström) and c = 10.30 Å (corresponding to b according to Carlström), again by X-ray diffraction analyses (Sikorski et al. 2009). In a rather recent work, using Fourier-transform infrared (FITR) and solid state cross-polarising/magic-angle spinning (CP-MAS) ^{13}C NMR spectrophotometers, the crystalline structure of α-, β- and γ-chitin were characterised in detail using crab shell, squid pen and '*Lucainade*' (probably *Lucanidae* beetles) as sources, respectively (Jang et al. 2004). It was confirmed that α- and γ-chitin have two types of hydrogen bonding, which stabilise the crystal unit in two directions, i.e. within the sheet of chitin fibres (intrasheet, between the carbonyl group of amide I and the amide II) and between sheets (intersheet, between the CH_2OH side chain and the carbonyl group). β-chitin, by contrast, has only the intrasheet hydrogen bonds. As determined by Differential Scanning Calorimetry (DSC), the evaporation of bound water occurs between 50 and 150 °C in all three types of chitin (endothermic peak). The maximum decomposition temperature of the crystalline structure (exothermic peak) depends on hydrogen bonding and is, therefore, highest in α-chitin (330 °C), lowest in β-chitin (230 °C) and intermediate in γ-chitin (310 °C). In extensive

X-ray diffraction assays, these authors also found that the crystalline structures of all three types of chitin are stable in the range of 25–250 °C.

The structural stability of chitin is illustrated by its occurrence in fossils. Palynological investigations identified Lepidopteran wing scales in sediments in the Trias more than 200 million years ago (mya) (van Eldijk et al. 2018; Zhang et al. 2018). These findings suggest that Lepidoptera are evolutionarily older than previously assumed, refuting the hypothesis of the co-evolution of Lepidoptera and flowering plants that rose maximum 160 mya in the Jurassic period (Katz 2018; Nepi et al. 2018). The oldest fossil containing chitin has been isolated in Burgess Shale in Canada. It is an artefact of the basal demosponge *Vauxia gracilenta* as old as 505 million years, the Middle Cambrian period (Ehrlich et al. 2013). Classical and modern methods were applied to detect fossilised chitin. Using Calcofluor White (CFW) that fluoresces upon binding to β-(1,4) and β-(1,3)-linked polysaccharides (Herth and Schnepf 1980), Ehrlich and colleagues identified chitin fibres in fossilised material.

Concise information on chitin fibre or microfibril length is scarce. After demineralisation, depigmentation and deproteinisation of the lobster (*Homarus americanus*) cuticle, chitin nanofibres with lengths around 1.000 nm (maximum 3.000 nm) and the average width of 4.0 nm were isolated (Mushi et al. 2014).

Chitin fibres are synthesised by chitin synthases, which are membrane-inserted glycosyltransferases with multiple transmembrane domains (see Chap. 3). At the cytoplasmic face of this enzyme, the activated chitin monomer UDP-GlcNAc—produced in the cytoplasm by the Leloir pathway (Merzendorfer and Zimoch 2003; Araujo et al. 2005; Schimmelpfeng et al. 2006; Tonning et al. 2006; Moussian 2008, 2013)—enters the active site of the enzyme, that offers an UDP-less GlcNAc as an acceptor. The second UDP-GlcNAc is bound to the first one releasing the UDP moiety. In a comprehensible model, the initial two GlcNAcs rotate by 180° thereby exposing the acceptor C4 of the second residue for the next reaction (Dorfmueller et al. 2014). In the following, rotation of every other GlcNAc residue allows stereotypic continuation of polymerisation. Thus, the presentation of NAc moieties on C2 alternates (Fig. 2.2).

Nonsense and missense mutations in chitin synthase coding genes cause a chitin-less cuticle in the fruit fly *Drosophila melanogaster* (Moussian et al. 2005). Likewise, downregulation of chitin synthase transcripts by RNA interference provokes a reduction in cuticular chitin in various insects (Arakane et al. 2005, 2008; Zhang et al. 2010). The cuticle without or with reduced chitin is, in all cases, lethal. Inhibition of chitin synthase activity by insecticides like Nikkomycin Z reduces the amounts of chitin at the same time interfering with the organisation of the chitin-containing extracellular matrix and the structure of the respective organ (Schonitzer and Weiss 2007; Gangishetti et al. 2009). This observation indicates that the chitin structure and organisation depend on chitin amounts.

Both in fungi and in arthropods, pure chitin does not seem to prevail. Some acetyl groups are removed during cell wall or cuticle differentiation by chitin deacetylases thereby converting chitin at least partially to chitosan (Neville 1975). This issue is dealt with in Chap. 5 of this book.

2.3 Chemistry to Structure

Chitin is present in most invertebrate animals including arthropods, molluscs, nematodes and Porifera, but also in fungi (Zhang et al. 2005; Ehrlich et al. 2007a, b). In Porifera, like in the Arthropod cuticle, the matrix consists of α-chitin, while in mollusc beaks β-chitin prevails and γ-chitin is found in insect cocoons.

Chitin is the major scaffolding component of the insect cuticle accounting for 20–50% of the weight of the cuticle (Chapman 2013). Generally, in terrestrial insects chitin amounts scale isometrically with body size (Lease and Wolf 2010). This suggests an optimised ratio between cuticle thickness and body volume. The chemistry of chitin, a polymer of an amino sugar, defines its physical properties. The asymmetry of the chitin crystallite, in any case, entails that stiffness or elasticity along the three axes a, b or c differs. The highest stiffness value is observed on the b-axis (after Carlström) along the chain of covalently bound sugar monomers. The two other axes that are stabilised by hydrogen bonds are probably more elastic.

In a series of simulation experiments, for a α-chitin unit consisting of four pairs of antiparallel chains with a diameter of 2×2 nm^2 and a length of 5–35 nm an elastic modulus of 92.26 GPa was determined (Yu and Lau 2015). Addition of proteins preferably with rich on β-folds and water, decreased the elastic modulus to 36.39 GPa. This is probably due to the decrease of hydrogen bonds in chitin–chitin interactions. In the model, ductility increased with the length of fibres but reached a plateau longer than 20 nm. Another work, applying X-ray diffraction, measured the elastic modulus of α-chitin parallel to the fibre axis to be 41 GPa (Nishino et al. 1999). These values are clearly below the estimated stiffness of 150 GPa for chitin nanofibres (Vincent and Wegst 2004), which is markedly higher than the value of pure copper (124 GPa) and 50 GPa lower than the value of steel (200 GPa) (Callister and Rethwisch 2013). Experimentally determined values for chitin-containing cuticle types largely vary depending on the prevalent component (Vincent and Wegst 2004). The Young modulus of a Resilin-rich cuticle is about 1 MPa (Resilin is a chitin-binding protein rich in elastic cuticles), of soft cuticles between 1 kPa and 50 MPa, of sclerotised cuticles between 1 and 20 GPa. Thus, in nature, the mechanical properties of chitin-containing structures differ due to the association of chitin with proteins and organic molecules (sclerotisation). These differences do not distinguish only different types of cuticles, but may also prevail within the same cuticle. In the American lobster, for instance the stiffness of the outer procuticle, i.e. exocuticle is around nine GPa, whereas in the inner endocuticle it is around 4 GPa (Raabe et al. 2005a, b).

In nature, according to Neville (Neville 1975), around 17 antiparallel running chitin fibres are bundled forming chitin microfibrils with a diameter of around 3 nm and a length ranging from few nanometres to micrometres. Usually, these microfibrils are arranged in parallel constituting two-dimensional horizontal sheets, the laminae. This arrangement is stabilised by hydrostatic interactions within a fibre and between fibres. The laminae, in turn, are stacked to establish the procuticle. Yves Bouligand showed in a series of ultrastructural experiments that stacking of these sheets is helicoidal in the cuticle of some crustaceans (Bouligand 1965). Subsequently, Neville and Luke showed that this organisation occurs also in insects (Neville and Luke 1969). Bouligand's model of chitin microfibril organisation ('twisted plywood') has been unanimously accepted to date. The helicoidal organisation of the procuticle, however, was already observed by the Italian entomologist Antonio Berlese in 1909 (Berlese 1909). In his book 'Gli Insetti', in Fig. 516 on page 468, he drew cuticle laminae that are stacked helicoidally. The underlying microscopic method of this scheme is unclear. Bouligand did not seem to know this work and did not cite it in his seminal article in 1965. Besides a helicoidal arrangement, in certain types of cuticles like the beetle elytral cuticle, laminae may also be stacked in a preferred direction. In any case, to what extent chitin microfibrils interact with each other in the vertical direction, is not known. The helicoidal organisation of the chitin laminae in the (pro)-cuticle according to Berlese and Bouligand is also found in Oenychophores but not in Tardigrades (Harrison and Rice 1993). This indicates that the molecular mechanisms of chitin laminae organisation have been evolved very early in arthropod evolution, but after the separation of Tardigrades from Arthropods and Oenychophores.

2.4 Structure to Biology

Chitin as a constituent of the cuticle plays an important role in defining the physical or mechanical properties of the arthropod cuticle (Fig. 2.3). In the arthropod cuticle, chitin is associated with proteins and adopts higher order organisation that is crucial for its function. Commonly, as revealed by genome sequences, arthropod species have hundreds of chitin-binding proteins (Cornman et al. 2008; Cornman and Willis 2008; Futahashi et al. 2008; Cornman 2009; Cornman and Willis 2009; Cornman 2010; Rosenfeld et al. 2016). These proteins evolve fast as the number of chitin-binding proteins differs considerably within insect orders. According to Hamodrakas and colleagues, chitin may be recognised and bound by especially proteins with antiparallel β-sheet half-barrel structure (Hamodrakas et al. 2002).

In some specialised cuticle, like in the head skeleton of dipteran (e.g. *Drosophila melanogaster*) larvae that is used for food grinding, it is not possible to see chitin micorfibrils at the ultrastructural level. By contrast, the body cuticle of *D. melanogaster* larvae, which is soft and flexible, chitin microfibrils are well visible by electron microscopy. Assuming, therefore, that chitin fibres are rather short (unstructured) in the head skeleton but long (visible) in the body cuticle, we propose that short chitin microfibrils are associated with hard cuticles, whereas long chitin microfibrils are present in soft cuticles. These observations may also suggest that chitin microfibril length negatively correlates with stiffness or softness of the cuticle in arthropods for forces perpendicular to the microfibril axis.

The structure and the mechanical properties of the chitin-based cuticle have been extensively studied in the lobster *Homarus americanus* (Raabe et al. 2005a, b, 2006, 2007). The main finding is that the b-axis of the α-chitin unit cell (the c-axis

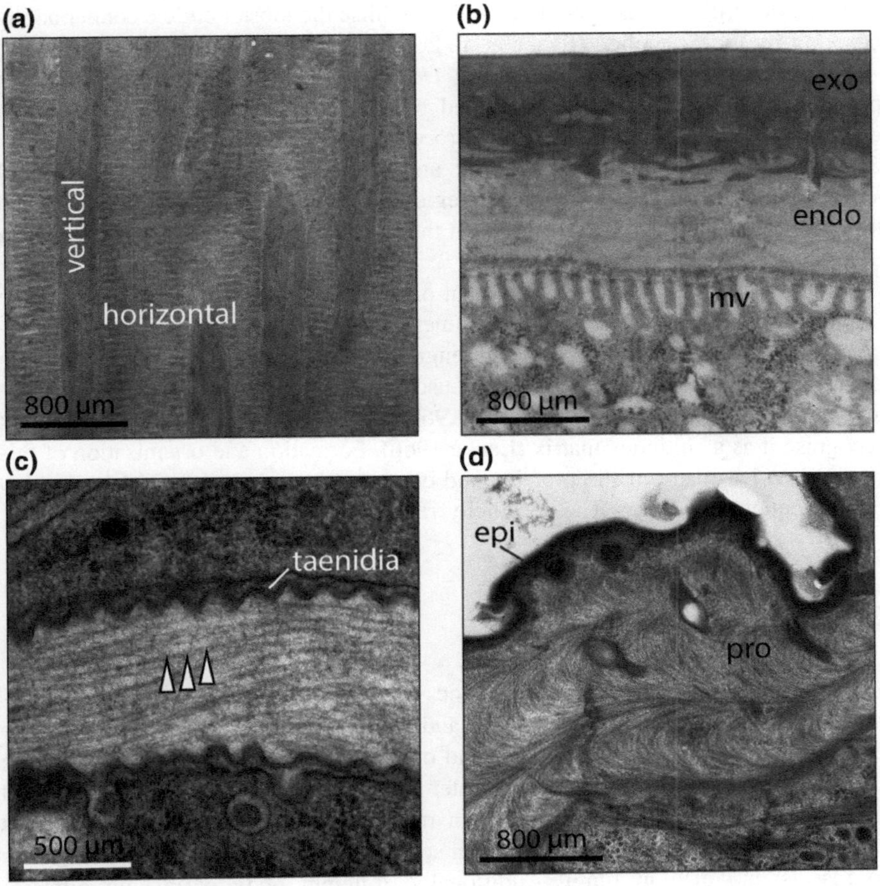

Fig. 2.3 Higher organisation of chitin. The cuticle of the beetle elytra (*Tribolium castaneum*) consists of vertical and horizontal chitin–protein matrices. **b** The procuticle of the *Tribolium castaneum* larval cuticle is subdivided into two regions with highly cross-linked (exocuticle, exo) and weakly cross-linked (endocuticle, endo) chitin adjacent to the microvilli (mv) that are involved in cuticle secretion. **c** The tracheal lumen of developing *Drosophila melanogaster* embryos contains a chitinous matrix (triangles) needed for tracheal morphogenesis. **d** Oblique sections through the procuticle (here a *Parhyale hawaiensis* juvenile) reveals the illusion arc-like chitin fibres indicative of a helicoidal arrangement of chitin laminae (see Chap. 9)

according to Carlström), has a preferred fibre texture normal (with some 10° of inclination) to the surface of the exoskeleton at any position of the body. The crystallographic texture of hard mineralised regions, as opposed to the texture of soft membranous regions, has, however, an additional albeit infrequent fibre orientation perpendicular to the normal. These fibres may represent chitin in the pore canals that run from the epidermal cells to the cuticle surface. Moreover, a twisted honeycomb structure has been described to stabilse the lobster cuticle counteracting cracking by forces applied (Raabe et al. 2005a, b).

In the beak of the Humboldt squid *Dosidicus gigas* β-chitin associates with proteins, which in turn are cross-linked by a variety of organic molecules that especially associate with histidines, a process named sclerotisation (Miserez et al. 2010). An opposite gradient of water and sclerotisation defines beak material stiffness and elasticity with the stiffest region with 5 GPa being at the beak tip and the softest region with 0.05 GPa being at the margin of the so-called wing (Miserez et al. 2008).

Chitin is not only a structural element of extracellular matrices but may also be implied in organ shape during development. In *D. melanogaster*, it was demonstrated that tracheal tube diameter and length regulation depend on a luminal chitin matrix (Fig. 2.3) (Tonning et al. 2005; Luschnig et al. 2006; Moussian et al. 2006). This matrix was already discovered in 1966 by M. Locke, who, however, did not recognise it as a chitinous matrix (Locke 1966). Formation and organisation of this matrix involve many of the proteins and enzymes that are also active during epidermal cuticle formation. The specificity remains to be investigated.

2.5 Concluding Remarks

In summary, the chemistry and biophysics of chitin alone, a versatile polysaccharide target for modifications in length and by deacetylation, are not sufficient to explain its use in a variety of tissues and organisms. Rather, associations of chitin with proteins, organic molecules and water are important modulators of mechanical and biophysical properties of the chitin matrix. The versatility of chitin is best illustrated by its presence in most animal species ranging from sponges, nematodes to insects and also in fungi. Additionally, different body parts with different functions are equipped with different, optimised types of chitin matrices. We can imagine that, with modern omics tools in hand, the molecular ecology of these structures will characterise the fourth era of chitin research.

Acknowledgements I am deeply thankful to Dr. Zhitao Yu, Department of Entomology, Kansas State University, Manhattan, USA, for her substantial contribution to the figures.

References

Arakane Y, Muthukrishnan S, Kramer KJ, Specht CA, Tomoyasu Y, Lorenzen MD, Kanost M, Beeman RW (2005) The Tribolium chitin synthase genes TcCHS1 and TcCHS2 are specialized for synthesis of epidermal cuticle and midgut peritrophic matrix. Insect Mol Biol 14(5):453–463

Arakane Y, Specht CA, Kramer KJ, Muthukrishnan S, Beeman RW (2008) Chitin synthases are required for survival, fecundity and egg hatch in the red flour beetle, Tribolium castaneum. Insect Biochem Mol Biol 38(10):959–962

Araujo SJ, Aslam H, Tear G, Casanova J (2005) Mummy/cystic encodes an enzyme required for chitin and glycan synthesis, involved in trachea, embryonic cuticle and CNS development–analysis of its role in Drosophila tracheal morphogenesis. Dev Biol 288(1):179–193

Berlese A (1909) Gli insetti – loro organizzazione, sviluppo, abitudini e rapporti coll'uomo. Milano, Società Editrice Libreria

Bouligand Y (1965) On a twisted fibrillar arrangement common to several biologic structures. C R Acad Sci Hebd Seances Acad Sci D 261(22):4864–4867

Braconnot H (1811) Sur la nature des champignons. Annales de chimie ou recueil de mémoires concernant la chimie et les arts qui en dépendent et spécialement la pharmacie 79:265–304

Brunner E, Ehrlich H, Schupp P, Hedrich R, Hunoldt S, Kammer M, Machill S, Paasch S, Bazhenov VV, Kurek DV, Arnold T, Brockmann S, Ruhnow M, Born R (2009) Chitin-based scaffolds are an integral part of the skeleton of the marine demosponge Ianthella basta. J Struct Biol 168(3):539–547

Callister WD, Rethwisch DG (2013) Materials science and engineering: an introduction. Wiley, New York

Carlstrom D (1957) The crystal structure of alpha-chitin (poly-N-acetyl-D-glucosamine). J Biophys Biochem Cytol 3(5):669–683

Chapman RF (2013) The Insects. Cambridge, Cambridge University Press, Structure and Function

Cornman RS (2009) Molecular evolution of Drosophila cuticular protein genes. PLoS ONE 4(12): e8345

Cornman RS (2010) The distribution of GYR- and YLP-like motifs in Drosophila suggests a general role in cuticle assembly and other protein-protein interactions. PLoS One 5(9)

Cornman RS, Togawa T, Dunn WA, He N, Emmons AC, Willis JH (2008) Annotation and analysis of a large cuticular protein family with the R&R Consensus in Anopheles gambiae. BMC Genom 9:22

Cornman RS, Willis JH (2008) Extensive gene amplification and concerted evolution within the CPR family of cuticular proteins in mosquitoes. Insect Biochem Mol Biol 38(6):661–676

Cornman RS, Willis JH (2009) Annotation and analysis of low-complexity protein families of Anopheles gambiae that are associated with cuticle. Insect Mol Biol 18(5):607–622

Crini G, Badot P-M, Guibal E (2007) Chitine et chitosane – Du biopolymère à l'application, Presse universitaire de Franche-Comté

Dauby P, Jeuniaux C (1986) Origine exogène de la chitine décelée chez les Spongiaires. Cah Biol Mar 28:121–129

Dorfmueller HC, Ferenbach AT, Borodkin VS, van Aalten DM (2014) A structural and biochemical model of processive chitin synthesis. J Biol Chem 289(33):23020–23028

Ehrlich H, Krautter M, Hanke T, Simon P, Knieb C, Heinemann S, Worch H (2007a) First evidence of the presence of chitin in skeletons of marine sponges. Part II. Glass sponges (Hexactinellida: Porifera). J Exp Zool B Mol Dev Evol 308(4):473–483

Ehrlich H, Maldonado M, Spindler KD, Eckert C, Hanke T, Born R, Goebel C, Simon P, Heinemann S, Worch H (2007b) First evidence of chitin as a component of the skeletal fibers of marine sponges. Part I. Verongidae (demospongia: Porifera). J Exp Zool B Mol Dev Evol 308(4):347–356

Ehrlich H, Rigby JK, Botting JP, Tsurkan MV, Werner C, Schwille P, Petrasek Z, Pisera A, Simon P, Sivkov VN, Vyalikh DV, Molodtsov SL, Kurek D, Kammer M, Hunoldt S, Born R, Stawski D, Steinhof A, Bazhenov VV, Geisler T (2013) Discovery of 505-million-year old chitin in the basal demosponge Vauxia gracilenta. Sci Rep 3:3497

Fränkel S, Kelly A (1901) Beiträge zur Constitution des Chitins. Monatsh Chem 23(2):123–132

Futahashi R, Okamoto S, Kawasaki H, Zhong YS, Iwanaga M, Mita K, Fujiwara H (2008) Genome-wide identification of cuticular protein genes in the silkworm, Bombyx mori. Insect Biochem Mol Biol 38(12):1138–1146

Gangishetti U, Breitenbach S, Zander M, Saheb SK, Muller U, Schwarz H, Moussian B (2009) Effects of benzoylphenylurea on chitin synthesis and orientation in the cuticle of the Drosophila larva. Eur J Cell Biol 88(3):167–180

Goncalves IR, Brouillet S, Soulie MC, Gribaldo S, Sirven C, Charron N, Boccara M, Choquer M (2016) Genome-wide analyses of chitin synthases identify horizontal gene transfers towards bacteria and allow a robust and unifying classification into fungi. BMC Evol Biol 16(1):252

Gonell HW (1926) Rontgenographische Studien an Chitin. Hoppe-Seyler's Zeitschrift fuer Physiologische Chemie Berlin 152:18–30

Hamodrakas SJ, Willis JH, Iconomidou VA (2002) A structural model of the chitin-binding domain of cuticle proteins. Insect Biochem Mol Biol 32(11):1577–1583

Harrison FW, Rice ME (1993) Onychophora, chilopoda, and lesser protostomata. John Wiley, New York

Herth W, Schnepf E (1980) The fluorochrome, calcofluor white, binds oriented to structural polysaccharide fibrils. Protoplasma 105(1–2):129–133

Jang MK, Kong BG, Jeong YI, Lee CH, Nah JW (2004) Physicochemical characterization of alpha-chitin, beta-chitin, and gamma-chitin separated from natural resources. J Polym Sci Part A Polym Chem 42(14):3423–3432

Jeuniaux C (1982) La chitine dans le regne animal. Bulletin de la Societe Zoologique de France 107(3):363–386

Katz O (2018) Extending the scope of Darwin's 'abominable mystery': integrative approaches to understanding angiosperm origins and species richness. Ann Bot 121(1):1–8

Lease HM, Wolf BO (2010) Exoskeletal chitin scales isometrically with body size in terrestrial insects. J Morphol 271(6):759–768

Ledderhose G (1876) Über salzsaures Glucosamin." Berichte der deutschen chemischen Gesellschaft: 1200–1201

Locke M (1966) The structure and formation of the cuticulin layer in the epicuticle of an insect, Calpodes ethlius (Lepidoptera, Hesperiidae). J Morphol 118(4):461–494

Lotmar W, Picken LER (1950) A new crystallographic modification of chitin and its distribution. Experientia 6(2):58–59

Luschnig S, Batz T, Armbruster K, Krasnow MA (2006) serpentine and vermiform encode matrix proteins with chitin binding and deacetylation domains that limit tracheal tube length in Drosophila. Curr Biol 16(2):186–194

Merzendorfer H, Zimoch L (2003) Chitin metabolism in insects: structure, function and regulation of chitin synthases and chitinases. J Exp Biol 206(Pt 24):4393–4412

Meyer KH, Pankow G (1935) Sur la constitution et la structure de la chitine. Helv Chim Acta 18 (1):589–598

Minke R, Blackwell J (1978) The structure of alpha-chitin. J Mol Biol 120(2):167–181

Miserez A, Rubin D, Waite JH (2010) Cross-linking chemistry of squid beak. J Biol Chem 285 (49):38115–38124

Miserez A, Schneberk T, Sun C, Zok FW, Waite JH (2008) The transition from stiff to compliant materials in squid beaks. Science 319(5871):1816–1819

Morgulis S (1916) The Chemical Constitution of Chitin. Science 44(1146):866–867

Moussian B (2008) The role of GlcNAc in formation and function of extracellular matrices. Comp Biochem Physiol B Biochem Mol Biol 149(2):215–226

Moussian B (2013) The Arthropod Cuticle. In: Minelli A, Boxshall G, Fusco G (eds) Arthropod biology and evolution. Springer, Berlin, Heidelberg, pp 171–196

Moussian B, Schwarz H, Bartoszewski S, Nusslein-Volhard C (2005) Involvement of chitin in exoskeleton morphogenesis in Drosophila melanogaster. J Morphol 264(1):117–130

Moussian B, Tang E, Tonning A, Helms S, Schwarz H, Nusslein-Volhard C, Uv AE (2006) Drosophila Knickkopf and Retroactive are needed for epithelial tube growth and cuticle differentiation through their specific requirement for chitin filament organization. Development 133(1):163–171

Mushi NE, Butchosa N, Salajkova M, Zhou Q, Berglund LA (2014) Nanostructured membranes based on native chitin nanofibers prepared by mild process. Carbohydr Polym 112:255–263

Nepi M, Grasso DA, Mancuso S (2018) Nectar in Plant-Insect Mutualistic Relationships: From Food Reward to Partner Manipulation. Front Plant Sci 9:1063
Neville AC (1975) Biology of the arthropod cuticle. Springer Verlag, Berlin Heidelberg New York
Neville AC, Luke BM (1969) A Two-system model for chitin-protein complexes in insect cuticles. Tissue Cell 1(4):689–707
Nishino T, Matsui R, Nakamae K (1999) Elastic modulus of the crystalline regions of chitin and chitosan. J Polym Sci B: Polym Phys 37(11):1191–1196
Odier A (1823) Mémoires sur la composition chimique des parties cornées des insectes. Mémoires de la Société d'Histoire Naturelle de Paris 1:29–42
Peters W (1972) Occurrence of chitin in Mollusca. Comparat Biochem Physiol B 41((3)):541–550
Raabe D, Al-Sawalmih A, Yi SB, Fabritius H (2007) Preferred crystallographic texture of alpha-chitin as a microscopic and macroscopic design principle of the exoskeleton of the lobster Homarus americanus. Acta Biomater 3(6):882–895
Raabe D, Romano P, Sachs C, Al-Sawalmih A, Brokmeier H-G, Yi S-B, Servos G, Hartwig HG (2005a) Discovery of a honeycomb structure in the twisted plywood patterns of fibrous biological nanocomposite tissue. J Cryst Growth 283(1–2):1–7
Raabe D, Romano P, Sachs C, Fabritius H, Al-Sawalmih A, Yi S, Servos G, Hartwig HG (2006) Microstructure and crystallographic texture of the chitin-protein network in the biological composite material of the exoskeleton of the lobster Homarus americanus. Materials Science and Engineering a-Structural Materials Properties Microstructure and Processing 421(1–2):143–153
Raabe D, Sachs C, Romano P (2005b) The crustacean exoskeleton as an example of a structurally and mechanically graded biological nanocomposite material. Acta Mater 53(15):4281–4292
Rosenfeld JA, Reeves D, Brugler MR, Narechania A, Simon S, Durrett R, Foox J, Shianna K, Schatz MC, Gandara J, Afshinnekoo E, Lam ET, Hastie AR, Chan S, Cao H, Saghbini M, Kentsis A, Planet PJ, Kholodovych V, Tessler M, Baker R, DeSalle R, Sorkin LN, Kolokotronis SO, Siddall ME, Amato G, Mason CE (2016) Genome assembly and geospatial phylogenomics of the bed bug Cimex lectularius. Nat Commun 7:10164
Rudall KM (1963) The chitin/protein complexes of insect cuticles. Treherne & Wiggles-worth, vol 1, Beament, Treherne & Wiggles-worth, vol 1, pp 257–313, 24 figs, pp 257–313
Schimmelpfeng K, Strunk M, Stork T, Klambt C (2006) Mummy encodes an UDP-N-acetylglucosamine-dipohosphorylase and is required during Drosophila dorsal closure and nervous system development. Mech Dev 123(6):487–499
Schonitzer V, Weiss IM (2007) The structure of mollusc larval shells formed in the presence of the chitin synthase inhibitor Nikkomycin Z. BMC Struct Biol 7:71
Semino CE, Allende ML (2000) Chitin oligosaccharides as candidate patterning agents in zebrafish embryogenesis. Int J Dev Biol 44(2):183–193
Semino CE, Specht CA, Raimondi A, Robbins PW (1996) Homologs of the Xenopus developmental gene DG42 are present in zebrafish and mouse and are involved in the synthesis of Nod-like chitin oligosaccharides during early embryogenesis. Proc Natl Acad Sci U S A 93(10):4548–4553
Sikorski P, Hori R, Wada M (2009) Revisit of alpha-chitin crystal structure using high resolution X-ray diffraction data. Biomacromol 10(5):1100–1105
Städeler G (1859) Untersuchungen über das Fibroin, Spongin und Chitin, nebst Bemerkungen über den thierischen Schleim. Justus Liebig Annalen der Chemie 111(1):12–28
Tang WJ, Fernandez J, Sohn JJ, Amemiya CT (2015) Chitin is endogenously produced in vertebrates. Curr Biol 25(7):897–900
Tonning A, Helms S, Schwarz H, Uv AE, Moussian B (2006) Hormonal regulation of mummy is needed for apical extracellular matrix formation and epithelial morphogenesis in Drosophila. Development 133(2):331–341
Tonning A, Hemphala J, Tang E, Nannmark U, Samakovlis C, Uv A (2005) A transient luminal chitinous matrix is required to model epithelial tube diameter in the Drosophila trachea. Dev Cell 9(3):423–430

van Eldijk TJB, Wappler T, Strother PK, van der Weijst CMH, Rajaei H, Visscher H, van de Schootbrugge B (2018) A Triassic-Jurassic window into the evolution of Lepidoptera. Sci Adv 4(1):e1701568

Vincent JF, Wegst UG (2004) Design and mechanical properties of insect cuticle. Arthropod Struct Dev 33(3):187–199

Watson BD (1965) The fine structure of the body-wall in a free-living nematode, Euchromadora vulgaris. Quart. J. micr. Sci. 106(1):75–81

Wester DH (1909) "Über die Verbreitung und Lokalisation des Chitins im Tierreich " Zool. Jahrb. Abt. Syst. 28:531–558

Wharton D (1980) Nematode egg-shells. Parasitology 81(2):447–463

Wharton DA, Jenkins T (1978) Structure and chemistry of the egg-shell of a nematode (Trichuris suis). Tissue Cell 10(3):427–440

Yu Z, Lau D (2015) molecular dynamics study on stiffness and ductility in chitin-protein composite. J Mater Sci 50:7149–7157

Zakrzewski AC, Weigert A, Helm C, Adamski M, Adamska M, Bleidorn C, Raible F, Hausen H (2014) Early divergence, broad distribution, and high diversity of animal chitin synthases. Genome Biol Evol 6(2):316–325

Zhang J, Liu X, Zhang J, Li D, Sun Y, Guo Y, Ma E, Zhu KY (2010) Silencing of two alternative splicing-derived mRNA variants of chitin synthase 1 gene by RNAi is lethal to the oriental migratory locust, Locusta migratoria manilensis (Meyen). Insect Biochem Mol Biol 40 (11):824–833

Zhang Q, Mey W, Ansorge J, Starkey TA, McDonald LT, McNamara ME, Jarzembowski EA, Wichard W, Kelly R, Ren X, Chen J, Zhang H, Wang B (2018) Fossil scales illuminate the early evolution of lepidopterans and structural colors. Sci Adv 4(4):e1700988

Zhang Y, Foster JM, Nelson LS, Ma D, Carlow CK (2005) The chitin synthase genes chs-1 and chs-2 are essential for C. elegans development and responsible for chitin deposition in the eggshell and pharynx, respectively. Dev Biol 285(2):330–339

Chapter 3
Chitin Prevalence and Function in Bacteria, Fungi and Protists

Lea Steinfeld, Ali Vafaei, Janin Rösner and Hans Merzendorfer

Abstract Chitin is an important structural polysaccharide, which supports and organizes extracellular matrices in a variety of taxonomic groups including bacteria, fungi, protists, and animals. Additionally, chitin has been recognized as a molecule that is required for *Rhizobia*-legume symbiosis and involved in arbuscular mycorrhizal signaling in the symbiotic interaction between terrestrial plants and fungi. Moreover, it serves as a unique molecular pattern in the plant defense system against pathogenic fungi and parasites, and in the innate and adaptive immune response of mammals and humans. In this review, we will focus on the prevalence and structural function of chitin in bacteria, fungi, and protists, with a particular focus on the evolution of chitin synthases and the function of chitin oligosaccharides as a signaling molecule in symbiosis and immunity.

Keywords *Rhizobia* · Fungi · Cell wall · Protists · Skeleton

3.1 Introduction

Extracellular matrices of bacteria, fungi, protists, algae, and animals are supported by various types of exopolysaccharides such as xanthans, alginate, glucans, cellulose, mannans, hyaluronan, or chitin. Chitin is a polysaccharide of *N*-acetylglucosamine (GlcNAc) units, which is produced by a variety of organisms belonging to different taxonomic groups. In common to all these organisms is that they possess a highly conserved biosynthetic machinery for chitin formation. The first step involves the formation of the activated sugar donors, UDP-GlcNAc, catalyzed by different enzymes of a side branch of the Leloir pathway (Muthukrishnan et al. 2012). In the second step, the sugar moieties of UDP-GlcNAc are consecutively transferred to the non-reducing end of the acceptor polysaccha-

L. Steinfeld · A. Vafaei · J. Rösner · H. Merzendorfer (✉)
Department of Chemistry and Biology – Molecular Biology, University of Siegen, 57076 Siegen, Germany
e-mail: hans.merzendorfer@uni-siegen.de

Q. Yang and T. Fukamizo (eds.), *Targeting Chitin-containing Organisms*, Advances in Experimental Medicine and Biology 1142,
https://doi.org/10.1007/978-981-13-7318-3_3

ride in a reaction catalyzed by the key enzyme chitin synthase, a processive, membrane-integral family 2 glycosyltransferase residing in the plasma membranes of chitin synthesizing cells (Dorfmueller et al. 2014; Zhu et al. 2016). Chitin can be considered a renewable resource, which receives increasing interest for various applications in the agricultural, chemical, and pharmaceutical industry. It is primarily found in ecdysozoan (including arthropods) and lophotrochozoan (including crustaceans) species, which accounts for large amounts of the total chitin biomass on earth. Actually, most chitin is extracted from waste shell material arising in crab and shrimp fishery.

Once deposited on the extracellular surfaces of cells and epithelia, chitin serves multiple functions. It is used as a fibrillar scaffold to support extracellular matrices such as fungal cell walls or arthropod epidermal and serosal cuticles (Merzendorfer 2011). Chitin is also found in the pharynx and eggshells of nematodes (Zhang et al. 2005), and is part of the particular complex and hard structures of mollusks radula teeth (Peters and Latka 1986) and annelids chaeta (Picken and Lotmar 1950; Tilic and Bartolomaeus 2016). It is also found as a component of invertebrate endoskeletons such as the squid pen of cephalopods (Stegemann 1963; Rudall and Kenchington 1973), as well as in mucous membranes termed as peritrophic matrices, which line the midgut of many insects (Zimoch and Merzendorfer 2002; Hegedus et al. 2009). Furthermore, chitin serves as a structural scaffold for various biocomposites produced by calcareous sponges (Ehrlich et al. 2007), corals (Goldberg 1978; Bo et al. 2012), various mussels and snails (Weiss and Schonitzer 2006), and diatoms, which produce chitin fibrils that extend from chitinous theca (Durkin et al. 2009). Chitin has also been detected in the lorica of heterotrich ciliates such as *Eufolliculina uhligi* (Mulisch and Hausmann 1983), and in the cysts of other unicellular protists including *Blepharisma undulans*, *Pseudomicrothorax dubius* (Mulisch and Hausmann 1989) and *Entamobae* species (Campos-Gongora et al. 2004). Notably, sequences for genuine chitin synthases have been identified in choanoflagellates, which are considered the sister group of metazoans, and in basal chordates, where chitin is found in the intestine and the epidermis. Finally, chitin oligosaccharide synthases, which are more closely related to hyaluronan synthases, have been reported in nitrogen-fixing rhizobial bacteria, as well in different vertebrates including amphibian and bony fishes (Debellé et al. 1992).

For a long time, chitin has been considered to be absent in chordates. Based on different spectroscopic, X-ray diffraction, and histochemical techniques (mainly lectin staining with wheat germ agglutinin), however, chitin has been postulated to be produced in some chordates (Wagner 1994). Accordingly, chitin was detected in peritrophic matrices of different ascidian tunicates (Peters 1966; Anno et al. 1974; Gowri et al. 1982), in the gill bars of the lancetfish, *Amphioxus lanceolatus* (Sannasi and Hermann 1970), in the epidermal cuticle of the blenny fish *Paralipophrys trigloides* (Wagner et al. 1993). Later on, *CHS*-like genes termed *DG42*, which are homologous to the rhizobial *nodC* encoding chitin oligosaccharides, have been found in the *Xenopus* frog (Semino and Robbins 1995), zebrafish, and mouse (Semino et al. 1996). In the beginning, it was discussed controversially whether the derived enzyme produces hyaluronan or chitin oligosaccharides, the latter of which could also act as a primers for hyaluronan synthesis (Meyer and Kreil 1996; Varki 1996). Subsequently, evidence accumulated that these enzymes initially produce chitin oligosaccharides that act as primers for the biosynthesis of hyaluronan, and that this process is required during early vertebrate embryogenesis (Semino and Allende 2000). Tang et al. (2015), however, discovered genuine *CHS* genes in the genomes of bony and cartilaginous fishes (*Osteichthyes* and *Chondrichthyes*), lampreys (*Petromyzon marinus*) lancelets (*Branchiostoma flordea*), frogs (*Xenopus Silurana tropicalis*), and axolotl (*Ambystoma mexicanum*). Frequently, two classes of *CHS* genes were recorded in these chordates, however, precise annotation and phylogenetic analysis are pending. Some chitin synthases encoded by these genes are presumably involved in the production of chitinous structures that have been discovered decades before. Bony fishes, whose genomes harbor genuine *CHS* genes, include zebrafish (*Danio rerio*), catfish (*Ictalurus punctatus*), Japanese eel (*Anguilla japonica*), spotted gar (*Lepisosteus oculatus*), Southern platyfish (*Xiphophorus maculatus*), sticklebacks (*Gasterosteus aculeatus*), and cartilaginous fishes comprise the elephant fish (*Callorhinchus milii*). In the zebrafish, CHS-1 is preferentially expressed during embryonic and larval development. In situ hybridization revealed, that *CHS-1* expression is observed in epithelial and mesenchymal cells of the larval gut. Furthermore, chitin detection with a fluorescent bacterial chitin-binding domain showed that chitin production starts 3 days after fertilization, and that most chitin produced later on is extracellular and distributed throughout the intestinal lumen (Tang et al. 2015). When *CHS-1* expression was silenced by RNA interference, the chitin signals observed in histochemistry were markedly reduced. In addition, RT-PCR and histochemical staining of chitin indicated that chitin is also produced in the scales of the zebrafish as well as the scales of juvenile Atlantic salmon, Salmo salar. Here, the authors report that chitin is accumulating intracellularly in epithelial goblet and club cells. *CHS* genes were also identified in two amphibian species, *Xenopus* and axolotl by this group. In *Xenopus*, chitin was detected mainly in superficial epidermal cells resembling squamous epithelia, Leydig epidermal cells, and mesenchymal (fibroblast-like) cells. In axolotl, chitin distribution was similar to that seen in fish scales. A further indication for the production of chitin in vertebrates was provided by preparing chemical extracts from fresh scales from the Atlantic salmon that were analyzed by microscopic Fourier transform infrared spectrometry. This experiment confirmed the presence of chitin in the scale extracts, although it was not really clear whether alpha, beta or gamma chitin was detected (Tang et al. 2015). However, despite the presence of chitin and the identification of putative genes encoding chitin synthases, unequivocal proof that these genes encode chitin synthases is pending. In his excellent review article, Robert Stern has critically discussed the chitin enigma in vertebrates (Stern 2017). His hypothesis that chitin is an evolutionary precursor of hyaluronan earns special attention.

Because chitin synthases are highly conserved enzymes, the presence of *CHS* genes in the genomes of organisms can be considered diagnostic for the capability of chitin biosynthesis. This approach may be even more reliable to predict the capability of chitin formation than using different histochemical staining techniques. Most of them have specificity issues, because the dyes frequently used to detect chitin, such as wheat germ agglutinin or calcofluor white, bind also to other polysaccharides and glycoproteins with various affinities. Fluorophore-coupled chitin-binding domains from *Bacillus circulans* chitinase A, which bind to this polymer more specifically, have been less widely used to detect chitin. Based on the increasing availability of genome sequences, an increasing number of genuine *CHS* genes has been identified over the past decades indicating that chitin is produced by a larger variety of organisms from different taxonomic groups than previously thought (Zakrzewski et al. 2014). In fact, chitin synthases show a wide-ranging distribution in metazoan taxa and exhibit extensive diversification. Homologs of genuine *CHS* genes have been detected in heterokont as well as in red and green algae, fungi, dinoflagellates, ciliates, and amoebozoans, basal metazoans such as sponges and corals, ecdysozoans including arthropods and nematodes, as well as lophotrochozoans including annelids and mollusks, and even in some chordates including amphioxiformes, tunicates, teleost and cartilaginous fishes, as well as amphibians. *CHS* genes have been classified into three divisions: fungal, diatome, and metazoan *CHS* genes (Zakrzewski et al. 2014). Fungal *CHS* genes are further grouped into seven classes (Roncero 2002; Horiuchi 2009; Merzendorfer 2011), and metazoan *CHS* genes into two large classes. Class I metazoan *CHS* include genes from filasterean species and choanoflagellates, sponges, cnidarians, annelids, mollusks, and lancelets. Class II metazoan genes can be further divided into the lophotrochozoan, ecdysozoan and deuterostome *CHS* groups. Notably, diversification of *CHS* genes is most pronounced in fungi and lophotrochozoans, of which some species have >20 and ≤ 10 paralogs. Chitin synthases from these two groups are frequently fused to different types of myosin motor domains, indicating that the interaction with the cytoskeleton is an important factor in controlling chitin secretion (Zakrzewski et al. 2014).

In this review, we will summarize current knowledge about the multiple functions of chitin in bacteria, fungi, and protist. Chitin synthesis in metazoans will be discussed elsewhere in this book.

3.2 Bacteria that Produce Chitin Oligosaccharides

Chitin synthesis has been considered characteristic for eukaryotic organisms. Hence, it came as a surprise when Atkinson and Long (1992) and Debelle et al. (1992) independently reported that the NodC protein from rhizobial bacteria (Alphaproteobacteria) have striking homologies to chitin synthase and related β-glycosyltransferases. *NodC* is one of several rhizobial nodulation genes required for the symbiosis between rhizobial bacteria and legumes. Stimulated by flavonoids released from the legume root, the bacteria produce signaling molecules called nodulation (Nod) factors, which induce differentiation of legume root cells into nodule structures. Actually, they trigger a division of cortical cells in the roots of leguminous plants. Upon this, the bacteria enter the cells of its host plant and differentiate into nitrogen-fixing bacteroids. The bacterial signaling factors are made of a chitin oligosaccharide backbone, which becomes chemically modified by acylation (see Fig. 3.1). Geremia et al. (1994) showed that the NodC protein from *Azorhizobium caulinodans* is a chitin oligosaccharide synthase, which polymerizes UDP-*N*-acetyl-D-glucosamine into chitin tetraose and chitin pentaose. This observation has been confirmed for NodC from *Rhizobium leguminosarum,* which was heterologously expressed in *Escherichia coli*. The recombinant protein directs the synthesis of chitin pentaose, chitin tetraose, chitin triose, and two unidentified modified chitin oligosaccharides (Kamst et al. 1995).

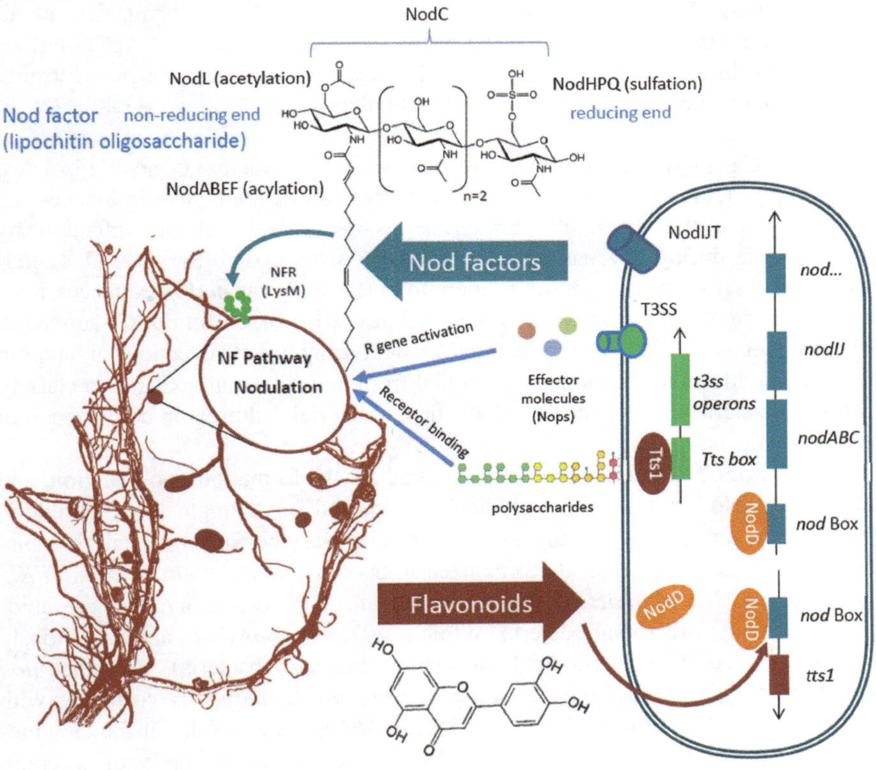

Fig. 3.1 Determinants of host specificity in *Rhizobia*-legume symbiosis involving lipochitin oligosaccharides

3.2.1 *Nodulation Genes and Their Functions in Establishing* Rhizobia-*Legume Symbiosis*

Dorfmueller et al. (2014) demonstrated that the rhizobial NodC protein from *Sinorhizobium meliloti* possesses structural, kinetic, and mechanistic characteristics commonly found in chitin synthases. They further showed that NodC is competitively inhibited by the fungal chitin synthase inhibitor Nikkomycin Z. Using the structure of a bacterial cellulose synthase BcsA as a template for homology modeling (Morgan et al. 2013), they mapped the membrane topology of NodC and identified conserved catalytic residues for chitin oligosaccharide synthesis. The derived structural model proved helpful in revealing general insight into the molecular mechanism of chitin synthesis. However, NodC lacks those transmembrane domains forming the polysaccharide-conducting channel which is present in BcsA, as well as in the structural models predicted by homology modeling for chitin synthase 2 from *Saccharomyces cerevisiae*, ScChs2 (Dorfmueller et al. 2014), and chitin synthase 1 from the beetle *Tribolium castaneum*, TcChs1 (Zhu et al. 2016). In the yeast chitin synthase ScChs3, only two of these pore-forming transmembrane helices could be modeled to the BcsA structure (Gohlke et al. 2017).

The *nodC* gene was used to examine the community structure of nodule isolates. Interestingly, temperature-dependent gene expression of *nodC* also influences the composition of the rhizobial microbiome as determined for the infection by bradyrhizobia during soybean nodulation (Shiro et al. 2016). Sarita et al. (2005) developed a set of group-specific primers to detect different *nodC* sequences from rhizobial bacteria in various root nodule isolates. Also based on *nodC* sequences, Bontemps et al. (2005) have generated a microarray-based detection and typing method that allows to discriminate rhizobial from non-rhizobial taxa and to classify rhizobial bacteria using genomic DNA from bacterial cultures or colonized root nodules.

Next to *nodC*, additional genes are required for the formation, modification, and secretion of Nod factors, most of them are located on a single large symbiotic plasmid. The number and organization of the genes vary between different rhizobial bacteria. The genes are typically organized in operons: *nodD, nodABCIJ, nodFEL, nodMNT, nodO, and nodHPQ*. However, not all of these operon/genes are ubiquitously found in rhizobial bacteria. While *nodD and nodABCIJ* are ubiquitously present in rhizobial bacteria (and nitrogen-fixing actinobacteria), the other *nod* operons/genes are present only in some species, which frequently correlates with host specificity (Carlson et al. 1994). The *nodABC* genes encode all the enzymes that are necessary to form the core Nod factor structure made of a chitin oligosaccharide backbone modified by deacetylation at the reducing end and acylation at the non-reducing end resulting in lipochitin oligosaccharides (LCO). While NodC produces the chitin oligosaccharide backbone, NodB is a chitin deacetylase, which removes the terminal N-acetylglucosamine at the non-reducing end of these chitin oligosaccharides (John et al. 1993), and the NodA protein, in turn, is an

N-acyltransferase that attaches a fatty acyl moiety to the deacetylated chitin oligosaccharide (Atkinson et al. 1994). LCOs undergo a series of additional chemical substitutions catalyzed by the products of further *nod* genes, which affect host specificity and rhizobial communities. These modifications, however, must not necessarily occur after the acylation step, as it has been assumed for a longer time, but occurs also prior acylation, which may even be the last step in LCO biosynthesis (Poinsot et al. 2016). The LCOs primarily determine host specificity in the *Rhizobium*-legume symbiosis, mediated by specific receptors, transporters or metabolic enzymes (Spaink and Carlson 1996). The lengths of both, the chitin oligosaccharide and the acyl chain, the saturation level of the acyl chain, and the substitutions of the reducing and non-reducing sugar residues are characteristic for the respective rhizobial species and strains (Wang et al. 2012). Actually, NodC contributes to host specificity by determining the length of the carbohydrate chain, which can strongly influence the activities on host plants (Kamst et al. 1997). In contrast to *nodABC*, the *nodD* gene and its homologues have important functions in establishing the *Rhizobium*-legume symbiosis by activating nodulation genes. The proteins encoded by *nodD* genes belong to the LysR family of transcriptional regulators, which are activated by flavonoids released into the rhizosphere by the respective host plant (Fig. 3.1). Interestingly, the flavonoids mediate increased binding of NodD to the promoter of *nod* genes triggered by DNA bending of conserved nod-boxes within the promoter sequence (Fisher et al. 1988; Feng et al. 2003). Notably, NodD proteins from different rhizobial species react to different sets of flavonoids, and this also accounts for rhizobial bacteria that express different *nodD* homologs (Broughton et al. 2000; Peck et al. 2006). The ability to respond to different types of plant-specific flavonoids via *nodD*-dependent activation of *nod* genes represents an early checkpoint in the symbiosis between legumes and rhizobial bacteria.

The ubiquitous *nodI* and *nodJ* genes encode proteins that are similar to capsular polysaccharide secretion proteins at the inner membrane from gram-negative bacteria (Vazquez et al. 1993), and have been shown to mediate efficient secretion of LCOs (Spaink et al. 1995). LCO variations depend on further modifications of the reduced and non-reduced ends, which are performed by proteins encoded by species-specific *nod* genes (Fig. 3.1). The roles of *nodFEL* genes are in the synthesis of the acyl moiety at the non-reducing end of Nod factors (as also *nodAB*). The functions of *nodEF* from *Rhizobium meliloti* have been examined in detail by Demont et al. (1993). NodE is a β-ketoacyl transferase, which is required for the synthesis of unusual polyunsaturated fatty acids at the non-reducing end. This modification has been shown to affect the host range between different *Rhizobium leguminosarum* biovars by changing the degree of hydrophobicity of the fatty acyl chain of the LCOs (Bloemberg et al. 1995). NodF, an acyl carrier protein, is also involved in this process, and likewise, it contributes to host specificity of LCO-mediated rhizobial symbiosis. NodG is homologous to FabG and both enzymes, when produced as recombinant proteins, exhibit 3-oxoacyl-acyl carrier protein reductase activity required for the first reduction step in each cycle of fatty acid elongation (Lopez-Lara and Geiger 2001). NodL is an *O*-acetyl transferase required for acetylation at the C-6 position of the non-reducing end (Bloemberg et al. 1994). Interestingly, there is an unrelated *O*-acetyl transferase in *R. leguminosarum*, which is encoded by the *nodX* gene and involved in the acetylation of pentameric LCOs (Firmin et al. 1993).

The *nodMNT* genes are commonly found in *Rhizobium* biovars. Nodulation and development of bacteroids are impaired in *nodM* and *nodN* deletion mutants used to inoculate *Medicago sativa* (Baev et al. 1992). The *nodM* gene encodes a glucosamine synthase (Baev et al. 1991), which is required for the formation of UDP-GlcNAc. Accordingly, galactosamine or *N*-acetylglucosamine substitutions are able to abolish impaired nodulation in *nodM* deletion mutants. The *nodN* gene codes for a protein containing a MaoC domain. Several bacterial proteins that are composed solely of this domain have (R)-specific enoyl-CoA hydratase activity, which might be involved in fatty acid synthesis. Finally, the *nodT* gene appears to be involved in secretion of LCOs and presumably encodes an outer membrane transport protein (Rivilla et al. 1995).

The *nodHPQ* genes are required for *O*-sulfation at the C-6 position of the reducing end (Roche et al. 1991). The *nodH* gene encodes a sulfotransferase, the *nodP* gene an ATP sulfurylase and the *nodQ* gene an APS kinase. Notably, the *nodH* is present in two copies in the metagenome of non-rhizobial but nitrogen-fixing *Frankia* actinobacteria (together with *nodABC* genes) and affects their host specificity and their stability (Nguyen et al. 2016).

NodO is a hydrophobic Ca^{2+} binding protein partially homologous to hemolysin, which can form a cation-specific channel in planar lipid bilayers (Sutton et al. 1994). It is only found in specific *R. leguminosarum* biovars and it can be secreted by hemolysin type I secretion systems (Scheu et al. 1992). NodO appears to play a significant role in promoting infection thread development. Recently, Yan et al. (2017) confirmed the impact of the *nodO* gene on the *Rhizobia*-legume symbiosis by analyzing a wildtype strain and a *nodO* deletion mutant in different legumes. In contrast to the wildtype, nodulation was impaired on an ultrastructural level in *Astragalus membranaceus* and *Caragana intermedia* infected with the in *nodO* deletion mutant. In promiscuous *Sophora flavescens*, however, no signs of altered nodulation could be observed when infected with the *nodO* deletion mutant.

Two further genes that are under the control of a nod-box and determine the host range are the *nodSU* genes (Lewin et al. 1990), which are part of a single *nodYABCSUIJ* operon in *Bradyrhizobium japonicum* (Gottfert et al. 1990). NodS functions as S-adenosyl-L-methionine-dependent methyltransferase, which is required for methylation of chitin oligosaccharides that are deacetylated at the non-reducing end (Geelen et al. 1995). In contrast, *nodU* is involved in 6-*O*-carbamoylation of nod factors (Jabbouri et al. 1995). The *nodK* and *nodY* genes are related to each other and found in some bradyrhizobial species preceding the *nodABC* genes (Dobert et al. 1994). Their function has not been elucidated so far. The host-specific nodulation gene nodZ encodes a LCO fucosyltransferase (Quinto et al. 1997), which is expressed in Bradyrhizobium *japonicum* (Stacey et al. 1994), *Sinorhizobium* species (Quesada-Vincens et al. 1997), and *Mesorhizobium loti* and in six *Mesorhizobium* strains infecting *A. membranaceus* and *C. intermedia* (Rodpothong et al. 2009).

In addition to Nod factors, some but not all rhizobial bacteria produce nodulation outer proteins (Nop), which they release via a type III secretion system (T3SS) (Deakin and Broughton 2009). The synthesis of both, the Nops and the T3SS components are controlled by plant flavonoids and the bacterial NodD, which activates the expression of the transcription factor TtsI. This transcription factor binds to promoters containing a tts-box, which is found also in those genes that control Nop and T3SS gene expression. The Nop effector molecules can stimulate the expression of resistance genes (R genes), which provoke effector-triggered immunity in some leguminous plant and restrict the nodulation with specific rhizobacterial strains. Also, polysaccharides on the surface of Rhizobacteria, which are recognized by plant lectins, may be determinants required for the establishment of symbiosis, but their precise role is currently uncertain (De Hoff et al. 2009).

3.3 Recognition of LCOs by Host Receptors

Rhizobial LCOs are recognized by legumes via heterodimeric, membrane-integral Nod factor Ser/Thr receptor kinases (NFRs) that exhibit extracellular LysM motifs frequently found in carbohydrate-binding proteins (Rodpothong et al. 2009; Wang et al. 2012). Nod factor binding triggers a signal transduction pathway, which has been partially elucidated in recent years (Kelly et al. 2017). This signaling induces a variety of symbiotic genes in the host plant leading to root cell differentiation and nodulation. Notably, LysM-type receptors such as CERK1 are also involved in arbuscular mycorrhizal signaling and in the immune defense against pathogenic fungi (Shinya et al. 2015). Mycorrhizal fungi produce lipochitin oligosaccharides (Myc factors) similar to Nod factors, as well as short-chain chitin oligomers, implying commonalities in signaling during mycorrhizal and rhizobial associations. On the other side, plants are also capable of recognizing fungal infections by detecting chitin oligosaccharides. In both cases, lysin motif receptors such as CERK1 are involved in the perception of chitin oligosaccharides, which trigger downstream signaling events to initiate defense reactions (Miya et al. 2007). However, fungal pathogens have evolved escape strategies by using effector molecules and/or altering the composition of their cell walls. The discovery of the dual function of CERK1 in both rice plant immunity and mycorrhizal symbiosis revealed exciting insight into the evolutionary relationships between immune and symbiotic systems in plants (Zhang et al. 2015). Obviously, there is a direct connection between plants chitin-mediated defense system and rhizobial symbiosis, which is reflected by mutual structural characteristics between chitin oligomers and Nod/Myc factors as well as the corresponding receptors.

Next to α-proteobacteria (Rhizobiales) and terrabacteria (Actinobacteria), whose *nodC* genes are more distantly related to eukaryotic chitin synthases, genes encoding chitin synthases that are more closely related to eukaryotic chitin synthases have been detected in a variety of γ-proteobacteria including Enterobacteriaceae (*Brenneria*, *Pectobacterium*, *Dickeya*, *Cedecea*, *Teredinidae*), Cellvibrionaceae (*Cellvibrio*), and Pseudomonadaceae (*Pseudomonas*) (Gonçalves et al. 2016). The fact, that all these genes form a monophyletic group, which is homologous to eukaryotic division 1 *CHS* genes, suggests a unique transfer of a eukaryotic gene to a bacterial genome and subsequent spreading by horizontal gene transfer between bacteria (Gonçalves et al. 2016). Some of these genes have been recognized as factors required for bacteria–plant interactions. Many of the bacteria that harbor division 1 *CHS* genes possess other β-glycosyltransferase genes to produce different types of bacterial exopolysaccharides such as alginate, curdlan, cellulose or hyaluronan, which are components of extracellular biofilms (Schmid et al. 2015). Whether division 1 chitin synthases of γ-proteobacteria are involved in the formation of longer chitin fibers to organize the extracellular matrix of biofilms, or in chito–oligosaccharide synthesis similar to NodC has to be determined in future experiments.

3.4 Fungi that Deposit Chitin (and Chitosan) as a Component of Cell (and Spore) Walls

Chitin is widely distributed in fungi such as basidiomycetes, ascomycetes, and phycomycetes, where it is a component of cell walls and other structural matrices of mycelia, stalks, and spores (Peter 2005). It has important functions, because it helps to withstand the cell's turgor pressure and stabilizes the cell wall during hyphal growth, budding, and cell division. The chitin nanofibrils produced are found predominantly in the α-configuration, which conveys high tensile strength due to the formation of numerous intra- and intermolecular hydrogen bonds within the chitin nanofibrils (Deringer et al. 2016; Kameda et al. 2005). The majority of chitin synthesis occurs at the sites of growth, or at regions where the cell wall has to be repaired, rebuild or restructured. In Baker's yeast, *S. cerevisiae*, chitin is mainly produced in the bud tip while growing and bud neck at the time of cytokinesis (Bowman and Free 2006). In addition, it is an intermediate in the formation of the spore wall and is synthesized during the mating process, where it is deposited at the subapical region of the shmoo tip (Cid et al. 1995). In filamentous fungi such as *Neurospora crassa* or *Aspergillus fumigatus*, chitin is mainly deposited at the hyphal tips, uniformly along the mycelia and at the septa. Although chitin has important functions, it accounts only for a smaller portion of the cell wall's total dry mass, which amounts to 1–2% (w/w) in yeast, 4% (w/w) in *N. crassa*, 2–6% (w/w) in *Candida albicans*, and 7–15% (w/w) in *A. fumigatus* (Free 2013).

Taken together, chitin and its derivative chitosan are major fibrillary components that function as a scaffold for the assembly of different components of fungal cell walls providing skeletal support essential for cell survival but also interactions with fungal surroundings (Roncero 2002).

3.5 Cell and Spore Wall Structure in Fungi

Although the fungal cell wall appears to be a static structure at a first glance, it is a highly dynamic extracellular matrix, whose basic structure is similar in all fungi. However, there are taxon-specific variations of a common theme. The structure and composition of the cell wall influence the function and interaction of the fungus with its environment, for example, by mediating adhesion or by the activation of signaling cascades. The fungal cell wall consists of a combination of fibrous and gel-like carbohydrate polymers forming a core scaffold to which other components are added. The rigidity of the scaffold helps to maintain the shape of the fungus while providing enough flexibility for cell growth. It also enables fungi to resist harsh environmental conditions such as high osmotic pressure or mechanical stress. About 90% (w/w) of the fungal cell wall consists of polysaccharides like glucans (α-1,3, β-1,3, and β-1,6 glucans, depending on the fungal species), mannans, and chitin, whereas glycoproteins make up only a small percentage of all cell wall components. Different fungal groups possess additional characteristic polysaccharides in their cell and spore walls such as β-1,4 glucans, galactomannans, galactosaminoglycans, as well as capsule polysaccharides such as glucuronoxylomannan and galactoxylomannan (Gow et al. 2017). In yeast, which has been extensively studied, underneath an outer fibrillar layer of mannoproteins the cell wall has an inner matrix of cross-linked β-(1,3)- and β-(1,6)-glucans and chitin (Fig. 3.2).

In most fungal species, such an inner core layer exists. To this fibrillary scaffold species-specific cell wall proteins (CWPs) and other proteins and polysaccharides are added by noncovalent or covalent binding. Noncovalently bound proteins are adhering to the cell wall by hydrogen bonding or hydrophobic and/or ionic interactions. Three types of covalently bound proteins have been recognized:

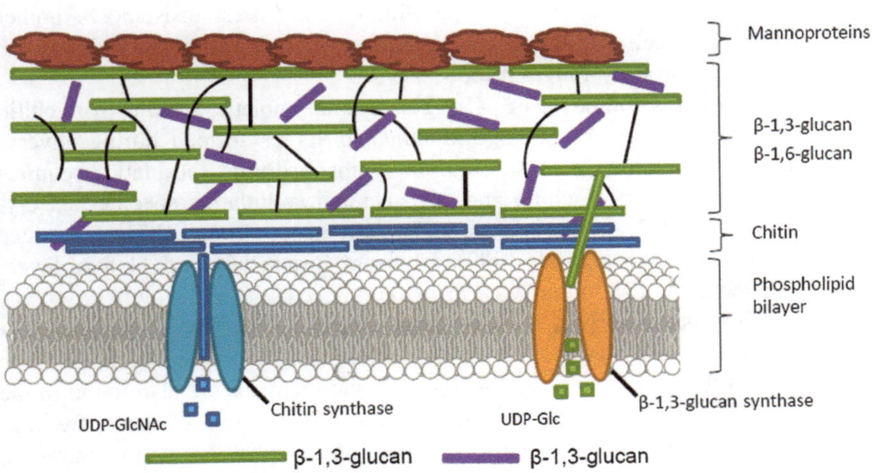

Fig. 3.2 Structure of the yeast cell wall

glycosylphosphatidylinositol (GPI) proteins that are linked to β-(1,6)-glucan by a glycosidic linkage; Pir proteins bound to β-(1,3)-glucan by an alkali–labile linkage and proteins bound to other proteins by a disulfide linkage (Orlean 2012).

The cell wall composition changes continuously to adapt fungal growth to different environmental conditions. Characterization of the molecular mechanisms of cell wall assembly and remodeling is essential, not only for a better understanding of fungal cell biology, but also for the identification of specific fungal–host interactions and the design of novel antifungal strategies. For a pathogenic fungus, cell wall remodeling is also an important mechanism to escape the host's defense system during colonization and infection. Generally, cell wall remodeling during growth and morphogenesis involves elongation, branching and cross-linking of newly synthesized polysaccharides to the existing cell wall. This process is mediated by transglycosylation, in which glycosidic linkages are broken up and reestablished to generate new cross-linkages between polysaccharides, particularly between chitin and β-glucans. The underlying mechanism for this process is now beginning to be deciphered and involves transglycosylases of the CRH family, which are unique to fungi and highly conserved (Arroyo et al. 2016). In yeast, Crh1 and Crh2 have redundant functions during vegetative growth. Next to their function in catalyzing cross-link formation, they also exhibit some hydrolase activity, which helps to interconnect existing polysaccharides. Another yeast transglycosylase, Crr1, takes part in spore wall biogenesis. Crh orthologs have been found in the genomes of *C. albicans, A. fumigatus* and *N. crassa* where they also appear to be involved in establishing cross-links between chitin and other cell wall polysaccharides. Induction of transglycosylase activity seems to be one mechanism that allows coping with cell wall stress. In a comprehensive study performed in *C. albicans*, Ene et al. (2015) examined the enzymes being responsible for cell wall remodeling. In particular, they showed that inactivation of Crh enzymes leads to increased cell wall elasticity, while their overexpression protects from osmotic stress. Hence, Crh enzymes help in controlling osmotic stress resistance. Another mechanism to resist cell wall stress is the induction of chitin synthesis, which is one of the main compensatory fungal responses.

Chitosan, which is composed of β-1,4-glucosamine moieties, results from chitin deacetylation and is present in large amounts in spores formed during adverse environmental conditions. In yeast, chitosan synthesis during sporulation requires Chs3 for the production of chitin and two secreted sporulation-specific deacetylases, Cda1 and Cda2, which remove the acetyl group from chitin to produce chitosan (Christodoulidou et al. 1999). Chitosan forms the third of four layers directly underneath the outermost layer composed of bisformyl dityrosine cross-linked by the activity of a P450 family protein, Dit2. The dityrosine layer together with the chitosan layer are unique components of the spore wall, while components of the inner two layers of β-glucan and mannans are also found in the cell wall (Briza et al. 1988) (Fig. 3.3). The outer two layers of the spore wall contribute significantly to the overall robustness of the spore, which in contrast to vegetative cells are less susceptible to digestive environments, heat stress, extreme pH values and high salt concentrations, as tested for yeast spores that, for instance,

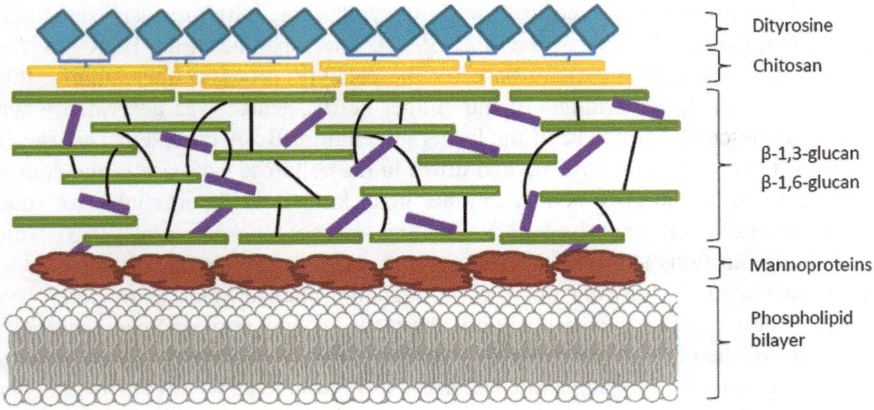

Fig. 3.3 Structure of the spore cell wall from *S. cerevisiae*

survive the passage through the intestinal tract of *Drosophila melanogaster* (Coluccio et al. 2008). From the finding that the spores from *Schizosaccharomyces pombe*, which lack the dityrosine layer, were yet highly tolerant to stress, it was concluded that chitosan is the crucial component that confers resistance to the spore.

However, chitosan is not only involved in protecting the spore from environmental stress. The outer two layers of the spore wall also contribute to the adherence between neighboring spores resulting in the formation of stable spore tetrads. It has also been shown that chitosan produced by zygomycetes has antimicrobial activity against pathogenic microorganisms like *Staphylococcus aureus, Escherichia coli, C. albicans,* and *Fusarium oxysporum* (Gharieb et al. 2015), and the formation of chitosan has been additionally considered to prevent hydrolysis of the polymer by chitinases.

3.5.1 *Structure and Regulation of the Chitin Synthase in Fungi*

Chitin is produced by an enzyme named chitin synthase (Chs), which belongs to family 2 of membrane-integral glycosyltransferases (GTF2). These enzymes produce chitin by catalyzing the transfer of the activated sugar donor UDP-*N*-acetylglucosamine to the non-reducing end of the growing chains. The catalytic core (CON1) of the Chs harbors conserved sequence motifs that are thought to be involved UDP-binding, donor and acceptor saccharide binding, and product binding (Merzendorfer 2011). Among others, these include the CHS signature sequence QRRRW, as well as the DxD and EDR motifs involved in product binding and catalysis, respectively. Actually, mutational studies performed in the 1990s have shown that these motifs are critical for chitin synthase activity in yeast (Nagahashi et al. 1995; Cos et al. 1998; Yabe et al. 1998). The aspartate in the EDR motif is thought to function as a general base for the catalytic process. Interestingly, this motif resides at the *N*-terminus of the "finger helix", which was described when analyzing the crystal structure of the bacterial cellulose BcsA complex. This finger helix has been shown to move up and down in the cellulose synthase in the course of the catalytic cycle governed by another helix known as "interfacial helix" that lies at the base of the carbohydrate-conducting channel (Morgan et al. 2013). This and other indications led to the hypothesis that catalysis is energetically coupled to the extrusion of the nascent polymer through the channel. Based on homology models using the BcsA as a structural template, a similar mechanism has been proposed for the chitooligosaccharide synthase NodC from *Sinorhizobium meliloti* and for the yeast chitin synthases Chs2 and Chs3 (Gohlke et al. 2017; Dorfmueller et al. 2014). Next to a conserved catalytic core, which faces the cytosolic site of the membrane, CHS enzymes possess several transmembrane helices that likely form a channel to translocate single chitin chains to the extracellular side, where they assemble into nanofibrils. However, there are still uncertainties how the chitin nanofiber form and how interconnections with other cell wall components are created (Gonçalves et al. 2016).

Regulation of chitin synthesis has been extensively studied in yeast, which has three different CHS enzymes that are differentially regulated on the levels of transcription, intracellular trafficking and posttranslational modifications (Merzendorfer 2011). This is because they adopt different functions during growth and cytokinesis. While Chs1 is involved in general cell wall repair, Chs2 is essential for primary septum formation, and Chs3 synthesizes the chitin ring at bud emergence, as well as the chitin in the lateral cell wall (Merzendorfer 2011). While the *CHS3* gene is constitutively expressed, intracellular trafficking of Chs3 and its regulatory subunit Chs4 are tightly controlled and the intracellular localization of both protein varies significantly, depending on which phase the cell has reached in the cell cycle (Gohlke et al. 2018). The exit of Chs3 from the endoplasmic reticulum (ER) depends on the ER chaperone Chs7 and the palmitoyltransferase Pfa4 (Lam et al. 2006; Trilla et al. 1999). The transport from the trans Golgi network (TGN) to the plasma membrane requires the exomer complex (Sanchatjate and Schekman 2006). During the cell cycle, Chs3 also shuttles between the chitosomes, which are considered to be endosomal reservoir vesicles, and the plasma membrane (Bartnicki-Garcia 2006). Refilling this reservoir with Chs3 is facilitated by AP-1-dependant endosomal recycling, but involves also the retromer complex (Arcones et al. 2016). In addition, proper localization of Chs4, which tethers Chs3 to the bud neck by interacting with the septin-binding protein Bni4 (Sacristan et al. 2012), depends on the septin-dependent kinase Gin4 (Gohlke et al. 2018). The intracellular trafficking of Chs1, which also shuttles between chitosomes and the plasma membrane, depends on constitutive secretory and endocytic pathways (Ziman et al. 1996).

In contrast, *chs2* expression oscillates during the budding cycle (Choi et al. 1994), and transcriptional control is mediated by the Mcm1-Ndd1-Fkh2 transcription factor complex (Chen et al. 2009). In the late anaphase, Chs2 is transported from the ER to the bud neck via the secretory pathway (Chuang and Schekman 1996). ER exit is inhibited by Cdk1-dependent phosphorylation of Chs2 (Teh et al. 2009). This block is relieved toward the end of mitosis leading to Cdc14-dependent dephosphorylation of Chs2 and its subsequent transport in COPII vesicles to the bud neck (Chin et al. 2012). Following endocytosis, Chs2 is directed to the vacuole for degradation by the major Pep4 protease. Other mechanisms of posttranslational regulation that may exist include proteolytic activation of zymogenic Chs3 forms (Choi et al. 1994), as well as oligomerization of Chs3 (Sacristan et al. 2012; Gohlke et al. 2017).

3.5.2 Evolution and Classification of Fungal CHS *Genes*

Fungal genomes harbor more than one *CHS* gene, and the gene number varies from 3 *CHS* genes in *S. cerevisiae* to 5 in *Wangiella dermatitidis*, to 8 in *A. fumigatus* and *Cryptocossus neoformans*, to more than 20 in several mucoromycete species (Ruiz-Herrera and Ortiz-Castellanos 2010; Gow et al. 2017). Phylogenetic analysis indicate that GT2 enzymes have their origin in a common ancestral molecule. Different phylogenetic CHS classification systems have been published based on distinct assumptions regarding their origin, diversity, and evolution. The increasing complexity in the classification of fungal *CHS* genes makes it difficult to keep the overview. Therefore, we provide a short chronological survey on the phylogenetic analyses of fungal *CHS* genes.

First approaches to categorize *CHS* genes from available sequences comprising different sources established five to six fungal *CHS* classes that group into two divisions (Roncero 2002; Ruiz-Herrera et al. 2002; Munro and Gow 2001). Division 1 harbors class I, II and III *CHS* genes, which all encode proteins that have a common domain organization with a catalytic domain flanked by a hydrophilic *N*-terminal and a hydrophobic C-terminal region, as described by (Bowen et al. 1992). In contrast, division 2 comprises class IV, V and VI *CHS* genes, which encode proteins with the catalytic domain being preceded by a cytochrome b_5-like domain (cyt-b_5; PF00173) supposed to be involved in binding of lipid ligands. In case of class V and VI *CHS* genes, a myosin head-like domain (MMD, PF00063) is observed at the *N*-terminus, which helps to localize Chs enzymes to the hyphae by the interaction with the actin cytoskeleton, and facilitates polarized exocytosis (Tsuizaki et al. 2009; Schuster et al. 2012).

With increasing amounts of genomic data that have become available in subsequent years, the classification system was continuously modified. Using PCR and degenerated primers specifically designed for conserved sequence motifs, Choquer et al. (2004) analyzed *CHS* genes in the plant-pathogen *Botrytis cinerea*, a filamentous ascomycete. Eight distinct *CHS* genes were identified, which were thought to comprise the entire *CHS* gene family of this fungus. Genomic Southern blots revealed that all of them were single copy genes. After sequencing, they grouped the *CHS* genes into two divisions and seven classes (I–VII, with two subclasses IIIa and IIIb). Division I and II are each made up of three classes I–III, and four classes IV–VII, respectively.

Likewise, Mandel et al. (2006) grouped *CHS* genes from Coccidioides posadasii into a system of two divisions and seven defined classes of fungal chitin synthases (I–VII). Because *C. posadasii* contains single members for each of the seven classes of fungal *CHS* genes, they suggested that it is a good model to investigate the respective role of these genes in fungal growth and differentiation. Using RT-PCR they obtained a first insight into their putative functions. *CpCHS2*, *CpCHS3*, and *CpCHS6* were preferentially expressed in the saprobic phase, whereas *CpCHS1* and *CpCHS4* were found to be more highly expressed in the parasitic phase, while *CpCHS5* and *CpCHS7* did not differ in their expression in both phases.

To search for genes that are functionally related in the biogenesis of the cell wall, Ruiz-Herrera and Ortiz-Castellanos (2010) analyzed the homologies of the enzymes responsible for the synthesis of different cell wall polysaccharides, including Chs, and covalently bound cell wall proteins in members of different fungal phyla. They analyzed >300 *CHS* genes from >50 fungal species and grouped them into 2 divisions comprising 5 classes. When they compared protozoan and animal *CHS* genes with the fungal genes, they found that all of them have a chitin synthase motif, which is more similar to the CS2 domain from fungal enzymes. Further, none of these genes contained a MMD domain. This suggests that protozoan and animal *CHS* genes are more closely related to fungal class IV chitin synthases, which therefore may be the common ancestor of fungal *CHS* genes.

Pacheco-Arjona and Ramirez-Prado (2014) provided the first large-scale phylogenetic classification of fungal *CHS* genes and identified a putative cell wall metabolism gene cluster in genomes from *Aspergillus* by synteny analysis. For this purpose, they analyzed *CHS* sequences from 54 fungal genomes, encompassing 21 orders from five phyla (Ascomycota, Basidiomycota, Microsporidia, Mucoromycotina, Chytridiomycota), and they were able to classify 347 Chs proteins into 7 phylogenetic clades grouping into 2 divisions. Interestingly, four classes (III, V, VI, and VII) can specifically be assigned to filamentous fungi whereas classes I, II, and IV can be found in yeast and filamentous fungi (Fernandes et al. 2016).

In another approach, Li et al. (2016) examined the evolution of the *CHS* gene family within the fungal tree of life, covering 18 divergent fungal lineages. By examination of more than 900 *CHS* genes from more than 100 fungal species, they grouped fungal *CHS* genes into 3 divisions containing 8 classes (I–VIII, including novel subclasses IVa, IVb, VIa–c) according to the phylogenetic positions and domain architectures. For the analysis of the domain architecture, they included the conserved Chs domain type I (CS1, PF01644), type II (CS2, PF03142) and a chitin synthase *N*-terminal domain (CS1N, PF08407). While in division I only CS1, CS2, and CS1N are found, division III comprises only CS2. In contrast, most *CHS* genes of division II additionally contain the Cyt b_5-domain, which is absent from some members of classes IVa and IVb in division II. The MMD domain and a DEK C-terminal (DEK_C, PF08766) domain, which may mediate oligomerization, were identified in classes V and VII. The scientists found that the fungal *CHS* gene family is made of at least 10 ancestral orthologous clades, which have gone through multiple independent duplications and losses in different evolutionary tracks of the fungal lineages. Particularly, class III genes have expanded in plant- or animal-pathogenic fungi of different lineages. Moreover, they showed that the newly identified VIb and VIc genes are mainly present in pathogenic fungi from Sordariomycetes and Dothideomycetes.

Analyzing 231 fungal species from 9 phyla, Liu et al. (2017) addressed the evolution of fungal chitin synthases combining phylogenetic and domain structure analysis, while also considering the adaptation of fungi to ecological niches. They identified 20 domains, which fall into two groups (A, B) and exist in seven class-specific types of configurations along with PF03142 (=CON1) (types: A1–A3 and B1–B4). The majority of fungal *CHS* genes contained the domains: PF00063, PF00173, PF08766, PF01644, and PF08407. The other 14 domains could be identified in just a few fungal *CHS* genes. Group A contains PF01644 followed by the CON1 region (PF01644-CON1) and can be sub-classified into 3 types: Type A1 (PF01644-CON1), Type A2 (PF08407-PF01644-CON1), and Type A3 (PF08407-PF01644-CON1 followed by PF03142). In contrast to group A, group B does not contain PF01644, but PF03142 and can be divided into four types: Type B1 (contains only PF03142), Type B2 (PF03142 and PF00173), Type B3 (PF00063, PF03142 and PF08766), and Type B4 (PF00063, PF00173, PF03142 and PF08766). The seven *CHS* classes in Ascomycota are defined by the specific combination and arrangement of domains. CHSI, CHSII, and CHSIII, belonging to division 1, exhibit the Type A3 configuration. CHSV, CHSVII, and CHSIV, belonging to division 2, adopt the type B4 or B2 configurations. CHSs with a type B1 configuration make up division III just containing CHSVI. The study provided evidence that all fungal *CHS* genes have a common ancestor and that gene duplication, domain recombination, divergence, and accretion led to the diversification of fungal *CHS* genes resulting in a classification system of at least seven classes grouped into three divisions with different domain configurations. This is in contrast to a previous phylogeny suggested by Pacheco-Arjona and Ramirez-Prado (2014), which groups *CHS* genes into two divisions, one containing Dikarya and one early diverging fungi. Liu et al. (2017) found that the contraction of the *CHS* gene family is morphology-specific, with a pronounced gene loss in unicellular fungi, while expansion of the *CHS* gene family is lineage-specific, most obvious in early diverging fungi. Moreover, ClassV and ClassVII *CHS* genes share the same domain structure, which was accomplished by the recruitment of domains PF00063 and PF08766, and subsequent duplications. The emergence of class V and VII *CHS* genes appears to be important for the morphogenesis of filamentous fungi by supporting hyphal tip growth, for the development of pathogenicity in pathogenic fungi, and for tolerance to heat stress in Pezizomycotina fungi. In addition, 832 *CHS* gene sequences from the CON1 region of Ascomycota were used to construct a phylogenetic tree of 7 classes representing the evolutionary history of *CHS* genes in fungi.

Taken together, depending on the number of analyzed genes and the phylogenetic methods used in the different phylogenetic studies different CHS classification systems can be created, which occasionally group *CHS* genes into different classes. For example, class VI was grouped either into division I (Odenbach et al. 2009) or into division II in another study (Nino-Vega et al. 2004). Similarly, some *CHS* genes from basidiomycetes were classified as class I genes in one study (Ruiz-Herrera and Ortiz-Castellanos 2010), but as class II genes in another study (Munro and Gow 2001). Hence, the creation of a "master" classification system for fungal *CHS* genes would help to avoid confusion in the denomination of single *CHS* genes. The grouping of representative *CHS* genes from *A. fumigatus*, *C. albicans*, *S. cerevisiae,* and *S. pombe* into fungal CSH classes is given in Table 3.1.

3.5.3 Chitin as a Target for the Immune System and Its Function in Pathogenicity and Symbiosis

It is well known that the fungal cell wall is not only responsible for the growth or protection of the fungi, but there is also an interaction of its components with receptors of hosts being involved in the immune system. Consequently, the cell wall plays a major role in defense responses of the host and represents an ideal target to investigate host–pathogen interactions for the development of new treatment strategies against invasive fungal infections (Tada et al. 2013). The innate immune system senses fungal pathogens mostly by fungal-specific cell wall components (Latgé 2007). Cell wall polysaccharides of fungi are in the focus of interest, particularly because they do not exist in healthy humans. Their innate immune system, as the first line of defense against invaders, recognize microbe- or pathogen-associated molecular patterns (MAMPs or PAMPs, respectively) by several pattern recognition receptors (PPRs) detecting evolutionary conserved structures or molecules of the pathogen and thereby triggering innate immune response (Takeuchi and Akira 2010) (See Chap. 4).

The immune receptors involved in the recognition of cell wall polysaccharides are well known (reviewed in (Gow et al. 2017). Several studies showed that chitin has size-dependent effects on innate and adaptive immune responses including the ability to recruit and activate immune cells and induce cytokine and chemokine production via a variety of cell surface receptors including mannose receptor, TLR-2, and Dectin-1 of macrophages (reviewed in Elieh Ali Komi et al. 2018) (See Chap. 4).

Chitin functions as a recognition pattern for fungal pathogens also in plants and triggers the plant defense system (See Chap. 4). Fungi are able to prevent their recognition to prevent elimination by the host defense. For this purpose, they have evolved different mechanisms that contribute to immune evasion. These include the production of LysM domain containing chitin-binding scavengers, which sequester free chitin oligosaccharides (de Jonge and Thomma 2009), or the conversion of chitin to its deacetylated form chitosan (El Gueddari et al. 2002). Further, the expression of PAMPs can be varied to escape the recognition by PRRs, and the immune response can be prevented by masking PAMPs by less immuno-stimulatory structures (Seider et al. 2010). Other evasive mechanisms

Table 3.1 Genes encoding chitin synthases in different fungi

Organism	Name	Class	NCBI Ref #	Location	Aliases	References
Aspergillus fumigatus AF293	*CHS A*	I	XP_749322.1	Chr. 2, NC_007195.1 (465854..468816, complement)	AFUA_2G01870, Afu2g01870	Nierman et al. (2005)
	CHS B	II	XP_746604.1	Chr. 4, NC_007197.1 (1170719..1174520, complement)	AFUA_4G04180, Afu4g04180	Nierman et al. (2005)
	CHS C	III	XP_748263.1	Chr. 5, NC_007198.1 (211013..213795)	AFUA_5G00760, Afu5g00760	Nierman et al. (2005)
	CHS D	VI	XP_752630.1	Chr. 1, NC_007194.1 (3327430..3329666)	AFUA_1G12600, Afu1g12600	Nierman et al. (2005)
	CHS E	V	XP_755677.1	Chr. 2, NC_007195.1 (3474296..3480038)	AFUA_2G13440, Afu2g13440	Aufauvre-Brown et al. (1997)
	CHS F	III	XP_747364.1	Chr. 8, NC_007201.1 (1330219..1333902)	AFUA_8G05630, Afu8g05630	Nierman et al. (2005)
	CHS G	III	XP_754184.1	Chr. 3, NC_007196.1 (3831868..3834921)	AFUA_3G14420, Afu3g14420	Nierman et al. (2005)
	CHS, putative	VII	XP_755676.1	Chr. 2, NC_007195.1 (3465628..3471074, complement)		Nierman et al. (2005)
Candida albicans SC5314	*CHS 1*	II	XP_717009.1	Chr. 7, NC_032095.1 (593118..596198)	CAALFM_C702770WA	Muzzey et al. (2013), van het Hoog et al. (2007) and Jones et al. (2004)
	CHS 2	I	XP_716433.1	Chr. R, NC_032096.1 (1926708..1929737, complement)	CAALFM_CR09020CA	Muzzey et al. (2013), van het Hoog et al. (2007) and Jones et al. (2004)
	CHS 3	IV	XP_722148.2	Chr. 1, NC_032089.1 (2856152..2859793, complement)	CAALFM_C113110CA	Sudoh et al. (1993)
	CHS8	I	XP_717760.1	Chr. 3; NC_032091.1 (123503..126820)		Jones et al. (2004)

(continued)

Table 3.1 (continued)

Organism	Name	Class	NCBI Ref #	Location	Aliases	References
Saccharomyces cerevisiae S288C	*CHS1*	I	NP_014207.1	Chr. 14, NC_001146.8 (276502..279897)	YNL192 W, USA4	Philippsen et al. (1997) and Goffeau et al. (1996)
	CHS2	II	NP_009594.1	Chr. 2, NC_001134.8 (311898..314789)	YBR038 W	Goffeau et al. (1996) and Feldmann et al. (1994)
	CHS3	IV	NP_009579.1	Chr. 2, NC_001134.8 (284428..287925, complement)	YBR023C, CAL1, CSD2, DIT101, KTI2	Goffeau et al. (1996) and Feldmann et al. (1994)
Schizosaccharomyces pombe	*CHS 1*	I	NP_592838.1	Chr. 1, NC_003424.3 (195940..199051, complement)	SPAC13G6.12c, SPAC24B11.01c	Wood et al. (2002)

including the formation of biofilms or spores are known to support microbial persistence in animals and human (Brunke et al. 2016). Other pathogens like entomopathogenic fungi make use of proteolytic and chitinolytic enzymes that act as virulence factors to overcome physical barriers. For instance, *Beauveria bassiana* secretes chitinases and β-*N*-acetylglucosaminidases to penetrate the exoskeleton of insects that acts as physicochemical barriers against pathogens or other harmful environmental factors (Herrera-Estrella and Chet 1999).

3.6 Chitin in Protists

In recent years, the available information on chitin metabolism and function has massively expanded due to numerous studies performed mainly in fungal and insects. However, there is comparably little known on chitin biology in protists, though there are many chitin-producing protozoan parasites that pose threats to plant, animal, and human health. Also in these organisms, chitin has a role in protection against mechanical and chemical stress. It also contributes to morphogenesis and maintenance of the cellular shape, which relies on chitin biosynthesis, modification, deposition, and degradation, the letter of which allows remodeling and alteration of the shape.

3.7 Methods Used for Chitin Detection in Protists

Numerous histochemical strategies have been applied to detect chitin in protists. However, the utilized dyes are known to have issues with respect to their specificity. This applies also to Calcofluor White, which binds to many β-1,4-linked polysaccharides including cellulose (Pringle 1991), and likewise to the chloride–iodine–zinc staining, which was used by Herth et al. (1977) to stain chitinous fibrils in the lorica of the flagellate chrysophyte *Poterioochromonas stipitata* (syn. *Oehromonas malhamensis*).

More specific techniques used for the ultrastructural detection and localization of chitin involve chitin-binding proteins (lectins) such as wheat germ agglutinin (WGA). For fluorescence-microscopic detection of chitin, these chitin-binding proteins can be attached to fluorescent dyes such as fluorescein isothiocyanate (FITC) and applied to unfixed or fixed whole-mount, or sectioned specimens (Sengbusch and Müller 1983). WGA coupled to FITC has been used in several studied to detect chitin by fluorescence microscopy. For instance, Ward et al. (1985) suggested that the outer cyst wall of *Giardia lamblia* is largely composed of chitin, as this extracellular structure was strongly labeled by FITC-WGA. Also, Durkin et al. (2009) used FITC-WGA to detect chitin in a protozoan species. They were able to localize chitin at the girdle band region in various diatom species of the genus Thalassiosira, demonstrating that chitin biosynthesis is more widespread in diatoms than originally thought. Biancalana et al.

(2017) developed a combinatorial protocol based on FITC-WGA and catalyzed reporter deposition fluorescence in situ hybridization (CARD-FISH) staining of chitin-containing organisms including diatoms and dinoflagellates from brackish water samples. WGA was also used in ultrastructural studies for chitin detection in protists. In an electron microscopic study published by Arroyo-Begovich and Cárabez-Trejo (1982), the authors reported that chitin is situated in the cyst wall of *Entamoeba invadens* using gold-conjugated WGA. Similarly, Greco et al. (1990) analyzed the ultrastructural localization of chitin in the cyst walls of *Euplotes muscicola* using gold-conjugated WGA. In addition, the cyst wall of the ciliate *Hyalophysa chattoni* was examined by this technique (Landers 1991), showing that it is made essentially out of chitin and furthermore that it contains neutral and acidic polysaccharides. Finally, Mulisch and Hausmann (1989) demonstrated the localization of chitin using WGA-gold labeled ultrathin sections of cysts from two phylogenetically distant ciliates, *Blepharisma undulans* and *Pseudomicrothorax dubius*, suggesting that chitin synthesis is an ancestral characteristic of ciliates. According to Giraud-Guille et al. (1990), chitin crystals can be detected directly by diffraction contrast transmission electron microscopy (DCTEM) in biological specimens including arthropods and annelids (Shillito et al. 1997). However, so far this method was not used to detect chitin in protists to our knowledge.

Infrared spectroscopy and X-ray diffraction are considered the most trustable and reliable methods to detect chitin (Muzzarelli 1977). However, due to the specific technical equipment required, only a few researchers have applied these methods for the examination and analysis of extracellular secretions of protists. Infrared spectroscopy has been used to confirm the presence of chitin on the cell surface of the phytopathogenic protozoan *Phytomonas franfai*, which was suggested already by wheat germ agglutinin staining and chitinolytic hydrolysis before (Nakamura et al. 1993). One of the first approaches to use X-ray diffraction for chitin detection in protists was published by (Sachs 1956). He found that the diffraction patterns were similar to those found in the carapace of the crayfish and in the ovipositor of the grasshopper, and concluded that the cyst membranes of *Pelomyxa illinoisensis* (also known as Chaos illinoisensis) contain chitin. Herth et al. (1977) applied X-ray diffraction to detected chitin in the lorica of the chrysoflagellate *Poterioochromonas stipitata* obtaining diffraction patterns that were similar to those obtained for fungal cell wall preparations. Finally, this method was also used to detect chitin in the cyst walls of the amoeboid protist *Entamoeba invadens* (Arroyo-Begovich et al. 1980).

An enzymatic approach to identify chitin is through the specific hydrolysis by chitinases. Using alkali-stable extracts from protists and purified chitinases, scientists from France were able to detect chitin in various covering structures mainly from ciliates including (Bussers and Jeuniaux 1974; Bussers et al. 1977; Greco et al. 1990).

Although all three crystallographic types of chitin may be recognized in protists, mostly β-chitin composed of parallel-oriented sugar chains has been reported to occur in protists. For instance, β-chitin is found in the spines of centric diatoms (Herth 1978). Besides, it is present in the lorica, which is a shell-like protective outer covering of ciliates. In *Eufolliculina uhligi*, the lorica consists of ribbon-like alkali-resistant fibrils that exhibit an X-ray diffraction pattern typical of β-chitin. It seems that the lorica is produced within intracellular vesicles, which release their initially amorphous content through exocytosis into the extracellular space (Mulisch and Hausmann 1983). Once they have reached the extracellular site, fibrillogenesis occurs and the microfibrils become deposited on the cell surface. Assembly into microfibrils is impaired when the lorica components are secreted into solutions containing Calcofluor White or Congo Red, as it is generally known for chitin. The lorica maintains its shape after extraction with 20% (w/v) NaOH. However, a wide flattened network of fibrils with a diameter of about 20 nm turns visible (Mulisch et al. 1983). The dynamics of chitin biosynthesis and the impact of inhibitors such as diflubenzuron and nikkomycin Z were studied in *Eufolliculina uhligi* by Schermuly et al. (1996). They used fluorescence microscopy after staining chitinous structures with monoclonal anti-chitin and FITC-coupled secondary antibodies. To label chitin, they fed tritiated *N*-acetylglucosamine to trophont *E. uhligi* for around 2 h, prior to the initiation of cell division. Applying diflubenzuron of nikkomycin Z decreased incorporation of tritiated *N*-acetylglucosamine into chitin. However, it did not affect chitin deposition or accumulation when swarmers were treated. In contrast, the amount of chitin was drastically decreased in the lorica of swarmers when extracted from trophonts that were exposed to the inhibitors.

3.8 Occurrence of Chitin in Protists

As discussed above, unambiguous identification of chitin in extracellular layers of protists is challenging but not impossible. Evidence has accumulated that chitin is present in different extracellular structures such as spines, stalks, loricae, and cysts in a variety of protists belonging to different taxonomic groups (see Table 3.2).

As the presence of *CHS* genes may be considered diagnostic for the capability of chitin biosynthesis, *CHS* genes have been spotted in many eukaryotes including fungi, basal metazoans, and bilaterian animals and were phylogenetically analyzed (Zakrzewski et al. 2014). According to the phylogeny provided in this study, the *CHS* genes from choanoflagellates and metazoan species can be assembled into two major clades (I and II) next to fungal and diatom *CHS* gene classes. Clade I comprises *CHS* genes from sponges, cnidarians, choanoflagellates, and branchiostomata and a few lophotrochozoan species. Clade II includes all of the remaining metazoan *CHS* genes. Clade II is considered a paraphyletic class with a more complicated evolution. Fungal CHS genes are grouped into seven clades. In addition, the authors of this study identified a clade of diatom *CHS* genes, which is most closely related to fungal *CHS* classes IV, V, and VII.

Homologs of fungal and metazoan *CHS* genes have been identified in many phylogenetically unrelated protist groups. Based on sequence analysis, the *CHS* genes of choanoflagellate, filasterea, chromalveolates (or SAR including stramenopiles, alveolates and Rhizaria), chlorophyta, amoebozoa, and apusozoa form distinct clades (Goncalves et al. 2016). In particular, the *CHS* genes from

Table 3.2 Distribution of chitin in protists

	Supergroup		Phylum/ Division/Class	Order	Family	Species	Localization	References
Protista	Excavata		Matamonada	Diplomonadida	Hexamitidae	*Giardia lamblia*	Cyst wall	Manning et al. (1992), Ortega-Barria et al. (1990) and Ward et al. (1985)
				Trichomonadida	Trichomonadidae	*Trichomonas vaginalis*	Cell surface	Kneipp et al. (1998)
						Tritrichomonas foetus	Cell surface	Kneipp et al. (1998)
	SAR	Stramenopiles	Ochrophyta/ Bacillariophycaea	Thalassiosirales	Thalassiosiraceae	*Thalassiosira pseudonana*	Cell walls	Durkin et al. (2009), Brunner et al. (2009) and Tesson et al. (2008)
						Thalassiosira fluviatilis	Spines	Blackwell et al. (1967) and Morin et al. (1986)
			Ochrophyta/ Bacillariophycaea	Stephano-discaceae	Cyclotella	*Cyclotella cryptica*	Spines	Herth and Zugenmaier (1977), Blackwell et al. (1967) and Herth (1978)
			Ochrophyta/ Synurophyceae	Ochromona-dales	Ochromonadaceae	*Poterioo-chromonas stipitata*	Stalk	Herth et al. (1977) and Herth (1980)
		Alveolata	Ciliophora/ Heterotrichea	Heterotrichida	Folliculinidae	Folliculinopsis producta	Lorica	J.-C et al. (1977) and Mulisch et al. (1986)
						Parafolliculina violacea	Lorica	Bussers and Jeuniaux (1974) and Agatha and Simon (2012)
						Eufolliculina uhligi	Lorica	Mulisch et al. (1983) and Schermuly et al. (1996)
					Phacodiniidae	*Phacodinium metschnikoffi*	Cyst Wall	Bussers and Jeuniaux (1974)

(continued)

Table 3.2 (continued)

Supergroup	Phylum/ Division/Class	Order	Family	Species	Localization	References
	Ciliophora/ Spirotrichea	Euplotida	Euplotidae	*Euplotes muscicola*	Cystic layer	Greco et al. (1990)
Rhizaria	Cercozoa/ Phycomycota	Plasmodio-phorida	Plasmodiophoridae	*Plasmodiophora brassicae*	Cell walls	Schwelm et al. (2015)
Archaeplastida	Chlorophyta/ Ulvophyceae	Cladophorales	Pithophoraceae	*Pithophora oedogonia*	Cell walls	Kapaun and Reisser (1995)
	Chlorophyta	Chlorellales	Chlorellaceae	*Chlorella vulgaris*	Cell walls	Kapaun and Reisser (1995)
	Rhodophyta	Corallinales	Florideophyceae	*Clathromorphum compactum*	Skelet organic matrix	Rahman and Halfar (2014)
Opisthokonta	Holozoa/ Mesomycetozoea	Dermocystida	Rhinosporideaceae	*Rhinosporidium seeberi*	Cell walls	Mendoza et al. (2002)
		Ichthyophonida	incertae sedis	*Ichthyophonus hoferi*		Mendoza et al. (2002)
	Choanozoa/ Choanoflagellates	Craspedida	Salpingoecidae	*Salpingoeca sp.*	Theca	Buck (1990)
		Acanthoecida	Acanthoecidae	*Diaphanoeca sp.*		
Amoebozoa	Tubulinea	Tubulinida	Archamoebae	*Chaos illinoisensis*	Cyst walls	Sachs (1956)
	Archamoebae	Pelobiontida	Pelomyxidae	*Pelomyxa illinoisensis*		
	Rhizopoda/ Lobosea	Amoebida	Entamoebidae	*Entamoeba histolytica*		Arroyo-Begovich et al. (1980)
				Entamoeba invadens		

choanoflagellates and filastereans appear to be a sister group of fungal groups I-III genes lacking myosin motor domains and the metazoan *CHS* group.

Chitinous cell walls and extracellular linings are widespread in eukaryotes. In many protists, chitin has been considered a structural constituent of lorica and cyst walls, which occur in various protist lineages. Protist genera with chitinous loricae and cyst walls include *Giardia* (Diplomonads-Zoo/Sarcomastigophora), Nephromyces (Apicomplexa/Chromalveolata), Entamoeba (Amoebozoa), Trichomonas (Trichomonadida/Parabasalia), Poteriochromonas (syn. Ochromonas, Chrysophyta/ Chromalveolata), Thalassiosira (Heterokontophyta/Chromalveolata), and Euplotes (Ciliata-Chromalveolata) (Herth et al. 1977; Arroyo-Begovich and Cárabez-Trejo 1982; Ward et al. 1985; Saffo and Fultz 1986; Greco et al. 1990; Kneipp et al. 1998; Brunner et al. 2009).

Ciliates are defined as a monophyletic group of unicellular organisms, which belong to the Alveolata taxon. Various ciliate loricae and cysts appear to consist of chitin (Table 3.2). Its existence has been shown by chitinase digestion in numerous species of holotrichs such as *Pseudomicrothorax*, heterotrichs such as Parafolliculina and peritrichs such as *Cothurnia* and *Opisthonecta* (Bussers and Jeuniaux 1974; Mulisch and Hausmann 1989; Calvo et al. 2003). However, chitin appears to be absent in hypotrichs (except for *Euplotes*), *Colpoda*, *Bresslaua*, *Woodruffia*, *Didinium*, and *tintinnids* (Bussers and Jeuniaux 1974; Bussers 1976). Chitin was also reported in the ciliate *Didinium and Bursaria*, which are meanwhile grouped into the classes Litostomatea and Colpodea, respectively (Rieder 1973; Small and Lynn 1981; Lynn 2008). The loricae of heterotrich folliculinids contain 20 nm fibrils of β-chitin, which are embedded in a matrix of pigments, proteins, and mucopolysaccharides (Mulisch et al. 1983). The structure of the cyst walls of the heterotrichs *Stentor*, *Blepharisma*, *Climacostomum,* and *Fabrea* also contain chitin fibrils but in different arrangements (Repak and Anderson 1990). Furthermore, chitinous cysts have been observed in apostome ciliates. During its life cycle, *Hyalophysa chattoni*, a polymorphic symbiont of the estuarine grass shrimp *Palaemonetes pugio,* forms two fundamentally diverse cysts, the phoront (symbiotic) and the tomont (reproductive) cysts, which both contain chitin (Landers 1991).

The cyst walls are made of two layers. While the inner layer seems to be a chitinous coating produced in a non-exocytotic process, the outer layer does not contain carbohydrates but proteins (Landers 1991). Chitin has also been identified in the cyst walls of *Euplotes muscicola* (Greco et al. 1990).

Only a few studies reported chitin in heterotrophic flagellates. In two families of choanoflagellates, Salpingoecidae and Codonosigidae, chitin has been detected as a component of the theca, a secreted covering, which can also be composed of cellulose or mucopolysaccharides (Buck 1990). Our knowledge of chitin prevalence and function is also limited in diatoms. These protists belong to the eukaryotic lineage of stramenopiles, and their *CHS* genes fall into four different phylogenetic clades. Two diatom genera (Thalassiosira and Cyclotella) have been reported two synthesize long fibers of chitin that protrude from the theca through pores in the silica cell wall. The diatom *Thalassiosira pseudonana* possesses six *CHS* genes, which encode three types of chitin synthases suggesting multiple cellular roles of chitin-related processes (Durkin et al. 2009). Notably, the amount of transcripts increases when the cells start to divide after silicic acid and/or iron-depletion simulating environmental fluctuations known to affect the cell wall. It suggests that the production of chitin in the cell wall is enhanced when growth is no longer possible. This may mediate sinking out from the surface layer, which could be a survival mechanism in response to adverse conditions at the surface.

3.9 Chitin in Pathogenic Protists

Chitin is also a component of pathogenic protists causing serious health problems. *Entamoeba histolytica*, for instance, affects worldwide 50 million people and causes about 100,000 casualties per year (Pineda and Perdomo 2017). Thus, it is ranked as the third frequent reason for death by protozoan parasites after Malaria and trypanosomiasis. It causes diarrhea, dysentery, and hepatic liver abscess. The contagious and diagnostic form of *E. histolytica* is the quadrinucleate cyst. Chitin is a significant portion of the cyst wall of different *Entamoeba* species (Arroyo-Begovich et al. 1980), also supported by the fact that inhibitors of chitin synthesis prevent cyst formation by vegetative cells (Avron et al. 1982). Also the cyst wall of the reptilian parasite *E. invadensis* contains chitin fibrils as well as three groups of chitin-binding lectins, which are cross-linking with chitin fibrils (Frisardi et al. 2000 and van Dellen et al. 2006). A couple of *CHS* genes were identified in three different *Entamoeba* species, *E. histolytica, E. dispar* (another human pathogenic form) and *E. invadens* (Campos-Gongora et al. 2004). In vegetative cells of *Entamoeba* (trophozoites) *CHS-1* and *CHS-2* genes are not expressed, but considerable amounts of *CHS-1* and *CHS-2* mRNAs were detectable 4–8 h after initiation of cyst formation induced by glucose deprivation.

Moreover, the flagellated pathogens causing trichomoniasis, a urogenital infection of humans and cattle, produce chitinous coverings. Kneipp et al. (1998) demonstrated that the *Tritrichomonas vaginalis* and *T. foetus* parasites deposit chitin on the cell surface, as evidenced by decoration with different chitin-binding molecules. It was suggested that inhibitors of chitin synthesis could potentially prevent the infection. Because chitinases also appear to be involved in pathogenicity, they have been suggested as targets for therapeutic intervention as well (Loiseau et al. 2002).

Diplomonads are flagellates, of which many have a parasitic lifecycle. Among them, *Giardia lamblia* causes greater health problems, as it brings along giardiasis, a parasitosis of the human intestinal tract that causes diarrhea but is rarely precarious. The disease is transmitted by infective cysts, which are defecated by the parasitized host. The rigid cyst wall consists of chitin as an essential structural constituent (Ward et al. 1985). Illustrating the limitations of WGA-stainings, Ortega-Barria et al. (1990) has shown that WGA binding, which effectively inhibits growth, was not affected by chitinase treatment, suggesting that another *N*-acetylglucosamine containing component of the cyst wall was stained, and that chitin, which is yet present, may not even be accessible to WGA. Interestingly, combinations of silver and chitosan nanoparticles were reported to eradicate the parasite from stool and intestine (Said et al. 2012).

The life cycle of numerous protozoan parasites includes two phases: the first is the vegetative phase with cells that are called trophozoites. In this phase, the parasite can colonize the walls of the upper parts of the small intestine of the vertebrate host provoking different kinds of intestinal pathologies. The second phase is the cyst, which is resistant to harmful environmental conditions and highly contagious to the host. The cyst walls of *Entamoeba* and *Giardia* are made of a fibrillar matrix of polysaccharides and proteins, which form an outer layer of about 120–150 nm, as revealed by transmission electron microscopy (Chávez-Munguía et al. 2007). The polysaccharides are associated with different cyst wall proteins, some of which have carbohydrate-binding properties similar to lectins. The precise cyst wall structure varies between different species. In some parasites, the microfibrils form a single continuous layer on the surface of the cell membrane, in other it forms a double-layered covering with 1–3 pores (ostioles) surrounding the ectocyst and the endocyst (Chávez-Munguía et al. 2007). The structural characteristics of the cyst walls of *Entamoeba* and *Giardia* are summarized in Fig. 3.4.

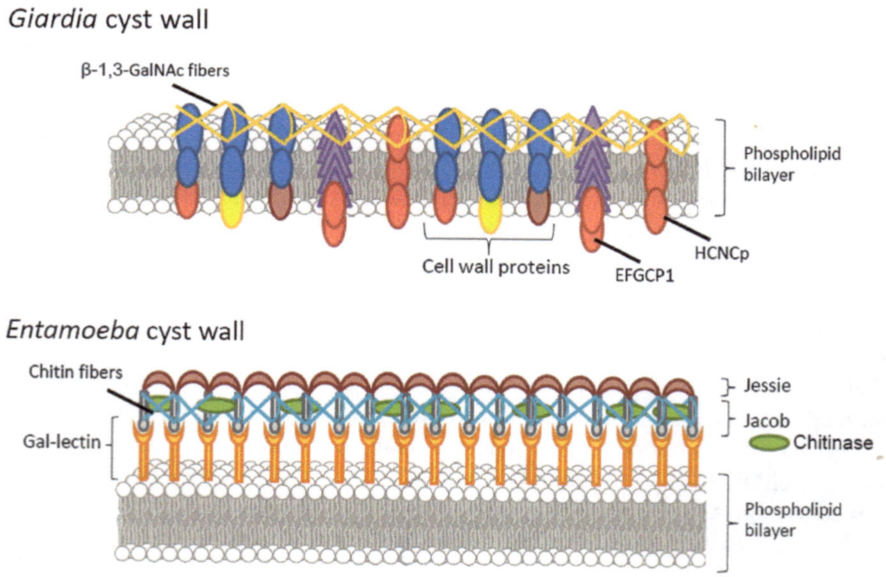

Fig. 3.4 Structure of cell and cyst walls in *Entamoeba* and *Giardia* parasites. During encystation, which is paralleled by decreasing uptake of glucose and oxygen, specific vesicles are formed and fuse with the plasma membrane to secrete components of the nascent cyst wall. The *Entamoeba* cyst wall is composed of chitin fibrils and three chitin-binding proteins (chitinase, Jessie, and Jacob). The *Giardia* cyst wall contains proteins including lectins that bind to curled fibrils of the β-1,3-GalNAc polymer, and the more recently discovered High Cysteine Non-variant Cyst proteins (HCNCp) and Epidermal Growth Factor (EGF)-like Cyst Proteins (EFGCP1) (Samuelson et al. 2013). Abbreviations: Gal, galactose; GalNAc, *N*-acetylgalactosamine; LR, leucine-rich; CR, cysteine-rich

3.10 Chitin Formation in Heterokont and Archaeplastidal Algae

Flagellated algae are polyphyletic and include organisms that appeared at various times during eukaryotic evolution. They can live either heterotrophic or phototrophic. In the latter case, they harbor chloroplasts, which allow them to profit from photosynthesis. In the case of Archaeplastida, which next to plants and glaucophytes (blue-green algae) comprise rhodophytes (red algae) and chlorophytes (green algae), the chloroplasts are derived from cyanobacteria that were obtained by primary endosymbiosis resulting in two plastidial membranes (Keeling 2010). Heterokont alga, ciliates, Apicomplexa, dinoflagellates, haptophytes, and cryptomonads have engulfed red algae, and euglenids engulfed green algae during evolution in a secondary endosymbiotic process. These processes usually result in three plastidial membranes. In addition, dinoflagellates have engulfed diatoms, haptophytes and cryptomonads resulting in tertiary endosymbiosis with varying numbers of plastidial membranes depending on how many membranes have been maintained during evolution. Dinoflagellates also internalized green algae by serial secondary endosymbiosis (Keeling 2010).

Most species of the Archaeplastida form cell walls that contain cellulose and glycoproteins. However, some members of the Chlorophyta have been shown to produce a chitinous cell wall (Muzzarelli 1977). In *Pithophora oedogonia*, chitin was detected with histochemical methods in cross-wall discs and smaller amounts in the longitudinal cell walls (Pearlmutter and Lembi 1978). In 2002, Kawasaki and colleagues reported chitin in unicellular green *Chlorella* algae that were infected with the chlorovirus CVK2 (Kawasaki et al. 2002). CVK2 is a large icosahedral virus with a double-stranded DNA genome, which infects specific *Chlorella* strains (Yamada et al. 2006). The genome of CVK2 includes a gene that encodes a GT2 family enzyme synthesizing hyaluronan and/or chitin, which is deposited on the surface of the *Chlorella* cell wall. Which polysaccharides are produced depends on genetic rearrangements in the *Chlorella* genome, resulting in functional genes encoding hyaluronan and/or chitin synthases (Mohammed Ali et al. 2005). Notably, chitin synthesis by *Chlorella* cells that were infected with slow-growing CVNF-1 viruses was even more efficient than in CVK2-infected algae. The developed *Chlorella*-virus system may be used as an environmentally safe procedure to synthesize useful material from light and CO_2 (Rakkhumkaew et al. 2018).

Recently, chitin was also reported in a calcified coralline red algae, *Clathromorphum compactum* (Rahman and Halfar 2014). *In C. compactum*, chitin serves as a template for nucleation to control biomineralization, and to increase skeletal strength.

Heterokonts (also known as stramenopiles) are a large infrakingdom, which includes bacillariophytes (diatoms), chrysophytes (golden algae), phaecophytes (brown algae), xanthophytes (yellow-green algae), some heterotrophic protists, as well as other associated organisms (Patterson 1989a, b). Chitin has been reported only in a few species of some of these classes, e.g., in chrysophytes. In the flagellate chrysophyte, *Poteriochromonas stipata*, helically ordered chitin fibrils were reported to be present in the wine-glass-formed lorica (Herth et al. 1977; Herth 1980; Sengbusch and Müller 1983). Also in diatoms, unicellular, yellow-brown colored algae, chitin biosynthesis appears to be required for the synthesis of the cell wall and extracellular spines. In particular, two diatom species, *Thalassiosira fluviatilis* (McLachlan et al. 1965) and *Cyclotella cryptica* (Herth 1978) have been studied in detail. In both species, β–chitin was detected in the extracellular spines using infrared spectrometry or X-ray diffraction analysis (Blackwell et al. 1967). The presence of chitin was further evidenced by chitinolytic degradation using transmission electron microscopy (Lindsay and Gooday 1985). In *T. fluviatilis*, chitinous extracellular thinner and thicker fibers originate from marginal and central pores in the silica valves, respectively. The thicker fibers form ropes that connect neighboring cells resulting in cell chains of variable length, which can levitate in the water (Aumeier and Menzel 2012).

Ultrastructural analysis suggests that the chitin fibers are formed at specialized membrane regions underneath each of the pores. The membranes at these regions are unusually thick in ultrathin cross-sections and have a high electron contrast, which suggests the presence of high amounts of transmembrane proteins including chitin synthases, which may secrete single chitin chains that then assemble into microfibrils on the extracellular site (Herth 1978). Genomic research performed in a variety of diatom genera led to the identification of hundreds of genes encoding putative chitin synthases, which fall into four phylogenetic clades (Durkin et al. 2009). Apart from *Thalassiosira spec.* and *Cyclotella spec.*, diatom species with *CHS* genes include *Skeletonema costatum*, *Chaetoceros socialis*, *Lithodesmium undulatum,* and *P. tricornutum.* Presumably, these *CHS* genes have different functions according to protein domain analysis and gene expression studies. In *Thalassiosira pseudonana*, six *CHS* genes encode three types of chitin synthases, and two of them are transcriptionally regulated depending on the nutritional conditions. Together, the results from this study suggest that the capability of chitin formation is widespread among diatoms and that chitin is associated not only with extracellular spines, but is also likely a component of the cell wall. The latter has been confirmed in a study by Brunner et al. (2009), who have shown that the cell walls of *T. pseudonana* contain a network-like chitinous scaffold. These biosilica-like scaffolds consist of interleaved 25 nm-fibers, which contain additional unidentified components. The chitin-skeleton may serve as a scaffold for biomineralization and mechanical stability. Similar results were obtained by Tesson et al. (2008), who showed by solid-state NMR that purified silica shells from *T. pseudonana* contain chitin next to proteins and lipids. Interestingly, chitin appears to be important for cell separation, resilience, and sedimentation. When Morin et al. (1986) analyzed the effect of the chitin synthesis inhibitors Polyoxin D in T. fluviatilis and C. cryptica, the chitin fiber-lacking cells grew at lower population density, and showed marked agglutination and higher sedimentation rates. Hence, the chitin fibers may have a function in regulating cellular adherence and buoyancy.

In addition to *CHS* genes, diatoms possess also genes encoding chitin-modifying enzymes such as chitinases or chitin-organizing proteins with chitin-binding domains (Durkin et al. 2009). One of the latter proteins discovered in *T. pseudonana* is p150, which has three potential *N*-glycosylation sites and three chitin-binding domains (Davis et al. 2005). Its expression is cell cycle regulated, and it is localized to the girdle band region. In copper-stressed cells, which have morphological abnormalities, it covers the elongated girdle band region. Hence, p150 may be involved in stabilizing the cells during the division process and environmental stress.

Two *CHS* genes of the *Thalassiosira* genome encode chitin synthases, which have an *N*-terminal myosin motor domain, as is was also reported for filamentous fungi and mollusks. This may indicate that these chitin synthases are involved in polarized secretion of chitin mediated by the interaction of the enzyme with the actin cytoskeleton.

3.11 Conclusions

Chitin serves different functions in most organisms capable of chitin biosynthesis. On the one side, it protects from mechanical stress and conveys resilience to the extracellular scaffold structure due to the polymers intrinsic tensile strength. On the other side, it is involved remodeling and regeneration of the coverings, and in morphogenesis defining the shape. Hence, chitin synthesis, deposition, modification, and degradation are important determinants of differentiation and morphogenesis. The latter accounts also for chitin oligosaccharides that act as symbiotic factors controlling the differentiation of root cells and nodulation in *Rhizobia*-legume symbiosis or arbuscular mycorrhizal symbiosis between plants and fungi. As chitin formation occurs in the life cycle of many fungi and protists that cause diseases in plants, mammals, and humans, which are not considered to produce chitin under healthy conditions, there is an increasing interest in the understanding of chitin synthesis and degradation, in order to design therapeutic drugs to cure infections. Furthermore, chitin oligosaccharides act as elicitors of innate and adaptive immune responses. In particular, it has been implicated to play an important role in allergic asthma mediated by TH2 cells. As the receptor that binds chitin oligomers have been identified, it might be possible to develop appropriate molecules that impair pattern recognition and hence help to stop the inflammatory processes.

References

Agatha S, Simon P (2012) On the nature of tintinnid loricae (Ciliophora: Spirotricha: Tintinnina): a histochemical, enzymatic, EDX, and high-resolution TEM study. Acta Protozool 51:1–19

Anno K, Otsuka K, Seno N (1974) A chitin sulfate-like polysaccharide from the test of the tunicate *Halocynthia roretzi*. Biochim Biophys Acta 362:215–219

Arcones I, Sacristan C, Roncero C (2016) Maintaining protein homeostasis: early and late endosomal dual recycling for the maintenance of intracellular pools of the plasma membrane protein Chs3. Mol Biol Cell 27:4021–4032

Arroyo-Begovich A, Cárabez-Trejo A (1982) Location of chitin in the cyst wall of *Entamoeba invadens* with colloidal gold tracers. J Parasitol 68:253–258

Arroyo-Begovich A, Carabez-Trejo A, Ruiz-Herrera J (1980) Identification of the structural component in the cyst wall of *Entamoeba invadens*. J Parasitol 66:735–741

Arroyo J, Farkas V, Sanz AB, Cabib E (2016) Strengthening the fungal cell wall through chitin-glucan cross-links: effects on morphogenesis and cell integrity. Cell Microbiol 18: 1239–1250

Atkinson EM, Long SR (1992) Homology of *Rhizobium meliloti* NodC to polysaccharide polymerizing enzymes. Mol Plant Microbe Interact 5:439–442

Atkinson EM, Palcic MM, Hindsgaul O, Long SR (1994) Biosynthesis of *Rhizobium meliloti* lipooligosaccharide Nod factors: NodA is required for an *N*-acyltransferase activity. Proc Natl Acad Sci USA 91:8418–8422

Aufauvre-Brown A, Mellado E, Gow NA, Holden DW (1997) *Aspergillus fumigatus* chsE: A gene related to chs3 of *Saccharomyces cerevisiae* and important for hyphal growth and conidiophore development but not pathogenicity. Fungal Genet Biol 21:141–152

Aumeier C, Menzel D (2012) Secretion in the diatoms. In: Vivanco JM, Baluska F (eds) Secretions and exudates in biological systems. Springer, Berlin, pp 221–250

Avron B, Deutsch RM, Mirelman D (1982) Chitin synthesis inhibitors prevent cyst formation by *Entamoeba trophozoites*. Biochem Biophysical Res Commun 108:815–821

Baev N, Endre G, Petrovics G, Banfalvi Z, Kondorosi A (1991) Six nodulation genes of nod box locus 4 in *Rhizobium meliloti* are involved in nodulation signal production: nodM codes for D-glucosamine synthetase. Mol Gen Genet 228(1–2):113–124

Baev N, Schultze M, Barlier I, Ha DC, Virelizier H, Kondorosi E, Kondorosi A (1992) *Rhizobium nodM* and nodN genes are common *nod* genes: *nodM* encodes functions for efficiency of Nod signal production and bacteroid maturation. J Bacteriol 174:7555–7565

Bartnicki-Garcia S (2006) Chitosomes: past, present and future. FEMS Yeast Res 6:957–965

Biancalana F, Kopprio GA, Lara RJ, Alonso C (2017) A protocol for the simultaneous identification of chitin-containing particles and their associated bacteria. Syst Appl Microbiol 40:314–320

Blackwell J, Parker K, Rudall K (1967) Chitin fibres of the diatoms *Thalassiosira fluviatilis* and *Cyclotella cryptica*. J Mol Biol 28:383–385

Bloemberg GV, Kamst E, Harteveld M, van der Drift KM, Haverkamp J, Thomas-Oates JE, Lugtenberg BJ, Spaink HP (1995) A central domain of *Rhizobium* NodE protein mediates host specificity by determining the hydrophobicity of fatty acyl moieties of nodulation factors. Mol Microbiol 16:1123–1136

Bloemberg GV, Thomas-Oates JE, Lugtenberg BJ, Spaink HP (1994) Nodulation protein NodL of Rhizobium leguminosarum *O*-acetylates lipo-oligosaccharides, chitin fragments and *N*-acetylglucosamine *in vitro*. Mol Microbiol 11:793–804

Bo M, Bavestrello G, Kurek D, Paasch S, Brunner E, Born R, Galli R, Stelling AL, Sivkov VN, Petrova OV, Vyalikh D, Kummer K, Molodtsov SL, Nowak D, Nowak J, Ehrlich H (2012) Isolation and identification of chitin in the black coral *Parantipathes larix* (Anthozoa: Cnidaria). Int J Biol Macromol 51(1–2):129–137

Bontemps C, Golfier G, Gris-Liebe C, Carrere S, Talini L, Boivin-Masson C (2005) Microarray-based detection and typing of the *Rhizobium* nodulation gene *nodC*: potential of DNA arrays to diagnose biological functions of interest. Appl Environ Microbiol 71:8042–8048

Bowen AR, Chen-Wu JL, Momany M, Young R, Szaniszlo PJ, Robbins PW (1992) Classification of fungal chitin synthases. Proc Nati Acad Sci USA 89:519–523

Bowman SM, Free SJ (2006) The structure and synthesis of the fungal cell wall. BioEssays 28:799–808

Briza P, Ellinger A, Winkler G, Breitenbach M (1988) Chemical composition of the yeast ascospore wall. The second outer layer consists of chitosan. J Biol Chem 263:11569–11574

Broughton WJ, Jabbouri S, Perret X (2000) Keys to symbiotic harmony. J Bacteriol 182:5641–5652

Brunke S, Mogavero S, Kasper L, Hube B (2016) Virulence factors in fungal pathogens of man. Curr Opin Microbiol 32:89–95

Brunner E, Richthammer P, Ehrlich H, Paasch S, Simon P, Ueberlein S, van Pée KH (2009) Chitin-based organic networks: an integral part of cell wall biosilica in the diatom *Thalassiosira pseudonana*. Angew Chem Int Ed Engl 48:9724–9727

Buck KR (1990) Choanomastigotes (choanoflagellates). In: Margulis L, Corliss JO, Melkonian M, Chapman DJ (eds) Handbook of the Protoctista: the structure, cultivation, habits and life histories of the eukaryotic microorganisms and their descendants exclusive of animals, plants and fungi. Jones and Bartlett Publishers, Boston, pp 194–199

Bussers JC, Jeuniaux C (1974) Recherche de la chitine dans les productions métaplasmatiques de quelques ciliés. Protistologica 10:43–46

Bussers JC, Voss-Foucart MF, Bouchez-Decloux N (1977) Ultrastructure and chemical composition of the lorica of *Folliculitis* products (Ciliata Heterotricha). Abstr Int Congr Protozool. 50:358

Bussers JC (1976) Structure et composition du kyste de résistance de 4 protozoaires ciliés. Protistologica 12:87–100

Calvo P, Fernandez-Aliseda MC, Garrido J, Torres A (2003) Ultrastructure, encystment and cyst wall composition of the resting cyst of the peritrich ciliate *Opisthonecta henneguyi*. J Eukaryot Microbiol 50:49–56

Campos-Gongora E, Ebert F, Willhoeft U, Said-Fernandez S, Tannich E (2004) Characterization of chitin synthases from *Entamoeba*. Protist 155:323–330

Carlson RW, Price NP, Stacey G (1994) The biosynthesis of rhizobial lipo-oligosaccharide nodulation signal molecules. Mol Plant Microbe Interact 7:684–695

Chávez-Munguía B, Omaña-Molina M, González-Lázaro M, González-Robles A, Cedillo-Rivera R, Bonilla P, Martínez-Palomo A (2007) Ultrastructure of cyst differentiation in parasitic protozoa. Parasitol Res 100:1169–1175

Chen SF, Juang YL, Chou WK, Lai JM, Huang CY, Kao CY, Wang FS (2009) Inferring a transcriptional regulatory network of the cytokinesis-related genes by network component analysis. BMC Syst Biol 3:110

Chin CF, Bennett AM, Ma WK, Hall MC, Yeong FM (2012) Dependence of Chs2 ER export on dephosphorylation by cytoplasmic Cdc14 ensures that septum formation follows mitosis. Mol Biol Cell 23:45–58

Choi WJ, Santos B, Duran A, Cabib E (1994) Are yeast chitin synthases regulated at the transcriptional or the posttranslational level? Mol Cell Biol 14:7685–7694

Choquer M, Boccara M, Goncalves IR, Soulie MC, Vidal-Cros A (2004) Survey of the *Botrytis cinerea* chitin synthase multigenic family through the analysis of six euascomycetes genomes. Eur J Biochem 271:2153–2164

Christodoulidou A, Briza P, Ellinger A, Bouriotis V (1999) Yeast ascospore wall assembly requires two chitin deacetylase isozymes. FEBS Lett 460:275–279

Chuang JS, Schekman RW (1996) Differential trafficking and timed localization of two chitin synthase proteins, Chs2p and Chs3p. J Cell Biol 135:597–610

Cid VJ, Durán A, del Rey F, Snyder MP, Nombela C, Sánchez M (1995) Molecular basis of cell integrity and morphogenesis in *Saccharomyces cerevisiae*. Microbiol Rev 59:345–386

Coluccio AE, Rodriguez RK, Kernan MJ, Neiman AM (2008) The yeast spore wall enables spores to survive passage through the digestive tract of *Drosophila*. PLoS One 3:e2873

Cos T, Ford RA, Trilla JA, Duran A, Cabib E, Roncero C (1998) Molecular analysis of Chs3p participation in chitin synthase III activity. Eur J Biochem 256:419–426

Davis AK, Hildebrand M, Palenik B (2005) A stress-induced protein associated with the girdle band region of the diatom *Thalassiosira pseudonana* (Bacillariophyta). J Phycol 41:577–589

De Hoff PL, Brill LM, Hirsch AM (2009) Plant lectins: the ties that bind in root symbiosis and plant defense. Mol Genet Genomics 282:1–15

de Jonge R, Thomma BP (2009) Fungal LysM effectors: extinguishers of host immunity? Trends Microbiol 17:151–157

Deakin WJ, Broughton WJ (2009) Symbiotic use of pathogenic strategies: rhizobial protein secretion systems. Nat Rev Microbiol 7:312–320

Debelle F, Rosenberg C, Denarie J (1992) The *Rhizobium*, *Bradyrhizobium*, and *Azorhizobium* NodC proteins are homologous to yeast chitin synthases. Mol Plant Microbe Interact 5: 443–446

Demont N, Debelle F, Aurelle H, Denarie J, Prome JC (1993) Role of the *Rhizobium meliloti nodF* and nodE genes in the biosynthesis of lipo-oligosaccharidic nodulation factors. J Biol Chem 268:20134–20142

Deringer VL, Englert U, Dronskowski R (2016) Nature, strength, and cooperativity of the hydrogen-bonding network in alpha-chitin. Biomacromol 17:996–1003

Dobert RC, Breil BT, Triplett EW (1994) DNA sequence of the common nodulation genes of *Bradyrhizobium elkanii* and their phylogenetic relationship to those of other nodulating bacteria. Mol Plant Microbe Interact 7:564–572

Dorfmueller HC, Ferenbach AT, Borodkin VS, van Aalten DM (2014) A structural and biochemical model of processive chitin synthesis. J Biol Chem 289:23020–23028

Durkin CA, Mock T, Armbrust EV (2009) Chitin in diatoms and its association with the cell wall. Eukaryot Cell 8:1038–1050

Ehrlich H, Maldonado M, Spindler KD, Eckert C, Hanke T, Born R, Goebel C, Simon P, Heinemann S, Worch H (2007) First evidence of chitin as a component of the skeletal fibers of marine sponges. Part I. *Verongidae* (demospongia: Porifera). J Exp Zool B Mol Dev Evo 308:347–356

El Gueddari NE, Rauchhaus U, Moerschbacher BM, Deising HB (2002) Developmentally regulated conversion of surface-exposed chitin to chitosan in cell walls of plant pathogenic fungi. New Phytol 156:103–112

Elieh Ali Komi D, Sharma L, Dela Cruz CS (2018) Chitin and Its effects on inflammatory and immune responses. Clin Rev Allergy Immunol 54:213–223

Ene IV, Walker LA, Schiavone M, Lee KK, Martin-Yken H, Dague E, Gow NA, Munro CA, Brown AJ (2015) Cell wall remodeling enzymes modulate fungal cell wall elasticity and osmotic stress resistance. MBio 6:e00986

Feldmann H, Aigle M, Aljinovic G, Andre B, Baclet M, Barthe C, Baur A, Becam A, Biteau N, Boles E (1994) Complete DNA sequence of yeast chromosome II. EMBO J 13:5795–5809

Feng J, Li Q, Hu HL, Chen XC, Hong GF (2003) Inactivation of the nod box distal half-site allows tetrameric *NodD* to activate *nodA* transcription in an inducer-independent manner. Nucleic Acids Res 31:3143–3156

Fernandes C, Gow NAR, Gonçalves T (2016) The importance of subclasses of chitin synthase enzymes with myosin-like domains for the fitness of fungi. Fungal Biol Rev 30:1–14

Firmin JL, Wilson KE, Carlson RW, Davies AE, Downie JA (1993) Resistance to nodulation of cv. Afghanistan peas is overcome by *nodX*, which mediates an *O*-acetylation of the *Rhizobium leguminosarum* lipo-oligosaccharide nodulation factor. Mol Microbiol 10:351–360

Fisher RF, Egelhoff TT, Mulligan JT, Long SR (1988) Specific binding of proteins from *Rhizobium meliloti* cell-free extracts containing NodD to DNA sequences upstream of inducible nodulation genes. Genes Dev 2:282–293

Free SJ (2013) Fungal cell wall organization and biosynthesis. Adv Genet 81:33–82

Frisardi M, Ghosh SK, Field J, Van Dellen K, Rogers R, Robbins P, Samuelson J (2000) The most abundant glycoprotein of amebic cyst walls (Jacob) is a lectin with five Cys-rich, chitin-binding domains. Infect Immun 68:4217–4224

Geelen D, Leyman B, Mergaert P, Klarskov K, Van Montagu M, Geremia R, Holsters M (1995) NodS is an S-adenosyl-L-methionine-dependent methyltransferase that methylates chitooligosaccharides deacetylated at the non-reducing end. Mol Microbiol 17:387–397

Geremia RA, Mergaert P, Geelen D, Van Montagu M, Holsters M (1994) The NodC protein of *Azorhizobium caulinodans* is an *N*-acetylglucosaminyltransferase. Proc Natl Acad Sci USA 91:2669–2673

Gharieb MM, El-Sabbagh SM, Shalaby MA, Darwesh OM (2015) Production of chitosan from different species of zygomycetes and its antimicrobial activity. Int J Sci Eng Res 6:123–130

Giraud-Guille M-M, Chanzy H, Vuong R (1990) Chitin crystals in arthropod cuticles revealed by diffraction contrast transmission electron microscopy. J Struct Biol 103:232–240

Goffeau A, Barrell BG, Bussey H, Davis R, Dujon B, Feldmann H, Galibert F, Hoheisel J, Jacq C, Johnston M (1996) Life with 6000 genes. Science 274:546–567

Gohlke S, Heine D, Schmitz HP, Merzendorfer H (2018) Septin-associated protein kinase Gin4 affects localization and phosphorylation of Chs4, the regulatory subunit of the Baker's yeast chitin synthase III complex. Fungal Genet Biol 117:11–20

Gohlke S, Muthukrishnan S, Merzendorfer H (2017) *In vitro* and *in vivo* studies on the structural organization of Chs3 from *Saccharomyces cerevisiae*. Int J Mol Sci 18 pii:E702

Goldberg WM (1978) Chemical changes accompanying maturation of the connective tissue skeletons of gorgonian and antipatharian corals. Marine Biol 49:203–210

Gonçalves IR, Brouillet S, Soulié MC, Gribaldo S, Sirven C, Charron N, Boccara M, Choquer M (2016) Genome-wide analyses of chitin synthases identify horizontal gene transfers towards bacteria and allow a robust and unifying classification into fungi. BMC Evol Biol 16:252

Gottfert M, Hitz S, Hennecke H (1990) Identification of *nodS* and *nodU*, two inducible genes inserted between the *Bradyrhizobium japonicum nodYABC* and *nodIJ* genes. Mol Plant Microbe Interact 3:308–316

Gow NA, Latge JP, Munro CA (2017) The fungal cell wall: structure, biosynthesis, and function. Microbiol Spectr 5. https://doi.org/10.1128/microbiolspec.funk-0035-2016

Gowri N, Sundara-Rajulu G, Aruchami M (1982) Presence of gamma-chitin in the peritrophic membrane of tunicates. In: Hirano S, Tokura S (eds) Second international conference on chitin and chitosan, Tottoni, Japan, 1982. J Eukaryot Microbiol, pp 77–81

Greco N, Bussers JC, Van Daele Y, Goffinet G (1990) Ultrastructural localization of chitin in the cystic wall of *Euplotes muscicola* Kahl (Ciliata, Hypotrichia). Eur J Protistol 26:75–80

Hegedus D, Erlandson M, Gillott C, Toprak U (2009) New insights into peritrophic matrix synthesis, architecture, and function. Ann Rev Entomol 54:285–302

Herrera-Estrella A, Chet I (1999) Chitinases in biological control. EXS 87:171–184

Herth W (1978) A special chitin-fibril-synthesizing apparatus in the centric diatom *Cyclotella*. Naturwissenschaften 65:260–261

Herth W (1980) Calcofluor white and Congo red inhibit chitin microfibril assembly of *Poterioochromonas*: evidence for a gap between polymerization and microfibril formation. J Cell Biol 87:442–450

Herth W, Kuppel A, Schnepf E (1977) Chitinous fibrils in the lorica of the flagellate chrysophyte *Poteriochromonas stipitata* (syn. *Ochromonas malhamensis*). J Cell Biol 73:311–321

Herth W, Zugenmaier P (1977) Ultrastructure of the chitin fibrils of the centric diatom *Cyclotella cryptica*. J Ultrastruct Res 61:230–239

Horiuchi H (2009) Functional diversity of chitin synthases of *Aspergillus nidulans* in hyphal growth, conidiophore development and septum formation. Med Mycol 47(Suppl 1):S47–S52

Jabbouri S, Fellay R, Talmont F, Kamalaprija P, Burger U, Relic B, Prome JC, Broughton WJ (1995) Involvement of nodS in N-methylation and nodU in 6-O-carbamoylation of *Rhizobium* sp. NGR234 nod factors. J Biol Chem 270:22968–22973

John M, Röhrig H, Schmidt J, Wieneke U, Schell J (1993) *Rhizobium* NodB protein involved in nodulation signal synthesis is a chitooligosaccharide deacetylase. Procd Natl Acad Sci USA 90:625–629

Jones T, Federspiel NA, Chibana H, Dungan J, Kalman S, Magee B, Newport G, Thorstenson YR, Agabian N, Magee P (2004) The diploid genome sequence of *Candida albicans*. Proc Natl Acad Sci USA 101:7329–7334

Kameda T, Miyazawa M, Ono H, Yoshida M (2005) Hydrogen bonding structure and stability of alpha-chitin studied by 13C solid-state NMR. Macromol Biosci 5:103–106

Kamst E, Pilling J, Raamsdonk LM, Lugtenberg BJ, Spaink HP (1997) *Rhizobium nodulation* protein NodC is an important determinant of chitin oligosaccharide chain length in Nod factor biosynthesis. J Bacteriol 179:2103–2108

Kamst E, van der Drift KM, Thomas-Oates JE, Lugtenberg BJ, Spaink HP (1995) Mass spectrometric analysis of chitin oligosaccharides produced by *Rhizobium* NodC protein in *Escherichia coli*. J Bacteriol 177:6282–6285

Kapaun E, Reisser W (1995) A chitin-like glycan in the cell wall of a *Chlorella* sp. (Chlorococcales, Chlorophyceae). Planta 197:577–582

Kawasaki T, Tanaka M, Fujie M, Usami S, Sakai K, Yamada T (2002) Chitin synthesis in chlorovirus CVK2-infected chlorella cells. Virology 302:123–131

Keeling PJ (2010) The endosymbiotic origin, diversification and fate of plastids. Philos Trans R Soc Lond B Biol Sci 365(1541):729–748

Kelly S, Radutoiu S, Stougaard J (2017) Legume LysM receptors mediate symbiotic and pathogenic signalling. Curr Opin Plant Biol 39:152–158

Kneipp LF, Andrade AF, de Souza W, Angluster J, Alviano CS, Travassos LR (1998) *Trichomonas vaginalis* and *Tritrichomonas foetus*: Expression of chitin at the cell surface. Exp Parasitol 89:195–204

Lam KK, Davey M, Sun B, Roth AF, Davis NG, Conibear E (2006) Palmitoylation by the DHHC protein Pfa4 regulates the ER exit of Chs3. J Cell Biol 174:19–25

Landers SC (1991) Secretion of the reproductive cyst wall by the apostome ciliate *Hyalophysa chattoni*. Eur J Parasitol 27:160–167

Latgé JP (2007) The cell wall: a carbohydrate armour for the fungal cell. Mol Microbiol 66:279–290

Lewin A, Cervantes E, Chee-Hoong W, Broughton WJ (1990) nodSU, two new nod genes of the broad host range *Rhizobium* strain NGR234 encode host-specific nodulation of the tropical tree *Leucaena leucocephala*. 3:317–326

Li M, Jiang C, Wang Q, Zhao Z, Jin Q, Xu JR, Liu H (2016) Evolution and functional insights of different ancestral orthologous clades of chitin synthase genes in the fungal tree of life. Front Plant Sci 7:37

Lindsay GJ, Gooday GW (1985) Action of chitinase on spines of the diatom *Thalassiosira fluviatilis*. Carbohydr Polym 5:131–140

Liu R, Xu C, Zhang Q, Wang S, Fang W (2017) Evolution of the chitin synthase gene family correlates with fungal morphogenesis and adaption to ecological niches. Sci Rep 7:44527

Loiseau PM, Bories C, Sanon A (2002) The chitinase system from *Trichomonas vaginalis* as a potential target for antimicrobial therapy of urogenital trichomoniasis. Biomed Pharmacother 56:503–510

Lopez-Lara IM, Geiger O (2001) The nodulation protein NodG shows the enzymatic activity of an 3-oxoacyl-acyl carrier protein reductase. Mol Plant Microbe Interact 14:349–357

Lynn D (2008) The ciliated protozoa: Characterization, classification and guide to the literature, 3rd edn. Springer, Dordrecht, p 605

Mandel MA, Galgiani JN, Kroken S, Orbach MJ (2006) *Coccidioides posadasii* contains single chitin synthase genes corresponding to classes I to VII. Fungal Genet Biol 43:775–788

Manning P, Erlandsen SL, Jarroll EL (1992) Carbohydrate and amino acid analyses of *Giardia muris* cysts. J Protozool 39:290–296

McLachlan J, McInnes A, Falk M (1965) Studies on the chitan (chitin: poly-*N*-acetylglucosamine) fibers of the diatom *Thalassiosira fluviatilis* Hustedt: I. Production and isolation of chitan fibers. Can J Bot 43:707–713

Mendoza L, Taylor JW, Ajello L (2002) The class mesomycetozoea: a heterogeneous group of microorganisms at the animal-fungal boundary. Ann Rev Microbiol 56:315–344

Merzendorfer H (2011) The cellular basis of chitin synthesis in fungi and insects: common principles and differences. Eur J Cell Biol 90:759–769

Meyer MF, Kreil G (1996) Cells expressing the DG42 gene from early *Xenopus* embryos synthesize hyaluronan. Proc Natl Acad Sci USA 93:4543–4547

Miya A, Albert P, Shinya T, Desaki Y, Ichimura K, Shirasu K, Narusaka Y, Kawakami N, Kaku H, Shibuya N (2007) CERK1, a LysM receptor kinase, is essential for chitin elicitor signaling in *Arabidopsis*. Proc Natl Acad Sci USA 104:19613–19618

Mohammed Ali AM, Kawasaki T, Yamada T (2005) Genetic rearrangements on the Chlorovirus genome that switch between hyaluronan synthesis and chitin synthesis. Virology 342:102–110

Morgan JL, Strumillo J, Zimmer J (2013) Crystallographic snapshot of cellulose synthesis and membrane translocation. Nature 493(7431):181–186

Morin LG, Smucker RA, Herth W (1986) Effects of two chitin synthesis inhibitors on *Thalassiosira fluviatilis* and *Cyclotella cryptica*. FEMS Microbiol Lett 37:263–268

Mulisch M, Harry O, Patterson D, Wyatt C (1986) Folliculinids (Ciliata: Heterotrichida) from Portaferry, Co., Down, including a new species of *Metafolliculina* Dons, 1924. Ir Nat J 22:1–7

Mulisch M, Hausmann K (1983) Lorica Construction in *Eufolliculina sp.* (Ciliophora, Heterotrichida). J Protozool 30:97–104

Mulisch M, Hausmann K (1989) Localization of chitin on ultrathin sections of cysts of two ciliated protozoa, *Blepharisma undulans* and *Pseudomicrothorax dubius*, using colloidal gold conjugated wheat germ agglutinin. Protoplasma 152:77–86

Mulisch M, Herth W, Zugenmaier P, Hausmann K (1983) Chitin fibrils in the lorica of the ciliate *Eufolliculina uhligi*: ultrastructure, extracellular assembly and experimental inhibition. Biol Cell 49:169–177

Munro CA, Gow NAR (2001) Chitin synthesis in human pathogenic fungi. Med Mycol 39:41–53

Muthukrishnan S, Merzendorfer H, Arakane Y, Kramer KJ (2012) Chitin Metabolism in Insects. In: Gilbert LI (ed) Insect Biochemistry and Molecular Biology. Elsevier, San Diego, pp 193–253

Muzzarelli RAA (1977) Chitin. Pergamon Press, Oxford

Muzzey D, Schwartz K, Weissman JS, Sherlock G (2013) Assembly of a phased diploid *Candida albicans* genome facilitates allele-specific measurements and provides a simple model for repeat and indel structure. Genome Biol 14:1

Nagahashi S, Sudoh M, Ono N, Sawada R, Yamaguchi E, Uchida Y, Mio T, Takagi M, Arisawa M, Yamada-Okabe H (1995) Characterization of chitin synthase 2 of *Saccharomyces cerevisiae*. Implication of two highly conserved domains as possible catalytic sites. J Biol Chem 270:13961–13967

Nakamura CV, Esteves MJ, Andrade AF, Alviano CS, de Souza W, Angluster J (1993) Chitin: a cell-surface component of *Phytomonas francai*. Parasitol Res 79:523–526

Nguyen TV, Wibberg D, Battenberg K, Blom J, Vanden Heuvel B, Berry AM, Kalinowski J, Pawlowski K (2016) An assemblage of *Frankia* Cluster II strains from California contains the canonical nod genes and also the sulfotransferase gene *nodH*. BMC Genom 17:796

Nierman WC, Pain A, Anderson MJ, Wortman JR, Kim HS, Arroyo J, Berriman M, Abe K, Archer DB, Bermejo C (2005) Genomic sequence of the pathogenic and allergenic filamentous fungus *Aspergillus fumigatus*. Nature 438:1151

Nino-Vega G, Carrero L, San-Blas G (2004) Isolation of the *CHS4* gene of *Paracoccidioides brasiliensis* and its accommodation in a new class of chitin synthases. Med Mycol 42:51–57

Odenbach D, Thines E, Anke H, Foster AJ (2009) The *Magnaporthe grisea* class VII chitin synthase is required for normal appressorial development and function. Mol Plant Pathol 10:81–94

Orlean P (2012) Architecture and biosynthesis of the *Saccharomyces cerevisiae* cell wall. Genetics 192:775–818

Ortega-Barria E, Ward HD, Evans JE, Pereira ME (1990) *N*-Acetyl-glucosamine is present in cysts and trophozoites of *Giardia lamblia* and serves as receptor for wheatgerm agglutinin. Mol Biochem Parasitol 43:151–165

Pacheco-Arjona JR, Ramirez-Prado JH (2014) Large-scale phylogenetic classification of fungal chitin synthases and identification of a putative cell-wall metabolism gene cluster in *Aspergillus* genomes. PLoS ONE 9:e104920

Patterson D (1989a) Stramenopiles: chromophytes from a protistan perspective. Chromophyte Algae Probl Perspect 357–379

Patterson D (1989b) Stramenopiles: chromophytes from a protistological perspective. In: Green JC, Leadbeater BSC, Diver WL (eds) The chromophyte algae: problems and perspectives. Clarendon Press, Oxford, pp 357–379

Pearlmutter NL, Lembi CA (1978) Localization of chitin in algal and fungal cell walls by light and electron microscopy. J Histochem Cytochem 26:782–791

Peck MC, Fisher RF, Long SR (2006) Diverse flavonoids stimulate NodD1 binding to nod gene promoters in *Sinorhizobium meliloti*. J Bacteriol 188:5417–5427

Peter MG (2005) Chitin and chitosan in fungi. Biopolymers 6:123–157

Peters W (1966) Chitin in tunicata. EXS 22:820–821

Peters W, Latka I (1986) Electron microscopic localization of chitin using colloidal gold labelled with wheat germ agglutinin. Histochem 84:155–160

Philippsen P, Kleine K, Pöhlmann R, Düsterhöft A, Hamberg K, Hegemann JH, Obermaier B, Urrestarazu L, Aert R, Albermann K (1997) The nucleotide sequence of *Saccharomyces cerevisiae* chromosome XIV and its evolutionary implications. Nature 387:93–98

Picken LE, Lotmar W (1950) Oriented protein in chitinous structures. Nature 165:599–600

Pineda E, Perdomo D (2017) *Entamoeba histolytica* under oxidative stress: What countermeasure mechanisms are in place? Cells 6:44

Poinsot V, Crook MB, Erdn S, Maillet F, Bascaules A, Ane JM (2016) New insights into Nod factor biosynthesis: Analyses of chitooligomers and lipo-chitooligomers of *Rhizobium* sp. IRBG74 mutants. Carbohydr Res 434:83–93

Pringle JR (1991) Staining of bud scars and other cell wall chitin with Calcofluor. Methods Enzymol 194:732–735

Quesada-Vincens D, Fellay R, Nasim T, Viprey V, Burger U, Prome JC, Broughton WJ, Jabbouri S (1997) *Rhizobium* sp. strain NGR234 NodZ protein is a fucosyltransferase. J Bacteriol 179:5087–5093

Quinto C, Wijfjes AH, Bloemberg GV, Blok-Tip L, Lopez-Lara IM, Lugtenberg BJ, Thomas-Oates JE, Spaink HP (1997) Bacterial nodulation protein NodZ is a chitin oligosaccharide fucosyltransferase which can also recognize related substrates of animal origin. Proc Natl Acad Sci USA 94:4336–4341

Rahman MA, Halfar J (2014) First evidence of chitin in calcified coralline algae: new insights into the calcification process of *Clathromorphum compactum*. Sci Rep 4:6162

Repak AJ, Anderson OR (1990) The fine structure of the encysting salt marsh heterotrich ciliate *Fabrea salina*. J Morphol 205:335–341

Rieder N (1973) Elektronenoptische und histochemische Untersuchungen an der Cystenhülle von *Didinium nasutum* OF Müller (Ciliata, Holotricha). Arch Protistenk 115:125–131

Rivilla R, Sutton JM, Downie JA (1995) *Rhizobium leguminosarum* NodT is related to a family of outer-membrane transport proteins that includes TolC, PrtF, CyaE and AprF. Gene 161:27–31

Roche P, Debelle F, Maillet F, Lerouge P, Faucher C, Truchet G, Denarie J, Prome JC (1991) Molecular basis of symbiotic host specificity in *Rhizobium meliloti*: *nodH* and *nodPQ* genes encode the sulfation of lipo-oligosaccharide signals. Cell 67:1131–1143

Rodpothong P, Sullivan JT, Songsrirote K, Sumpton D, Cheung KW, Thomas-Oates J, Radutoiu S, Stougaard J, Ronson CW (2009) Nodulation gene mutants of *Mesorhizobium loti* R7A-nodZ and nolL mutants have host-specific phenotypes on *Lotus* spp. Mol Plant Microbe Interact 22:1546–1554

Roncero C (2002) The genetic complexity of chitin synthesis in fungi. Curr Genet 41:367–378

Rudall KM, Kenchington W (1973) The chitin system. Biol Rev 48:597–633

Ruiz-Herrera J, Gonzalez-Prieto JM, Ruiz-Medrano R (2002) Evolution and phylogenetic relationships of chitin synthases from yeasts and fungi. FEMS Yeast Res 1:247–256

Ruiz-Herrera J, Ortiz-Castellanos L (2010) Analysis of the phylogenetic relationships and evolution of the cell walls from yeasts and fungi. FEMS Yeast Res 10:225–243

Sachs IB (1956) The chemical nature of the cyst membrane of *Pelomyxa illinoisensis*. Trans Am Microsc Soc 75:307–313

Sacristan C, Reyes A, Roncero C (2012) Neck compartmentalization as the molecular basis for the different endocytic behaviour of Chs3 during budding or hyperpolarized growth in yeast cells. Mol Microbiol 83:1124–1135

Saffo MB, Fultz S (1986) Chitin in the symbiotic protist *Nephromyces*. Can J Bot 64:1306–1310

Said DE, Elsamad LM, Gohar YM (2012) Validity of silver, chitosan, and curcumin nanoparticles as anti-*Giardia* agents. Parasitol Res 111:545–554

Samuelson J, Bushkin GG, Chatterjee A, Robbins PW (2013) Strategies to discover the structural components of cyst and oocyst walls. Eukaryot Cell 12:1578–1587

Sanchatjate S, Schekman R (2006) Chs5/6 complex: a multiprotein complex that interacts with and conveys chitin synthase III from the trans-Golgi network to the cell surface. Mol Biol Cell 17:4157–4166

Sannasi A, Hermann HR (1970) Chitin in the cephalochordata, *Branchisotoma floridae*. EXS 26:351–352

Sarita S, Sharma PK, Priefer UB, Prell J (2005) Direct amplification of rhizobial *nodC* sequences from soil total DNA and comparison to nodC diversity of root nodule isolates. FEMS Microbiol Ecol 54:1–11

Schermuly G, Markmann-Mulish U, Mulisch M (1996) *In vitro* studies of the pathway of chitin synthesis in the ciliated protozoon *Eufolliculina uhligi*. In: Domard A, Jeuniaux C, Muzzarelli RAA, Roberts G (eds) Advances in chitin science. Jaques Anrés, Lyon, pp 10–17

Scheu AK, Economou A, Hong GF, Ghelani S, Johnston AW, Downie JA (1992) Secretion of the *Rhizobium leguminosarum* nodulation protein NodO by haemolysin-type systems. Mol Microbiol 6:231–238

Schmid J, Sieber V, Rehm B (2015) Bacterial exopolysaccharides: Biosynthesis pathways and engineering strategies. Front Microbiol 6:496

Schuster M, Treitschke S, Kilaru S, Molloy J, Harmer NJ, Steinberg G (2012) Myosin-5, kinesin-1 and myosin-17 cooperate in secretion of fungal chitin synthase. EMBO J 31:214–227

Schwelm A, Fogelqvist J, Knaust A, Jülke S, Lilja T, Bonilla-Rosso G, Karlsson M, Shevchenko A, Dhandapani V, Choi SR (2015) The *Plasmodiophora brassicae* genome reveals insights in its life cycle and ancestry of chitin synthases. Sci Rep 5:11153

Seider K, Heyken A, Lüttich A, Miramón P, Hube B (2010) Interaction of pathogenic yeasts with phagocytes: survival, persistence and escape. Curr Opin Mircobiol 13:392–400

Semino CE, Allende ML (2000) Chitin oligosaccharides as candidate patterning agents in zebrafish embryogenesis. Int J Dev Biol 44:183–193

Semino CE, Robbins PW (1995) Synthesis of "Nod"-like chitin oligosaccharides by the *Xenopus* developmental protein DG42. Proc Natl Acad Sci U S A 92:3498–3501

Semino CE, Specht CA, Raimondi A, Robbins PW (1996) Homologs of the *Xenopus* developmental gene DG42 are present in zebrafish and mouse and are involved in the synthesis of Nod-like chitin oligosaccharides during early embryogenesis. Proc Natl Acad Sci USA 93:4548–4553

Pv Sengbusch, Müller U (1983) Distribution of glycoconjugates at algal cell surfaces as monitored by FITC-conjugated lectins. Studies on selected species from *Cyanophyta*, *Pyrrhophyta*, *Raphidophyta*, *Euglenophyta*, *Chromophyta*, and *Chlorophyta*. Protoplasma 114:103–113

Shillito B, Lechaire JP, Childress J, Gaill F (1997) Diffraction contrast imaging of extracellular matrix components using zero-loss filtering. J Struct Biol 120:85–92

Shinya T, Nakagawa T, Kaku H, Shibuya N (2015) Chitin-mediated plant–fungal interactions: catching, hiding and handshaking. Curr Opin Plant Biol 26:64–71

Shiro S, Kuranaga C, Yamamoto A, Sameshima-Saito R, Saeki Y (2016) Temperature-dependent expression of nodc and community structure of soybean-nodulating *Bradyrhizobia*. Microbes Environ 31:27–32

Small EB, Lynn DH (1981) A new macrosystem for the phylum *Ciliophora doflein*, 1901. Biosystems 14:387–401

Spaink HP, Carlson RW (1996) Regulation of plant morphogenesis by Lipo-Chitin oligosaccharides. Crit Rev Plant Sci 15:559–582

Spaink HP, Wijfjes AH, Lugtenberg BJ (1995) *Rhizobium* NodI and NodJ proteins play a role in the efficiency of secretion of lipochitin oligosaccharides. J Bacteriol 177:6276–6281

Stacey G, Luka S, Sanjuan J, Banfalvi Z, Nieuwkoop AJ, Chun JY, Forsberg LS, Carlson R (1994) *nodZ*, a unique host-specific nodulation gene, is involved in the fucosylation of the lipooligosaccharide nodulation signal of *Bradyrhizobium japonicum*. J Bacteriol 176:620–633

Stegemann H (1963) Protein (conchagen) and chitin in the supporting tissue of the cuttlefish. Hoppe Seylers Z Physiol Chem 331:269–279

Stern R (2017) Go fly a chitin: the mystery of chitin and chitinases in vertebrate tissues. Front Biosci (Landmark Ed) 22:580–595

Sudoh M, Nagahashi S, Doi M, Ohta A, Takagi M, Arisawa M (1993) Cloning of the chitin synthase 3 gene from *Candida albicans* and its expression during yeast-hyphal transition. Mol Gen Genet 241 (3–4):351-358

Sutton JM, Lea EJ, Downie JA (1994) The nodulation-signaling protein NodO from *Rhizobium leguminosarum* biovar viciae forms ion channels in membranes. Proc Natl Acad Sci USA 91:9990–9994

Tada R, Latge JP, Aimanianda V (2013) Undressing the fungal cell wall/cell membrane-the antifungal drug targets. Curr Pharm Des 19:3738–3747

Takeuchi O, Akira S (2010) Pattern recognition receptors and inflammation. Cell 140:805–820

Tang WJ, Fernandez J, Sohn JJ, Amemiya CT (2015) Chitin is endogenously produced in vertebrates. Curr Biol 25:897–900

Teh EM, Chai CC, Yeong FM (2009) Retention of Chs2p in the ER requires N-terminal CDK1-phosphorylation sites. Cell Cycle 8:2964–2974

Tesson B, Masse S, Laurent G, Maquet J, Livage J, Martin-Jézéquel V, Coradin T (2008) Contribution of multi-nuclear solid state NMR to the characterization of the *Thalassiosira pseudonana* diatom cell wall. Anal Bioanal Chem 390:1889–1898

Tilic E, Bartolomaeus T (2016) Structure, function and cell dynamics during chaetogenesis of abdominal uncini in *Sabellaria alveolata* (Sabellariidae, Annelida). Zoological Lett 2:1

Trilla JA, Duran A, Roncero C (1999) Chs7p, a new protein involved in the control of protein export from the endoplasmic reticulum that is specifically engaged in the regulation of chitin synthesis in *Saccharomyces cerevisiae*. J Cell Biol 145:1153–1163

Tsuizaki M, Takeshita N, Ohta A, Horiuchi H (2009) Myosin motor-like domain of the class VI chitin synthase CsmB is essential to its functions in *Aspergillus nidulans*. Biosci Biotechnol Biochem 73:1163–1167

van Dellen KL, Chatterjee A, Ratner DM, Magnelli PE, Cipollo JF, Steffen M, Robbins PW, Samuelson J (2006) Unique posttranslational modifications of chitin-binding lectins of *Entamoeba invadens* cyst walls. Eukaryot Cell 5:836–848

van het Hoog M, Rast TJ, Martchenko M, Grindle S, Dignard D, Hogues H, Cuomo C, Berriman M, Scherer S, Magee B (2007) Assembly of the *Candida albicans* genome into sixteen supercontigs aligned on the eight chromosomes. Genome Biol 8:R52

Varki A (1996) Does DG42 synthesize hyaluronan or chitin?: A controversy about oligosaccharides in vertebrate development. Proc Natl Acad Sci USA 93:4523–4525

Vazquez M, Santana O, Quinto C (1993) The NodL and NodJ proteins from *Rhizobium* and *Bradyrhizobium* strains are similar to capsular polysaccharide secretion proteins from gram-negative bacteria. Mol Microbiol 8:369–377

Wagner GP (1994) Evolution and multi-functionality of the chitin system. EXS 69:559–577

Wagner GP, Lo J, Laine R, Almeder M (1993) Chitin in the epidermal cuticle of a vertebrate (*Paralipophrys trigloides*, Blenniidae, Teleostei). EXS 49:317–319

Wang D, Yang S, Tang F, Zhu H (2012) Symbiosis specificity in the legume: rhizobial mutualism. Cell Microbiol 14:334–342

Ward HD, Alroy J, Lev BI, Keusch GT, Pereira ME (1985) Identification of chitin as a structural component of *Giardia cysts*. Infect Immun 49:629–634

Weiss IM, Schonitzer V (2006) The distribution of chitin in larval shells of the bivalve mollusk *Mytilus galloprovincialis*. J Struct Biol 153:264–277

Wood V, Gwilliam R, Rajandream MA, Lyne M, Lyne R, Stewart A, Sgouros J, Peat N, Hayles J, Baker S (2002) The genome sequence of *Schizosaccharomyces pombe*. Nature 415:871

Yabe T, Yamada-Okabe T, Nakajima T, Sudoh M, Arisawa M, Yamada-Okabe H (1998) Mutational analysis of chitin synthase 2 of *Saccharomyces cerevisiae*. Identification of additional amino acid residues involved in its catalytic activity. Eur J Biochem 258:941–947

Yamada T, Onimatsu H, Van Etten JL (2006) Chlorella viruses. Adv Virus Res 66:293–336

Yan H, Xie JB, Ji ZJ, Yuan N, Tian CF, Ji SK, Wu ZY, Zhong L, Chen WX, Du ZL, Wang ET, Chen WF (2017) Evolutionarily conserved *nodE*, *nodO*, T1SS, and hydrogenase system in Rhizobia of *Astragalus membranaceus* and *Caragana intermedia*. Front Microbiol 8:2282

Zakrzewski AC, Weigert A, Helm C, Adamski M, Adamska M, Bleidorn C, Raible F, Hausen H (2014) Early divergence, broad distribution, and high diversity of animal chitin synthases. Genome Biol Evol 6:316–325

Zhang X, Dong W, Sun J, Feng F, Deng Y, He Z, Oldroyd GE, Wang E (2015) The receptor kinase CERK1 has dual functions in symbiosis and immunity signalling. Plant J 81:258–267

Zhang Y, Foster JM, Nelson LS, Ma D, Carlow CK (2005) The chitin synthase genes *chs-1* and *chs-2* are essential for *C. elegans* development and responsible for chitin deposition in the eggshell and pharynx, respectively. Dev Biol 285:330–339

Zhu KY, Merzendorfer H, Zhang W, Zhang J, Muthukrishnan S (2016) Biosynthesis, turnover, and functions of chitin in insects. Annu Rev Entomol 61:177–196

Ziman M, Chuang JS, Schekman RW (1996) Chs1p and Chs3p, two proteins involved in chitin synthesis, populate a compartment of the *Saccharomyces cerevisiae* endocytic pathway. Mol Biol Cell 7:1909–1919

Zimoch L, Merzendorfer H (2002) Immunolocalization of chitin synthase in the tobacco hornworm. Cell Tissue Res 308:287–297

Chapter 4
Immune Responses of Mammals and Plants to Chitin-Containing Pathogens

Xi Jiang, Han Bao, Hans Merzendorfer and Qing Yang

Abstract Chitin-containing organisms, such as fungi and arthropods, use chitin as a structural component to protect themselves from harsh environmental conditions. Hosts such as mammals and plants, however, sense chitin to initiate innate and adaptive immunity and exclude chitin-containing organisms. A number of protein factors are then expressed, and several signaling pathways are triggered. In this chapter, we focus on the responses and signal transduction pathways that are activated in mammals and plants upon invasion by chitin-containing organisms. As host chitinases play important roles in the glycolytic processing of chitin, which is then recognized by pattern-recognition receptors, we also pay special attention to the chitinases that are involved in immune recognition.

Keywords Human chitinase · Chitin · Pattern-recognition receptors · Plant signal transduction pathways

X. Jiang · H. Bao · Q. Yang (✉)
School of Bioengineering, Dalian University of Technology, No. 2 Linggong Road, Dalian 116023, China
e-mail: qingyang@dlut.edu.cn

X. Jiang
e-mail: xijiang@mail.dlut.edu.cn

H. Bao
e-mail: baohan728@mail.dlut.edu.cn

H. Merzendorfer
Department of Chemistry and Biology – Molecular Biology, University of Siegen, 57076 Siegen, Germany
e-mail: hans.merzendorfer@uni-siegen.de

Q. Yang
State Laboratory of Biology for Plant Diseases and Insect Pests, Institute of Plant Protection at Chinese Academy of Agricultural Sciences, No. 2 Yuanmingyuan West Road, Beijing 100193, China

Q. Yang and T. Fukamizo (eds.), *Targeting Chitin-containing Organisms*, Advances in Experimental Medicine and Biology 1142,
https://doi.org/10.1007/978-981-13-7318-3_4

Abbreviations

Arg1	Arginase 1
CCL2	chemokine (C-C motif) ligand 2
CD11	cluster of differentiation 11
CXCL8	chemokine (C-X-C motif) ligand 8
ECD	ectodomain
EF-Tu	elongation factor thermo unstable
FIBCD1	fibrinogen C domain containing 1
Fizz1	found in inflammatory zone 1
flg22	flagelin 22
HDM	house dust mite
GlcNAc	2-acetamido-2-deoxy-D-glucopyranose
IL	interleukin
LRR	leucine-rich repeat
LTD4	leukotriene D4
MALT1	mucosa-associated lymphoid tissue lymphoma translocation protein 1
MAPK	mitogen-activated protein kinase
MPK3	mitogen-activated protein kinase 3
MyD88	myeloid differentiation factor 88
M2	alternatively activated macrophages
$(NAG)_n$	β-1,4-linked oligosaccharide of GlcNAc with a polymerization degree of n
NMR	nuclear magnetic resonance
NOD2	nucleotide-binding oligomerization domain-containing protein 2
PAMPs	pathogen-associated molecular patterns
PGD2	prostaglandin D2
PRRs	pattern-recognition receptors
RLCK	receptor-like cytoplasmic kinase
ROS	reactive oxygen species
TLR	toll-like receptor
TNF	tumor necrosis factor
TSLP	thymic stromal lymphopoietin
WRKY	transcription factor with a ~60-residue DNA-binding domain containing a highly conserved heptapeptide motif WRKYGQK

4.1 Introduction

Chitin-containing organisms, including fungi, nematodes, and arthropods, are potential pathogens of plants and mammals. The chitin of fungal cell walls and arthropod's exoskeletons plays a major role in defense responses of the host and represents an ideal target to investigate host–pathogen interactions for the development of new treatment strategies against invasive pathogen infections (Tada et al. 2013). Various receptors, cellular factors, and chitinases, the enzymes that degrade chitin, are thus produced by hosts, to sense and finally eliminate these pathogens (Di Rosa et al. 2016). For example, the expression levels of the acidic mammalian chitinase (AMCase) and chitotriosidase (CHIT1) increase dramatically in human upon pathogen invasion (Nair et al. 2005). Since mammals and plants do not contain endogenous chitin, the chitin in the cell walls of pathogens is recognized as an immune stimulator (Lee et al. 2008, 2011).

In humans, invasive fungal infections have evolved into an important healthcare issue, because especially in immunocompromised patients they often go ahead with very high mortality rates up to 95% and annually one million life-endangering infections (Brown et al. 2012). The fungal cell wall has also become important as a primary target to develop antifungal agents, vaccines, and immunotherapies, because it mostly consists of molecules that do not exist in humans (Gow et al. 2017; Tada et al. 2013). For instance, chitin synthesis is an ideal target, because chitin is required for structural integrity of the fungal cell. Several competitive inhibitors of the chitin synthase have been isolated from *Streptomycetes* spec., such as Nikkomycin Z or Polyoxins D, which are used as antifungal compounds (Bowman and Free 2006). There are other common antifungal drugs termed echinocandins such as caspofungin, anidulafungin, and micafungin that interfere with chitin levels by inhibiting β-1,3 glucan synthesis (Fernandes et al. 2016). The innate immune system senses fungal pathogens mostly by fungal-specific cell wall components (Latge 2007).

Mammalian pattern-recognition receptors (PRRs), including toll-like receptor 2 (TLR2) (Fuchs et al. 2018), dectin-1, mannose receptor, fibrinogen C domain containing 1 (FIBCD1), and the C-type lectin RegIIIγ, are able to bind chitin and induce the expression of various cytokines in innate immune cells to initiate immune responses. In lung tissue, chitin induces the expression of chemokine (C-C motif) ligand 2 (CCL2), interleukin-25 (IL-25), IL-33 and thymic stromal lymphopoietin (TLSP) by epithelial cells and of tumor necrosis factor (TNF), IL-12, and IL-18 by macrophages. In skin tissue, chitin induces chemokine (C-X-C motif) ligand 8 (CXCL8), IL-6 and TSLP expression in keratinocytes. In spleen cells, chitin increases the levels of interferon γ (IFN-γ) (Bourgeois et al. 2011) and IL-10. Mammalian chitinases are also involved in the immune response to chitin-containing pathogens, including chitin recognition, type 2 inflammation and pathogen exclusion.

Unlike mammals, every plant cell acts as a unit of the plant defense response in an uninterrupted arms race with pathogens. Plant PRRs, which are located at the plasma membrane, sense pathogens by recognizing chitin, peptidoglycan, and other conserved pathogen molecules known as pathogen-associated molecular patterns (PAMPs) (Dodds and Rathjen 2010). In *Arabidopsis*, *At*CERK1 and *At*LYK5 sense fungal chitin by forming receptor complexes (Cao et al. 2014), which then interact with intracellular LRR-receptor kinases to activate immune signaling pathways. In rice, *Os*CEBiP and *Os*CERK1 are the two major receptors forming a unique sandwich-type dimer that activates the chitin-related signal transduction pathway (Hayafune et al. 2014).

Humans and plants are the two predominantly studied organisms in terms of their defense mechanisms against pathogenic fungi and nematodes. In this chapter, we will provide a general description of the defense pathways, including recently discovered factors.

4.2 Human Defense Mechanism

4.2.1 Recognition of Chitin-Containing Pathogens

The host recognition of invading pathogens relies on PRRs, which recognize highly conserved invariable molecular patterns that are often essential for the survival of microorganisms (Bryant et al. 2015; Tschopp 2007). Chitin/chitosan are potential targets for recognition by the mammalian immune system, since mammals naturally lack these biopolymers (Chelsea et al. 2013). Pathogen recognition together with the release of proinflammatory cytokines induces the secretion of chitinases (e.g., CHIT1) from neutrophils and macrophages. AMCase and CHIT1 are expressed in humans to modify the size of chitin for PRR recognition. Several studies showed that chitin has size-dependent effects on innate and adaptive immune responses including the ability to recruit and activate immune cells and induce cytokine and chemokine production via a variety of cell surface receptors including TLR2, dectin-1, mannose receptor, FIBCD1 (Anders Schlosser et al. 2009), NKR-P1 (Tomas Semenuk et al. 2001), and RegIIIγ (Cash et al. 2006) (Fig. 4.1) have been identified as mammalian chitin-binding receptors (reviewed in (Komi et al. 2018)).

4.2.1.1 TLR2, Dectin-1, and Mannose Receptor

TLR2 (a receptor that binds various surface molecules of pathogens depending on the dimerization partner), dectin-1 (a β-glucan receptor that mediates T helper type 17 development and the subsequent recruitment of neutrophils) and mannose receptor (which binds to mannans and mannoproteins) participate in mediating immune responses to chitin (Dong et al. 2014; Dostert and Tschopp 2007). After binding to particulate chitin, dectin-1-induced signaling not only leads to the production of various cytokines and chemokines, including TNF, CXCL2, IL-2, IL-10, and IL-12, but also induces a respiratory burst and ligand uptake through phagocytosis (Da Silva et al. 2009). The signaling mediated by dectin-1 depends on the cytoplasmic part of dectin-1, which is constituted by a motif similar to the immunoreceptor tyrosine-based activation motif (ITAM). Once a ligand has bound to the extracellular part of dectin-1, the tyrosine of the ITAM-like motif becomes phosphorylated, and a pair of Syks are recruited. Syks are spleen tyrosine kinases that contain Src homology 2 domains (Becker et al. 2016).

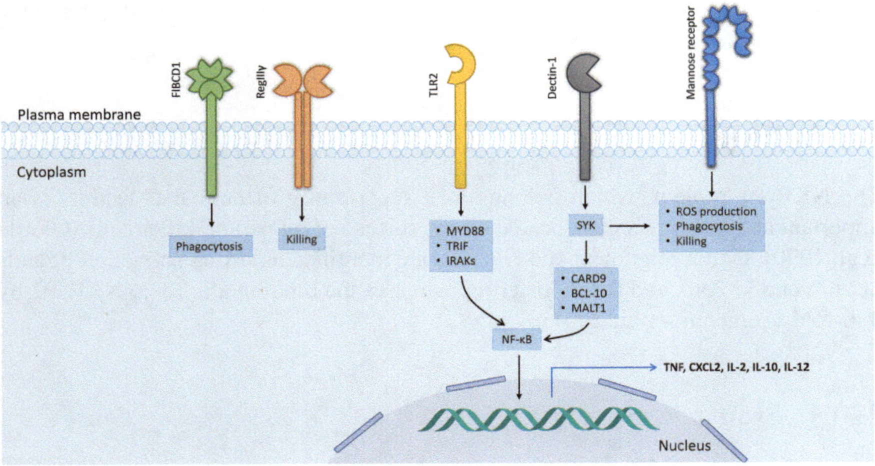

Fig. 4.1 Chitin-binding receptors from various superfamilies that sense chitin fragments through different pathways and activate signaling. FIBCD1 binds chitin and directs acetylated components for degradation in the cytoplasmic endosomes through endocytosis. Dectin-1 signaling induces respiratory burst and phagocytosis. TLR2 signaling mediated by the TIR domain induces IL-12 and TNF production. Mannose receptors participate in the endocytosis of chitinous materials by the formation of endosomes. Mannose receptors dissociate from ligands in a pH-dependent manner and recycle into the plasma membrane

In dendritic cells, caspase-recruitment domain protein 9 (CARD9) is involved in linking the activation of dectin-1/Syk to the activation of the adaptor protein Bcl-10/MALT1-dependent transcription factor NF-κB. The production of cytokines is then induced (Fig. 4.1). In chitin-induced macrophages, the IL-17A and TNF-α secretion are mediated by the TLR-2/MyD88 pathway and dectin-1/TLR2 pathway, respectively. The activation of dectin-1 in macrophages promotes phagocytosis, proinflammatory cytokine production, and reactive oxygen species (ROS) production.

4.2.1.2 FIBCD1

FIBCD1, a homotetrameric 55-kDa type II transmembrane protein, is expressed at high levels in the gastrointestinal tract (Thomsen et al. 2011). The ectodomain of FIBCD1, which comprises a coiled-coil region, a polycationic region, and C-terminal fibrinogen-related domain, is assembled into tetramers by disulfide linkages. According to functional analysis, the fibrinogen region is a calcium-dependent domain that binds acetylated components with high affinity. Screening for ligands revealed that FIBCD1 is a high-affinity receptor for chitin and chitin fragments. FIBCD1 facilitates the endocytosis of acetylated components and directs them for endosomal degradation. FIBCD1 may play an important role in controlling the exposure of chitin and chitin fragments, which is important for the immune defense against parasites and fungi and for modulating the immune response (Schlosser et al. 2009).

4.2.1.3 NKR-P1

The NKR-P1 protein, which belongs to a superfamily of animal C-lectins, is an important activating receptor located on the surface of rat natural killer cells (Giorda et al. 1990). Chitin oligomers and GlcNAc are identified as strong activation ligands in vitro and in vivo, and their clustering increases the binding affinity to NKR-P1 by 3–6-fold (Semeňuk et al. 2001).

4.2.1.4 RegIIIγ

RegIIIγ, which is expressed in intestinal epithelial cells, is a secretory C-type lectin. RegIIIγ binds intestinal bacteria but lacks the complement recruitment domains present in other microbe-binding mammalian C-type lectins. RegIIIγ expression is triggered by increased microbial epithelial contact at mucosal surfaces. RegIIIγ and its human counterpart, HIP/PAP, are direct-acting antimicrobial proteins that bind their bacterial targets through interactions with chitin and peptidoglycan.

4.2.2 Immune Responses

Various molecular signaling cascades are triggered to alter the cytokine profiles and cellular phenotypes after chitin is sensed by innate immune cells as a PAMP through specific membrane-bound receptors (Klauser et al. 2013). Here, we describe the immune responses in lung epithelial cells, macrophages, keratinocytes, and spleen cells after sensing chitin (Fig. 4.2).

4.2.2.1 Lung Epithelial Cells

Chitin exposure increases the expression of CCL2, IL-25, IL-33 and TLSP in lung epithelial cells to induce type 2 innate lymphoid cells (Van Dyken et al. 2014) (ILC2) to secrete cytokines IL-5 and IL-13, which are essential for the accumulation of eosinophils and alternatively activated macrophages. When chitin-bearing nematode parasites such as *Strongyloides venezuelensis* infect the lung, the type II alveolar epithelial (ATII) cells, which repair the injured alveolus by differentiating into alveolar epithelial type I cells (Roy et al. 2012), express IL-33 to induce ILC2 proliferation and the subsequent production of IL-5 and IL-13. Furthermore, mast

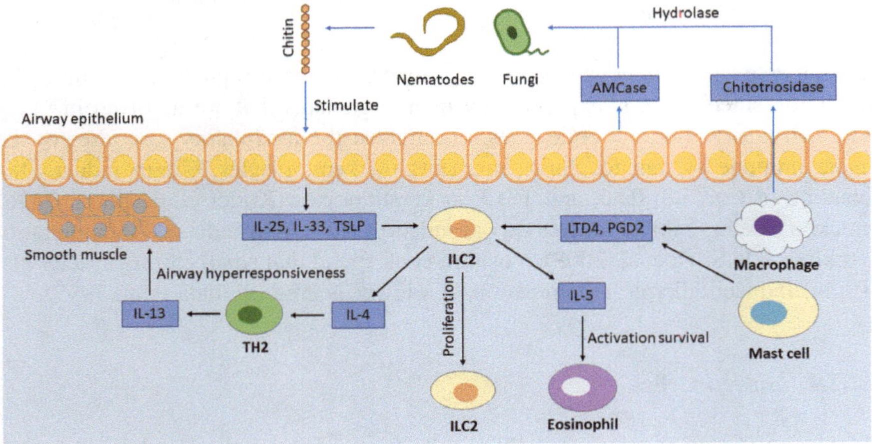

Fig. 4.2 Immune response against infection by chitin-containing nematodes and fungi. Infection by chitin-containing nematodes results in the release of IL-33, IL-25, and TSLP from the airway epithelium, which activates ILC2 cells. The ILC2 cells produce the Th2 cytokines IL-4, IL-5, and IL-13, which activate Th2 cells and eosinophils. Macrophages and mast cells contribute to further ILC2 activation by producing PGD2 and LTD4

cells and macrophages release PGD2 and LTD4 and stimulate additional ILC2 cells. IL-5 stimulates eosinophil activation and survival, whereas IL-13 induces airway hyperresponsiveness and, in concert with IL-9, promotes mucus production (Lund et al. 2013).

4.2.2.2 Macrophages

The intranasal administration of chitin particles induces the expression of cytokines, including TNF-α, IL-12, and IL-18, in alveolar macrophages. Upon chitin exposure, macrophages express Arg1, Ym1, Fizz1, mannose receptor, IL-10, the chemokines CCL17 and CCL24, and the eosinophil chemoattractant leukotriene B4 (Satoh et al. 2010).

Chitin also mediates alternative macrophage activation and enhances T cell function, NK cell activity and IFN-γ production (Cuesta et al. 2003). As shown in the study by Reese et al. a large number of alternatively activated macrophages (M2), namely, $CD11b^+$, $CD11c^-$, and $Gr1^-$ macrophages, are present in the lungs and peritoneum on the 9th day after infection by *Nippostrongylus brasiliensis*. Thus, chitin exposure during parasite infection induces the accumulation of alternatively activated macrophages (Reese et al. 2007). Chitin exposure can induce M2 polarization in vivo; however, macrophages fail to acquire an M2 phenotype in vitro. According to Roy et al. CCL2, which is secreted by $CD326^+$ airway epithelial cells in response to chitin, is a key factor involved in the alternative activation of macrophages and allergic inflammation in vivo (Roy et al. 2012).

4.2.2.3 Keratinocytes

Keratinocyte–chitin interactions may be important in the regulation of epidermal immunity, since chitin is expressed by microorganisms that are involved in many skin allergies. Koller and colleagues investigated the modulatory responses of the innate immune system to chitin fragments on keratinocytes. Chitin induces the secretion of CXCL8, IL-6, and TSLP in keratinocytes (Koller et al. 2011). TLR2 knockout cells exhibit decreased production of chitin-induced inflammatory cytokines (Da Silva et al. 2008). Yoshimi et al. found that small chitin particles (1–10 μm) induced allergic inflammation to a lower degree (Shibata et al. 1997).

4.2.2.4 Spleen Cells

Chitin was shown to stimulate spleen cells from ragweed-sensitized mice to produce IFN-γ and IL-10 but not IL-4 or IL-5, resulting in a shift in T helper cell responses in favor of Th1 responses. Chitin-treated mice exhibited significantly reduced IgE levels and inhibited inflammation of the peribronchial tissue, perivascular tissue and total lung (Shibata et al. 2000). As shown in the study by Wagener et al. chitin derived from *Candida albicans* is recognized by the mannose receptors NOD2 and TLR9, which participate in mediating an anti-inflammatory response through the secretion of cytokine IL-10 (Wagener et al. 2014).

4.2.3 AMCase and CHIT1

Both AMCase and CHIT1 are chitinases that belong to the glycosyl hydrolase family 18 (GH18). GH18 chitinases are expressed in a wide range of organisms from bacteria to humans. Chitin is the only documented substrate for these enzymes and is widely expressed in lower life forms. Although no endogenous substrate for these true chitinases has been identified in mammals, recent studies suggest that they play significant roles in inflammation (Mack et al. 2015), tissue injury responses, and the pathogenesis of human diseases (Lee et al. 2011).

AMCase is produced by lung epithelial cells, macrophages, and eosinophils and is distinguished from other chitinases by several unique features, including a low optimum pH and a preference for the β-anomer of the substrate (Boot et al. 2001). The crystal structure of AMCase reveals three second shell amino acid residues (His208, His269, and Arg145) that modulate the highly conserved chitinase active site and establish the low optimum pH of this enzyme (Bussink et al. 2008) (Olland et al. 2009). The indole ring of Try218 exhibits an unfavorable interaction with the terminal hydroxyl group of alpha-pentose but not with the terminal hydroxyl group of beta-pentose, substantiating the preference of the enzyme for the beta-anomer of the substrate. The exact physiological role of AMCase has not yet been clarified. AMCase is expressed at very high levels in patients with asthma and is induced by a

Th2-specific IL-13-mediated pathway in a chitin or ovalbumin (OVA) allergy model (Elias 2004; Shen et al. 2015; Shuhui et al. 2009). Knock-in mice with an inactive AMCase enzyme show enhanced pathological type 2 immune responses to inhaled house dust mites (HDM) (Kim et al. 2015). AMCase-deficient mice show defects in type 2 immunity against chitin-containing gastrointestinal nematodes (Vannella et al. 2016).

CHIT1, which is produced by mature monocyte-derived macrophages, Gaucher's cells, lung epithelial cells and lung macrophages, play an important role in the innate immune response to pathogens (Hollak et al. 1994; Aerts and Hollak 1997; Malaguarnera 2006). Two enzymatically active isoforms of CHIT1 (50 kDa and 39 kDa) and an alternatively spliced 40 kDa variant have been identified. The 50-kDa isoform is the predominantly secreted isoform, whereas the 39-kDa and 40-kDa isoforms are expressed and stored in intracellular lysosomes and lysosome-related organelles (LROs) (Boot et al. 1995). The biological significance of this chitinase as a component of lysosomes and LROs in disease pathogenesis is not completely understood. The degradation of fungal chitin by CHIT1 promotes Th2 cell accumulation and aggravates the disease in a pulmonary mycosis model. CHIT1-deficient mice are resistant to Th2-mediated fungal disease and survive much longer than the wild-type mice (Wiesner et al. 2015). Elevated levels of this enzyme have also been observed in subjects with a variety of other diseases, including infections (fungal and bacterial infections and malaria), chronic inflammation, liver diseases, and neurodegenerative diseases (Kzhyshkowska et al. 2007; Malaguarnera et al. 2006). More details of human chitinases structures, inhibitors and the roles in diseases are summarized in Chap. 11.

4.3 Plant Defense Mechanism

4.3.1 Immune Response

Chitin in the fungal cell wall is recognized by plant PRRs as a PAMP. The addition of fungal chitin to plant cell cultures results in cell membrane depolarization, medium alkalinization, and a rapid accumulation of phosphatidic acid (PA) and diacylglycerol pyrophosphate (DGPP), the phosphorylated derivative of PA (Felix et al. 1999; van der Luit et al. 2000). After PRRs bind chitin, plants produce and secrete antimicrobial effectors, e.g., chitinases or chitinase-like proteins (Sanchez-Vallet et al. 2015; Shinya et al. 2015). Chitinases catalyze the hydrolysis of chitin to produce chitin oligomers that trigger plant defense responses. These responses include Ca^{2+} spiking, ROS production, activation of the MAPK cascade, upregulation of defense gene expression, callose deposition, and molecular flux via plasmodesmata (Medzhitov 2007; Spoel and Dong 2012; Wan et al. 2008; Faulkner et al. 2013; Luna et al. 2011).

4.3.2 PRRs

Plant PRRs are ligand-binding receptors located on the cell surface. Receptor-like kinases (RLKs) are an important type of PRR. RLKs contain three domains, i.e., an extracellular domain for ligand recognition, a transmembrane domain and an intracellular kinase domain for triggering signal transduction. Receptor-like proteins (RLPs) are another important type of PRR that, in contrast to RLKs, contain a cytoplasmic C-terminal tail rather than a kinase domain at its C-terminus (Dodds and Rathjen 2010).

The perception of ligands such as chitin and flg22 is initiated by the formation of PRR complexes. Each PRR complex contains at least one RLK to activate signaling pathways. Plant PRRs with lysin motif (LysM)-containing ectodomains also form receptor complexes to recognize GlcNAc-containing ligands, such as fungal chitin, bacterial peptidoglycan (PGN), bacterial nodulation factors, and symbiotic fungal mycorrhiza factors (Zipfel and Oldroyd 2017).

Although the pathway is conserved, the composition and mechanism of PPRs vary among different species (Fig. 4.3). Here, we compare the PRRs from two extensively studied model organisms, *Arabidopsis* and rice.

4.3.2.1 PRRs in *Arabidopsis*

In *Arabidopsis thaliana*, two independent receptor complexes are induced by chitin. One complex consists of the LysM receptor kinase *At*CERK1 (chitin elicitor receptor kinase 1) and the *At*LYK5 (lysine motif receptor kinase 5) (or the LYK5 paralogue LYK4) (Cao et al. 2014). The other complex consists of *At*LYM2 (lysine motif domain-containing glycosylphosphatidylinositol-anchored protein 2) and an unknown PRR (Fig. 4.3) (Faulkner et al. 2013).

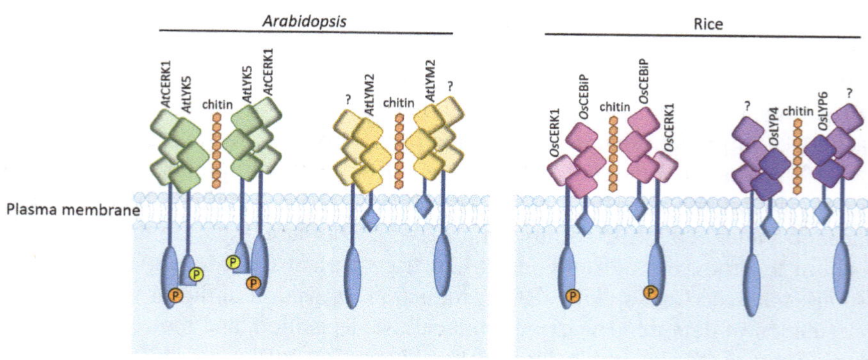

Fig. 4.3 LysM PRR complexes involved in chitin recognition. In plant cells, PRRs are located at the plasma membrane, forming complexes to sense chitin. In *Arabidopsis* (left), the LysM-receptor *At*CERK1 recognizes chitin with *At*LYK5; however, *At*LYM2 is another receptor to sense chitin by an unknown mechanism. In rice (right), *Os*CERK1 and *Os*CEBiP form a sandwich homodimer in the presence of $(GlcNAc)_8$. *Os*LYP4 and *Os*LYP6 are also involved in sensing chitin via an unknown mechanism

*At*CERK1

*At*CERK1 plays a key role in the responses to chitin oligosaccharides. These responses include MAPK activation, ROS generation, and the expression of defense-related genes (Miya et al. 2007; Petutschnig et al. 2010; Wan et al. 2008). The binding of $(NAG)_8$ induces the dimerization of the ectodomain of *At*CERK1 and subsequently activates the phosphorylation of the kinase domain of *At*CERK1. The phosphorylation of this kinase domain is inhibited by $(NAG)_4$ and $(NAG)_5$, contributing to the negative regulation of *At*CERK1. *At*CERK1 binds with high affinity to chitin, because chitin might simultaneously bind multiple *At*CERK1 receptors. Although the extracellular part of *At*CERK1 consists of three LysM domains that pack tightly against each other and form a spherical structure, only the second LysM domain is capable of chitin-binding (Liu et al. 2012b).

*At*LYK5

*At*LYK5, an RLK, binds to chitin with higher affinity than does *At*CERK1. *At*LYK5 exists as a homodimer in vivo but the homodimerization is independent of chitin. However, it is critical for chitin-induced homodimerization and autophosphorylation of *At*CERK1. The phosphorylated *At*CERK1 phosphorylates *At*LYK5 and then the phosphorylated *At*LYK5 is internalized into the late endocytic compartments. Endocytosis of *At*LYK5 not only strictly regulates signaling but also facilitates the assembly of receptor complexes with newly synthesized *At*LYK5 (Fig. 4.4) (Erwig et al. 2017). Due to the lack of essential catalytic residues, the kinase domain of *At*LYK5 is enzymatically inactive yet still being capable of interacting with *At*CERK1 (Erwig et al. 2017).

According to phylogenetic analysis, *At*LYK5 and its paralogue *At*LYK4 are located on the same branch. As functional redundancy between *At*LYK5 and *At*LYK4 has been observed in chitin-related responses, *At*LYK4 is also proposed to be associated with the formation of the chitin-sensing receptor complex (Cao et al. 2014).

*At*LYM2

*At*LYMs are *Arabidopsis* homologs of the rice receptor-like protein *Os*CEBiP. Among all the three *At*LYMs, *At*LYM2 exhibits the highest chitin binding affinity (Shinya et al. 2012). *At*LYM2 mediates a chitin-induced reduction in molecular flux via plasmodesmata (Faulkner et al. 2013), but it is not associated with AtCERK1-mediated defense responses, which indicates that there may exist two independent pathways responding to chitin in *Arabidopsis* (Faulkner et al. 2013; Shinya et al. 2012).

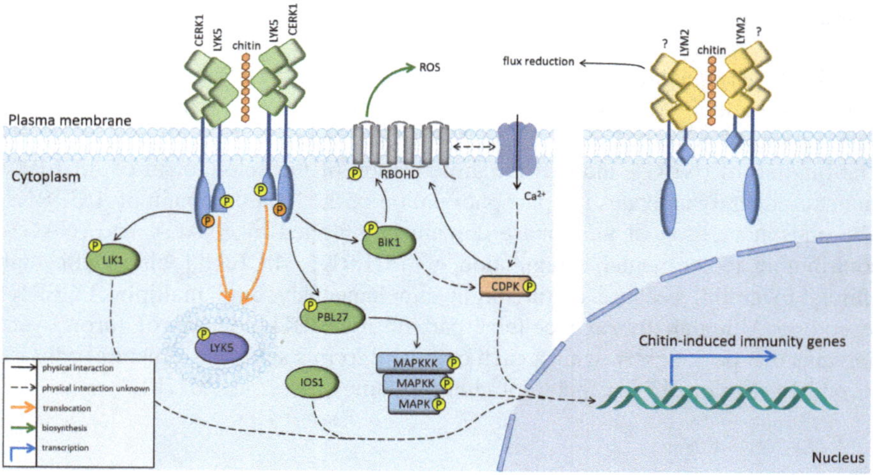

Fig. 4.4 The signaling pathway induced by chitin in *Arabidopsis*. Chitin induces CERK1-ECD dimerization and then activates its kinase domain's phosphorylation. CERK1 phosphorylates LYK5, and then the latter is then sorted into intracellular vesicles. LYM2 limits the molecular flux via plasmodesmata induced by chitin. Intracellular LRR-receptor kinases IOS1 and LIK1 also regulate chitin responses. Apart from phosphorylating cytoplasmic kinases PBL27 which, in turn, activates the MAPK cascade, CERK1 also activates BIK1 for ROS bursting

4.3.2.2 PRRs in Rice

The two major receptor complexes in rice include a complex composed of *Os*CEBiP (chitin elicitor binding protein) and *Os*CERK1, and a complex composed of *Os*LYP4, *Os*LYP6 (lysine motif containing proteins) and a third unknown PRR which may function like *Os*CEBiP (Fig. 4.3).

*Os*CEBiP and *Os*CERK1

*Os*CEBiP is the first chitin receptor identified in rice. It is an RLP anchored to the plasma membrane via a glycophosphatidylinositol (GPI) (Kaku et al. 2006) and binds chitin with high affinity (Shinya et al. 2012).

As *Os*CEBiP lacks the intracellular kinase domain, another type of PRR is needed for intracellular signal transduction. The RLK function is provided by *Os*CERK1, containing an additionally intracellular Ser/Thr kinase domain, *Os*CERK1 has been reported to activate chitin elicitor signaling by Kouzai et al. (2014). Unlike its *Arabidopsis* homologue, *Os*CERK1 contains only one extracellular LysM domain and does not demonstrate any binding affinity to chitin (Shimizu et al. 2010).

Both, *Os*CEBiP and *Os*CERK1, are capable of forming homo- or hetero-oligomers. NMR spectroscopy and molecular modeling studies demonstrated that

*Os*CEBiP and *Os*CERK1 form a unique sandwich-type dimer in the presence of $(NAG)_8$. This dimerization enables the central LysM domain in the *Os*CEBiP ectodomain to recognize chitin oligomers. In a modeled structure of the sandwich-type dimer, the central LysM motifs from the two *Os*CEBiPs interact with each other by binding one molecular $(NAG)_8$ from opposite directions. According to Hayafune et al. the *N*-acetyl group of $(NAG)_8$ is essential because deacetylated ligands can bind only with *Os*CEBiP monomers (Hayafune et al. 2014). As there are only three sugar rings bound to one *Os*CEBiP, a "sliding model" is proposed. In this model, two CEBiP-ECDs slide along a long chitin chain for optimal interaction with $(NAG)_8$ (Liu et al. 2016).

*Os*CERK1 also has been recognized to be necessary for the perception of short-chain chitin oligomers in arbuscular mycorrhizal signaling (Carotenuto et al. 2017) This finding suggests dual function of CERK1 in both plant immunity and mycorrhizal symbiosis, which brings new insight into the evolutionary interconnections of plant defense systems and plant-fungal symbiosis.

*Os*LYP4 and *Os*LYP6

*Os*LYP4 and *Os*LYP6 are lysine motif containing proteins with two characteristic LysM domains and a GPI anchor signal sequence. Similar to *Os*CEBiP, they are located at the plasma membrane, and bind peptidoglycans and chitin. Silencing of *Os*LYP4 and *Os*LYP6 specifically impaired peptidoglycan- or chitin-induced defense responses, suggesting their involvement in rice's susceptibility towards bacteria and fungi. In this regard, *Os*LYP4 and *Os*LYP6 are dual-function PRRs in rice innate immunity. (Kouzai et al. 2014; Liu et al. 2012a).

4.3.2.3 PRRs in Other Species

In other species, LysM-containing proteins are also involved in chitin sensing. Silencing *Nb*Cerk1 in *Nicotiana benthamiana* abolishes the ROS burst (Gimenez-Ibanez et al. 2009). In the instance that *Ta*CERK1 or *Ta*CEBiP is silenced in wheat, a mutant of the normally nonpathogenic fungus *Mycosphaerella graminicola*, *Mg3*LysM, colonizes leaf (Lee et al. 2014). Likewise, *Lotus japonicus* and *Medicago truncatula* possess very similar LysM-containing receptors, and mutants of the RLKs *Lj*LYS6, *Mt*LYK9, and RLP *Mt*LYR4 are unable to respond to chitin (Bozsoki et al. 2017).

4.3.3 Relevant Signal Transduction Pathways Involved in Chitin Perception

The pathways responsible for chitin perception vary in different species, but the downstream pathway may overlap with those triggered by some other PAMPs, such as flagellin or EF-Tu. For example, the regulatory receptor kinase malectin-like LRR-receptor kinase IOS1 in *Arabidopsis* is a positive regulator of the response to flagellin, and IOS1 also regulates chitin response (Yeh et al. 2016). The response to chitin and flg22 are negatively regulated by the LRR-receptor kinase LIK1, which is directly phosphorylated by *At*CERK1 (Le et al. 2014).

However, some genes also selectively respond to chitin oligosaccharides, such as MPK3, WRKY22, WRKY33, and WRKY53 (Wan et al. 2004, 2008).

4.3.3.1 Chitin-Induced Signal Transduction in *Arabidopsis*

*At*PBL27

In *Arabidopsis*, *At*CERK1 can phosphorylate the cytoplasmic kinase PBL27. The phosphorylated PBL27 is required for downstream signaling in consequence of chitin exposure. This downstream signaling includes MPK3/6 activation and callose deposition (Shinya et al. 2014). PBL27 positively regulates MAPKKK5 to specifically mediate chitin-induced MAPK activation.

*At*BIK1

Additionally, BIK1 (botrytis-induced kinase1) is an essential downstream element of pattern triggered immune (PTI) responses. It can interact with many PRRs, such as CERK1, FLS2, and ERF1 et al. The phosphorylated BIK1 for ROS generation may derive from a pathway dependent on chitin-induced *At*CERK1 activation (Fig. 4.4) (Laluk et al. 2011; Zhang et al. 2010).

4.3.3.2 Chitin-Induced Signal Transduction in Rice

*Os*RLCK185

*Os*RLCK185, the orthologue of *Arabidopsis* PBL27, belongs to the rice RLCK VII family. It is targeted by the *Xanthomonas oryzae* type III effector *Xoo*1488 that interacts with *Os*CERK1 to suppress chitin-induced defense responses. After chitin treatment, *Os*CERK1 phosphorylates *Os*RLCK185 and then the latter is released from the receptor complex to promote MAPK activation. To meet the requirement of chitin-induced MAPK activation and defense gene expression, *Os*RLCK185 acts upstream of the MAPK cascade (Fig. 4.5) (Yamaguchi et al. 2013).

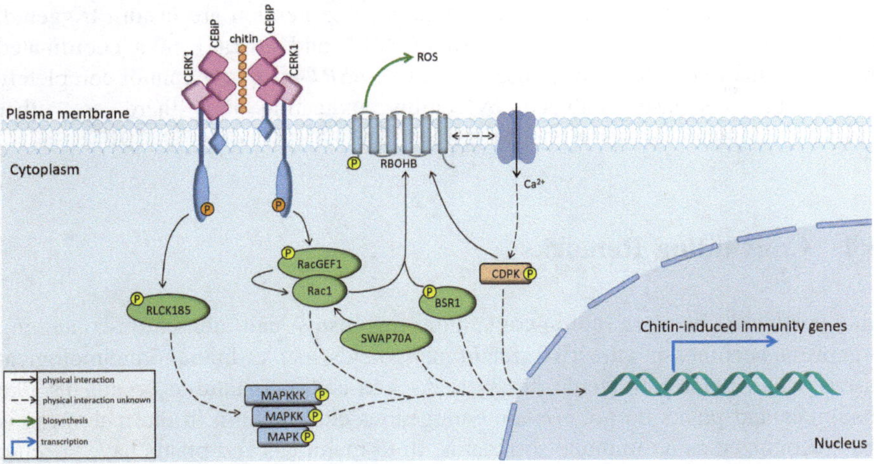

Fig. 4.5 The signaling pathway induced by chitin in rice. Two major PRRs, CEBiP and CERK1, spot chitin in rice. Dimerization of CERK1 and CEBiP activates immune signaling by means of a unique sandwich-type dimerization induced by chitin. CERK1 activates RacGEF1 by phosphorylation, and then RacGEF1 triggers the activation of Rac1. Rac1 interacts with RBOHB to activate ROS burst, and also activates the MAPK cascade. SWAP70A is the other GEF to activate Rac1. BSR1 participates in ROS production as well. CERK1 also phosphorylates RLCK185 to activate the MAPK cascade. BSR1 in conjunction with RLCK185 and Rac1 may function in a coordinated manner in the rice defense response

*Os*Rac1

*Os*Rac1 (Rac-like GTP-binding protein 1, a small GTPase from rice) is an important immune signaling element that can be activated by two GEFs (guanine nucleotide exchange factors), *Os*SWAP70A and *Os*RacGEF1. The GEF *Os*SWAP70A contains a Dbl (diffuse B-cell lymphoma)-homology domain, playing a role in ROS production and chitin-induced defense gene expression (Fig. 4.5) (Yamaguchi et al. 2012).

Downstream of *Os*CERK1/*Os*CEBiP-mediated signaling, *Os*Rac1 and *Os*RacGEF1 (guanine nucleotide exchange factor) jointly adjust chitin-induced immunity. Chitin activates *Os*CERK1 to phosphorylate *Os*RacGEF1, which consequently activates *Os*Rac1 (Akamatsu et al. 2013). In summary, the interaction of *Os*RacGEF1/*Os*Rac1 module with the *Os*CERK1/*Os*CEBiP complex, is a vital process for the hosts' resistance to fungal pathogens (Fig. 4.5) (Akamatsu et al. 2013).

*Os*BSR1

*Os*BSR1 (broad-spectrum resistance 1) is a receptor-like cytoplasmic kinase. In response to fungal invasion, *Os*BSR1 participates in a cytoplasmic phosphorylation-mediated signaling cascade to increase the hosts' susceptibility.

ROS burst and elevation of defense-related gene expression are in turn triggered. *Os*BSR1 may function along with *Os*RLCK185 and *Os*Racl in a coordinated manner in the rice defense response. However, *BSR1*-knockout cannot completely abolish the responses induced by chitin, indicating that there are other function-redundant factors produced in rice (Fig. 4.5) (Kanda et al. 2017).

4.4 Concluding Remarks

The cross-talk between chitin-containing organisms and non-chitin-containing organisms becomes an attractive area in many fields such as human immunological diseases and plant pathogenic diseases. As non-chitin-containing organisms like mammals and plants do not contain endogenous chitin, chitin in the pathogens is thus recognized as an immune stimulator. Both mammals and plants have specific PRRs to sense chitin. Cytoplasmic factors are activated once PRRs bind to chitin and then downstream immune responses will be set in motion.

References

Aerts JM, Hollak CE (1997) Plasma and metabolic abnormalities in Gaucher's disease. Bailliere's clinical haematology. 10(4):691–709

Akamatsu A, Wong HL, Fujiwara M, Okuda J, Nishide K, Uno K, Imai K, Umemura K, Kawasaki T, Kawano Y et al (2013) An OsCEBiP/OsCERK1-OsRacGEF1-OsRac1 module is an essential early component of chitin-induced rice immunity. Cell Host Microbe 13(4):465–476. https://doi.org/10.1016/j.chom.2013.03.007

Schlosser A, Thomsen T, Moeller JB, Nielsen O, Tornøe I, Mollenhauer J, Moestrup SK, Holmskov U (2009) Characterization of FIBCD1 as an acetyl group-binding receptor that binds chitin. J Immunol 183:3800–3809

Becker KL, Aimanianda V, Wang X, Gresnigt MS, Ammerdorffer A, Jacobs CW, Gazendam RP, Joosten LAB, Netea MG, Latge JP et al (2016) Aspergillus cell wall chitin induces anti- and proinflammatory cytokines in Human PBMCs via the Fc-gamma Receptor/Syk/PI3 K Pathway. Mbio 7(3). https://doi.org/10.1128/mBio:01823-15

Boot RG, Blommaart EF, Swart E, Ghauharali-van der Vlugt K, Bijl N, Moe C, Place A, Aerts JM (2001) Identification of a novel acidic mammalian chitinase distinct from chitotriosidase. J Biol Chem 276(9):6770–6778. https://doi.org/10.1074/jbc.M009886200

Boot RG, Renkema GH, Strijland A, van Zonneveld AJ, Aerts JM (1995) Cloning of a cDNA encoding chitotriosidase, a human chitinase produced by macrophages. J Biol Chem 270 (44):26252–26256

Bourgeois C, Majer O, Frohner IE, Lesiak-Markowicz I, Hildering K-S, Glaser W, Stockinger S, Decker T, Akira S, Mueller M et al (2011) Conventional dendritic cells mount a Type I IFN response against Candida spp. requiring novel phagosomal TLR7-mediated IFN-beta signaling. J Immunol 186(5):3104–3112. https://doi.org/10.4049/jimmunol.1002599

Bowman SM, Free SJ (2006) The structure and synthesis of the fungal cell wall. BioEssays 28(8):799–808. https://doi.org/10.1002/bies.20441

Bozsoki Z, Cheng J, Feng F, Gysel K, Vinther M, Andersen KR, Oldroyd G, Blaise M, Radutoiu S, Stougaard J (2017) Receptor-mediated chitin perception in legume roots is functionally separable from Nod factor perception. Proc Natl Acad Sci USA 114(38):E8118–E8127. https://doi.org/10.1073/pnas.1706795114

Brown GD, Denning DW, Gow NAR, Levitz SM, Netea MG, White TC (2012) Hidden killers: human fungal infections. Sci Transl Med 4(165). https://doi.org/10.1126/scitranslmed.3004404

Bryant CE, Orr S, Ferguson B, Symmons MF, Boyle JP, Monie TP (2015) International Union of Basic and Clinical Pharmacology. XCVI. Pattern recognition receptors in health and disease. Pharmacol Rev 67(2):462–504. https://doi.org/10.1124/pr.114.009928

Bussink AP, Vreede J, Aerts JM, Boot RG (2008) A single histidine residue modulates enzymatic activity in acidic mammalian chitinase. FEBS Lett 582(6):931–935. https://doi.org/10.1016/j.febslet.2008.02.032

Cao Y, Liang Y, Tanaka K, Nguyen CT, Jedrzejczak RP, Joachimiak A, Stacey G (2014) The kinase LYK5 is a major chitin receptor in Arabidopsis and forms a chitin-induced complex with related kinase CERK1. Elife. 3. https://doi.org/10.7554/elife.03766

Carotenuto G, Chabaud M, Miyata K, Capozzi M, Takeda N, Kaku H, Shibuya N, Nakagawa T, Barker DG, Genre A (2017) The rice LysM receptor-like kinase OsCERK1 is required for the perception of short-chain chitin oligomers in arbuscular mycorrhizal signaling. New Phytol 214 (4):1440–1446. https://doi.org/10.1111/nph.14539

Cash HL, Whitham CV, Behrendt CL, Hooper LV (2006) Symbiotic Bacteria direct expression of an intestinal bactericidal lectin. Science 313(5790):1126–1130

Chelsea L, Bueter CAS, Stuart M, Levitz (2013) Innate sensing of chitin and chitosan. PLoS Pathog 9(1):e1003080

Cuesta A, Esteban MÁ, Meseguer J (2003) In vitro effect of chitin particles on the innate cellular immune system of gilthead seabream (Sparus aurata L.). Fish Shellfish Immunol 15(1):1–11

Da Silva CA, Chalouni C, Williams A, Hartl D, Lee CG, Elias JA (2009) Chitin is a size-dependent regulator of macrophage TNF and IL-10 production. J Immunol 182(6):3573–3582. https://doi.org/10.4049/jimmunol.0802113

Da Silva CA, Hartl D, Liu W, Lee CG, Elias JA (2008) TLR-2 and IL-17A in chitin-induced macrophage activation and acute inflammation. J Immunol 181(6):4279–4286. https://doi.org/10.4049/jimmunol.181.6.4279

Di Rosa M, Distefano G, Zorena K, Malaguarnera L (2016) Chitinases and immunity: ancestral molecules with new functions. Immunobiol 221(3):399–411. https://doi.org/10.1016/j.imbio.2015.11.014

Dodds PN, Rathjen JP (2010) Plant immunity: towards an integrated view of plant-pathogen interactions. Nat Rev Genet 11(8):539–548. https://doi.org/10.1038/nrg2812

Dong B, Li D, Li R, Chen SC-A, Liu W, Liu W, Chen L, Chen Y, Zhang X, Tong Z (2014) A chitin-like component on sclerotic cells of Fonsecaea pedrosoi inhibits dectin-1-mediated murine Th17 development by masking β-glucans. PLoS ONE 9(12):e114113

Dostert C, Tschopp J (2007) DEteCTINg fungal pathogens. Nat Immunol 8(1):17

Elias JA (2004) Acidic mammalian chitinase in Asthmatic Th2 inflammation and IL-13 pathway activation. Science 304:1678–1682

Erwig J, Ghareeb H, Kopischke M, Hacke R, Matei A, Petutschnig E, Lipka V (2017) Chitin-induced and chitin elicitor receptor kinase1 (CERK1) phosphorylation-dependent endocytosis of Arabidopsis thaliana lysin motif-containing receptor-like kinase5 (LYK5). New Phytol 215(1):382–396. https://doi.org/10.1111/nph.14592

Faulkner C, Petutschnig E, Benitez-Alfonso Y, Beck M, Robatzek S, Lipka V, Maule AJ (2013) LYM2-dependent chitin perception limits molecular flux via plasmodesmata. Proc Natl Acad Sci USA 110(22):9166–9170. https://doi.org/10.1073/pnas.1203458110

Felix G, Duran JD, Volko S, Boller T (1999) Plants have a sensitive perception system for the most conserved domain of bacterial flagellin. Plant J 18(3):265–276. https://doi.org/10.1046/j.1365-313X.1999.00265.x

Fernandes C, Gow NAR, Goncalves T (2016) The importance of subclasses of chitin synthase enzymes with myosin-like domains for the fitness of fungi. Fungal Biol Rev 30(1):1–14. https://doi.org/10.1016/j.fbr.2016.03.002

Fuchs K, Cardona Gloria Y, Wolz O-O, Herster F, Sharma L, Dillen CA, Taumer C, Dickhofer S, Bittner Z, Dang T-M et al (2018) The fungal ligand chitin directly binds TLR2 and triggers inflammation dependent on oligomer size. EMBO reports 19(12). https://doi.org/10.15252/embr.201846065

Gimenez-Ibanez S, Hann DR, Ntoukakls V, Petutschnig E, Lipka V, Rathjen JP (2009) AvrPtoB targets the LysM receptor kinase CERK1 to promote bacterial virulence on plants. Curr Biol 19 (5):423–429. https://doi.org/10.1016/j.cub.2009.01.054

Giorda R, Rudert WA, Vavassori C, Chambers WH, Hiserodt JC, Trucco M (1990) NKR-P1, a signal transduction molecule on natural killer cells. Science 249(4974):1298–1300

Gow NAR, Latge J-P, Munro CA (2017) The fungal cell wall: structure, biosynthesis, and function. Microbiol Spectr 5(3). https://doi.org/10.1128/microbiolspec.funk-0035-2016

Hayafune M, Berisio R, Marchetti R, Silipo A, Kayama M, Desaki Y, Arima S, Squeglia F, Ruggiero A, Tokuyasu K et al (2014) Chitin-induced activation of immune signaling by the rice receptor CEBiP relies on a unique sandwich-type dimerization. Proc Natl Acad Sci USA 111(3):E404–E413. https://doi.org/10.1073/pnas.1312099111

Hollak C, van Weely S, Van Oers M, Aerts J (1994) Marked elevation of plasma chitotriosidase activity. A novel hallmark of Gaucher disease. J Clinic Investig 93(3):1288–1292

Kaku H, Nishizawa Y, Ishii-Minami N, Akimoto-Tomiyama C, Dohmae N, Takio K, Minami E, Shibuya N (2006) Plant cells recognize chitin fragments for defense signaling through a plasma membrane receptor. Proc Natl Acad Sci USA 103(29):11086–11091. https://doi.org/10.1073/pnas.0508882103

Kanda Y, Yokotani N, Maeda S, Nishizawa Y, Kamakura T, Mori M (2017) The receptor-like cytoplasmic kinase BSR1 mediates chitin-induced defense signaling in rice cells. Biosci Biotech Biochem 81(8):1497–1502

Kim LK, Morita R, Kobayashi Y, Eisenbarth SC, Lee CG, Elias J, Eynon EE, Flavell RA (2015) AMCase is a crucial regulator of type 2 immune responses to inhaled house dust mites. Proc Natl Acad Sci USA 112(22):E2891–2899. https://doi.org/10.1073/pnas.1507393112

Klauser D, Flury P, Boller T, Bartels S (2013) Several MAMPs, including chitin fragments, enhance at Pep-triggered oxidative burst independently of wounding. Plant Signal Behav 8(9): e25346

Koller B, Müller-Wiefel AS, Rupec R, Korting HC, Ruzicka T (2011) Chitin modulates innate immune responses of keratinocytes. PLoS ONE 6(2):e16594

Komi DEA, Sharma L, Dela Cruz CS (2018) Chitin and its effects on inflammatory and immune responses. Clin Rev Allergy Immunol 54(2):213–223. https://doi.org/10.1007/s12016-017-8600-0

Kouzai Y, Mochizuki S, Nakajima K, Desaki Y, Hayafune M, Miyazaki H, Yokotani N, Ozawa K, Minami E, Kaku H et al (2014) Targeted Gene Disruption of OsCERK1 Reveals Its Indispensable Role in Chitin Perception and Involvement in the Peptidoglycan Response and Immunity in Rice. Mol Plant-Microbe Interact 27(9):975–982. https://doi.org/10.1094/mpmi-03-14-0068-r

Kzhyshkowska JG, Goerdt A, Sergij (2007) Human chitinases and chitinase-like proteins as indicators for inflammation and cancer. Biomark Insights 2:117727190700200023

Laluk K, Luo H, Chai M, Dhawan R, Lai Z, Mengiste T (2011) Biochemical and genetic requirements for function of the immune response regulator botrytis-induced kinase1 in plant growth, ethylene signaling, and PAMP-triggered immunity in arabidopsis. Plant Cell 23(8): 2831–2849. https://doi.org/10.1105/tpc.111.087122

Latge J-P (2007) The cell wall: a carbohydrate armour for the fungal cell. Molec Microbiol 66 (2):279–290. https://doi.org/10.1111/j.1365-2958.2007.05872.x

Le MH, Cao Y, Zhang X-C, Stacey G (2014) LIK1, A CERK1-interacting kinase, regulates plant immune responses in arabidopsis. Plos One 9(7). https://doi.org/10.1371/journal.pone.0102245

Lee CG, Da Silva CA, Dela Cruz CS, Ahangari F, Ma B, Kang M-J, He C-H, Takyar S, Elias JA (2011) Role of chitin and chitinase/chitinase-like proteins in inflammation, tissue remodeling, and injury. Ann Rev Physiol 73:479–501

Lee CG, Da Silva CA, Lee J-Y, Hartl D, Elias JA (2008) Chitin regulation of immune responses: an old molecule with new roles. Curr Opinion Immunol 20(6):684–689

Lee W-S, Rudd JJ, Hammond-Kosack KE, Kanyuka K (2014) Mycosphaerella graminicola LysM effector-mediated stealth pathogenesis subverts recognition through both CERK1 and CEBiP homologues in wheat. Molec Plant-Microbe Interact 27(3):236–243. https://doi.org/10.1094/mpmi-07-13-0201-r

Liu B, Li J-F, Ao Y, Qu J, Li Z, Su J, Zhang Y, Liu J, Feng D, Qi K et al (2012a) Lysin Motif-containing Proteins LYP4 and LYP6 play dual roles in peptidoglycan and chitin perception in rice innate immunity. Plant Cell 24(8):3406–3419. https://doi.org/10.1105/tpc.112.102475

Liu T, Liu Z, Song C, Hu Y, Han Z, She J, Fan F, Wang J, Jin C, Chang J et al (2012b) Chitin-induced dimerization activates a plant immune receptor. Science 336(6085):1160–1164. https://doi.org/10.1126/science.1218867

Liu S, Wang J, Han Z, Gong X, Zhang H, Chai J (2016) Molecular mechanism for fungal cell wall recognition by rice chitin receptor OsCEBiP. Structure 24(7):1192–1200. https://doi.org/10.1016/j.str.2016.04.014

Luna E, Pastor V, Robert J, Flors V, Mauch-Mani B, Ton J (2011) Callose deposition: a multifaceted plant defense response. Molec Plant-Microbe Interact 24(2):183–193. https://doi.org/10.1094/mpmi-07-10-0149

Lund S, Walford HH, Doherty TA (2013). Type 2 innate lymphoid cells in allergic disease. Curr Immunol Rev 9(4):214-221

Mack I, Hector A, Ballbach M, Kohlhaufl J, Fuchs KJ, Weber A, Mall MA, Hartl D (2015) The role of chitin, chitinases, and chitinase-like proteins in pediatric lung diseases. Molec Cell Pediatr 2(1):3–3. https://doi.org/10.1186/s40348-015-0014-6

Malaguarnera L (2006) Chitotriosidase: the yin and yang. Cell Mol Life Sci CMLS 63(24):3018–3029

Malaguarnera L, Di Rosa M, Zambito AM, dell'Ombra N, Nicoletti F, Malaguarnera M (2006) Chitotriosidase gene expression in Kupffer cells from patients with non-alcoholic fatty liver disease. Gut 55(9):1313–1320

Medzhitov R (2007) Recognition of microorganisms and activation of the immune response. Nature 449(7164):819–826. https://doi.org/10.1038/nature06246

Miya A, Albert P, Shinya T, Desaki Y, Ichimura K, Shirasu K, Narusaka Y, Kawakami N, Kaku H, Shibuya N (2007) CERK1, a LysM receptor kinase, is essential for chitin elicitor signaling in Arabidopsis. Proc Natl Acad Sci USA 104(49):19613–19618. https://doi.org/10.1073/pnas.0705147104

Nair MG, Gallagher IJ, Taylor MD, Loke Pn, Coulson PS, Wilson R, Maizels RM, Allen JE (2005) Chitinase and Fizz family members are a generalized feature of nematode infection with selective upregulation of Ym1 and Fizz1 by antigen-presenting cells. Infect Immun 73(1):385–394

Olland AM, Strand J, Presman E, Czerwinski R, Joseph-McCarthy D, Krykbaev R, Schlingmann G, Chopra R, Lin L, Fleming M et al (2009) Triad of polar residues implicated in pH specificity of acidic mammalian chitinase. Protein Sci: Publ Protein Soc 18(3):569–578. https://doi.org/10.1002/pro.63

Petutschnig EK, Jones AME, Serazetdinova L, Lipka U, Lipka V (2010) The lysin motif receptor-like kinase (LysM-RLK) CERK1 is a major chitin-binding protein in arabidopsis thaliana and subject to chitin-induced phosphorylation. J Biol Chem 285(37):28902–28911. https://doi.org/10.1074/jbc.M110.116657

Reese TA, Liang H-E, Tager AM, Luster AD, Van Rooijen N, Voehringer D, Locksley RM (2007) Chitin induces accumulation in tissue of innate immune cells associated with allergy. Nature 447(7140):92

Roy RM, Wüthrich M, Klein BS (2012) Chitin elicits CCL2 from airway epithelial cells and induces CCR62-dependent innate allergic inflammation in the lung. J Immunol 189(5):2545–2552

Sanchez-Vallet A, Mesters JR, Thomma BPHJ (2015) The battle for chitin recognition in plant-microbe interactions. FEMS Microbiol Rev 39(2):171–183. https://doi.org/10.1093/femsre/fuu003

Satoh T, Takeuchi O, Vandenbon A, Yasuda K, Tanaka Y, Kumagai Y, Miyake T, Matsushita K, Okazaki T, Saitoh T (2010) The Jmjd3-Irf4 axis regulates M2 macrophage polarization and host responses against helminth infection. Nat Immunol 11(10):936

Schlosser A, Thomsen T, Moeller JB, Nielsen O, Tornøe I, Mollenhauer J, Moestrup SK, Holmskov U (2009) Characterization of FIBCD1 as an acetyl group-binding receptor that binds chitin. J Immunol 183(6):3800–3809

Semeňuk T, Krist P, Pavlíček J, Bezouška K, Kuzma M, Novák P, Křen V (2001) Synthesis of chitooligomer-based glycoconjugates and their binding to the rat natural killer cell activation receptor NKR-P1. Glycoconj J 18(10):817–826

Shen C-R, Juang H-H, Chen H-S, Yang C-J, Wu C-J, Lee M-H, Hwang Y-S, Kuo M-L, Chen Y-S, Chen J-K et al (2015) The correlation between chitin and acidic mammalian chitinase in animal models of allergic asthma. Int J Mol Sci 16(11):27371–27377. https://doi.org/10.3390/ijms161126033

Shibata Y, Foster LA, Bradfield JF, Myrvik QN (2000) Oral administration of chitin down-regulates serum IgE levels and lung eosinophilia in the allergic mouse. J Immunol 164(3):1314–1321

Shibata Y, Metzger WJ, Myrvik QN (1997) Chitin particle-induced cell-mediated immunity is inhibited by soluble mannan - Mannose receptor-mediated phagocytosis initiates IL-12 production. J Immunol 159(5):2462–2467

Shimizu T, Nakano T, Takamizawa D, Desaki Y, Ishii-Minami N, Nishizawa Y, Minami E, Okada K, Yamane H, Kaku H et al (2010) Two LysM receptor molecules, CEBiP and OsCERK1, cooperatively regulate chitin elicitor signaling in rice. Plant J 64(2):204–214. https://doi.org/10.1111/j.1365-313X.2010.04324.x

Shinya T, Motoyama N, Ikeda A, Wada M, Kamiya K, Hayafune M, Kaku H, Shibuya N (2012) Functional characterization of CEBiP and CERK1 homologs in arabidopsis and rice reveals the presence of different chitin receptor systems in plants. Plant Cell Physiol 53(10):1696–1706. https://doi.org/10.1093/pcp/pcs113

Shinya T, Nakagawa T, Kaku H, Shibuya N (2015) Chitin-mediated plant-fungal interactions: catching, hiding and handshaking. Curr Opin Plant Biol 26:64–71. https://doi.org/10.1016/j.pbi.2015.05.032

Shinya T, Yamaguchi K, Desaki Y, Yamada K, Narisawa T, Kobayashi Y, Maeda K, Suzuki M, Tanimoto T, Takeda J et al (2014) Selective regulation of the chitin-induced defense response by the Arabidopsis receptor-like cytoplasmic kinase PBL27. Plant J 79(1):56–66. https://doi.org/10.1111/tpj.12535

Shuhui L, Mok Y-K, Wong WSF (2009) Role of mammalian chitinases in asthma. Int Arch Allergy Immunol 149(4):369–377. https://doi.org/10.1159/000205583

Spoel SH, Dong X (2012) How do plants achieve immunity? Defence without specialized immune cells. Nat Rev Immunol 12(2):89–100. https://doi.org/10.1038/nri3141

Tada R, Latge J-P, Aimanianda V (2013) Undressing the fungal cell wall/cell membrane - the antifungal drug targets. Curr Pharm Design 19(20):3738–3747. https://doi.org/10.2174/1381612811319200012

Thomsen T, Schlosser A, Holmskov U, Sorensen GL (2011) Ficolins and FIBCD1: soluble and membrane bound pattern recognition molecules with acetyl group selectivity. Mol Immunol 48(4):369–381

Tomas Semenuk PK, Pavlıcek Jiri, Bezouska Karel, Kuzma Marek, Novak Petr, Kren Vladimır (2001) Synthesis of chitooligomer-based glycoconjugates and their binding to the rat natural killer cell activation receptor NKR-P1. Glycoconj J 18:817–826

Tschopp CDJ (2007) DEteCTINg fungal pathogens. J Biol Chem 8:17–18. https://doi.org/10.1074/jbc

van der Luit AH, Piatti T, van Doorn A, Musgrave A, Felix G, Boller T, Munnik T (2000) Elicitation of suspension-cultured tomato cells triggers the formation of phosphatidic acid and diacylglycerol pyrophosphate. Plant Physiol 123(4):1507–1515. https://doi.org/10.1104/pp.123.4.1507

Van Dyken SJ, Mohapatra A, Nussbaum JC, Molofsky AB, Thornton EE, Ziegler SF, McKenzie AN, Krummel MF, Liang H-E, Locksley RM (2014) Chitin activates parallel immune modules that direct distinct inflammatory responses via innate lymphoid type 2 and γδ T cells. Immunity 40(3):414–424

Vannella KM, Ramalingam TR, Hart KM, de Queiroz Prado R, Sciurba J, Barron L, Borthwick LA, Smith AD, Mentink-Kane M, White S et al (2016) Acidic chitinase primes the protective immune response to gastrointestinal nematodes. Nat Immunol 17(5):538–544. https://doi.org/10.1038/ni.3417

Wagener J, Malireddi RS, Lenardon MD, Köberle M, Vautier S, MacCallum DM, Biedermann T, Schaller M, Netea MG, Kanneganti T-D (2014) Fungal chitin dampens inflammation through IL-10 induction mediated by NOD2 and TLR9 activation. PLoS Pathog 10(4):e1004050

Wan J, Zhang X-C, Neece D, Ramonell KM, Clough S, Kim S-Y, Stacey MG, Stacey G (2008) A LysM receptor-like kinase plays a critical role in chitin signaling and fungal resistance in Arabidopsis. Plant Cell 20(2):471–481. https://doi.org/10.1105/tpc.107.056754

Wan JR, Zhang SQ, Stacey G (2004) Activation of a mitogen-activated protein kinase pathway in Arabidopsis by chitin. Mol Plant Pathol 5(2):125–135. https://doi.org/10.1111/j.1364-3703.2004.00215.x

Wiesner DL, Specht CA, Lee CK, Smith KD, Mukaremera L, Lee ST, Lee CG, Elias JA, Nielsen JN, Boulware DR (2015) Chitin recognition via chitotriosidase promotes pathologic type-2 helper T cell responses to cryptococcal infection. PLoS Pathog 11(3):e1004701

Yamaguchi K, Imai K, Akamatsu A, Mihashi M, Hayashi N, Shimamoto K, Kawasaki T (2012) SWAP70 functions as a Rac/Rop guanine nucleotide-exchange factor in rice. Plant J 70(3): 389–397. https://doi.org/10.1111/j.1365-313X.2011.04874.x

Yamaguchi K, Yamada K, Ishikawa K, Yoshimura S, Hayashi N, Uchihashi K, Ishihama N, Kishi-Kaboshi M, Takahashi A, Tsuge S et al (2013) A receptor-like cytoplasmic kinase targeted by a plant pathogen effector is directly phosphorylated by the chitin receptor and mediates rice immunity. Cell Host Microbe 13(3):347–357. https://doi.org/10.1016/j.chom.2013.02.007

Yeh Y-H, Panzeri D, Kadota Y, Huang Y-C, Huang P-Y, Tao C-N, Roux M, Chien H-C, Chin T-C, Chu P-W et al (2016) The Arabidopsis Malectin-like/LRR-RLK IOS1 is critical for BAK1-dependent and BAK1-independent pattern-triggered immunity. Plant Cell 28(7):1701–1721. https://doi.org/10.1105/tpc.16.00313

Zhang J, Li W, Xiang T, Liu Z, Laluk K, Ding X, Zou Y, Gao M, Zhang X, Chen S et al (2010) Receptor-like cytoplasmic kinases integrate signaling from multiple plant immune receptors and are targeted by a pseudomonas syringae effector. Cell Host Microbe 7(4):290–301. https://doi.org/10.1016/j.chom.2010.03.007

Zipfel C, Oldroyd GED (2017) Plant signalling in symbiosis and immunity. Nature 543(7645): 328–336. https://doi.org/10.1038/nature22009

Chapter 5
Chitin Organizing and Modifying Enzymes and Proteins Involved In Remodeling of the Insect Cuticle

Subbaratnam Muthukrishnan, Hans Merzendorfer, Yasuyuki Arakane and Qing Yang

Abstract Chitin, the extracellular matrix polysaccharide of insects and arthropods is widely distributed in nature in all kingdoms of life and serves a variety of functions. After synthesis by membrane-bound chitin synthases, it is extensively remodeled before incorporation into divergent matrices with wide-ranging physical and biological properties. This chapter discusses the properties of a variety of insect enzymes and proteins involved in this process. Chitin remodeling involves chitin synthases, which make the nascent chitin chains, and chitin deacetylases that partially deacetylate some of the *N*-acetylglucosamine residues either randomly or sequentially to yield local chitosan-like regions. Other proteins secreted into the procuticle or the midgut help in the assembly of single chitin chains into larger crystalline aggregates that measure in a few 100 nanometers. They are further embedded in a complex matrix of cuticular proteins or become associated with proteins containing chitin-binding domains to constitute the laminar procuticle or the lattice-like peritrophic matrix. During molting, previously formed laminar cuticle or PM are decrystallized/depolymerized to unmask the chitin chains, which then are degraded by a mixture of chitinolytic enzymes consisting of chitinases and *N*-acetylglucosaminidases present in molting fluid or in gut secretions. Some of the degradation products may be recycled for the synthesis of new matrices. We present a model of chitin synthesis, assembly, and degradation and the roles of these chitin-remodeling enzymes in this overall process.

S. Muthukrishnan (✉)
Department of Biochemistry and Molecular Biophysics, Kansas State University, Manhattan, KS 66503, USA
e-mail: smk@ksu.edu

H. Merzendorfer
Institute of Biology, University of Siegen, 57076 Siegen, Germany

Y. Arakane
Department of Applied Biology, Chonnam National University, Gwangju 500-757, South Korea

Q. Yang
School of Bioengineering, Dalian University of Technology, No. 2 Linggong Road, Dalian 116023, China

Q. Yang and T. Fukamizo (eds.), *Targeting Chitin-containing Organisms*, Advances in Experimental Medicine and Biology 1142,
https://doi.org/10.1007/978-981-13-7318-3_5

Keywords Chitin · Chitin remodeling · Chitin synthase · Chitinase · Chitin deacetylase · *N*-acetylglucosaminidase · Knickkopf · Cuticular proteins · Peritrophic matrix

5.1 Introduction

Chitin (along with cellulose and hyaluronan) is one of the three major matrix polysaccharides found in nature that provides structural support to cell walls or extracellular matrices of multicellular organisms (Muzzarelli 1973; Cohen 2010; Muthukrishnan et al. 2016). All three are β-1-4 linked linear sugar polymers that provide shape, rigidity and mechanical strength to the organism's body plan and protect it against predation, infection, and mechanical injury. The capability to synthesize chitin, which is found in diverse organisms including fungi, sponges, annelids, and arthropods, is thought to have been pivotal in the evolution of arthropods, as it allows survival in diverse aquatic and terrestrial environments. Several chapters in this book deal with the structural properties and modifications of chitin–protein composites that account for the diverse properties of the arthropod/insect cuticle or other chitin-containing structures. In this chapter, the focus will be exclusively on the enzymes involved in chitin remodeling that contribute to its unique structural properties. In particular, the structures, catalytic properties and enzyme mechanisms will be discussed, with a strong bias toward insect enzymes with which the authors are most familiar.

5.2 Chitin Synthases, the Enzymes that Polymerize Precursors of Chitin

Chitin synthases (CHS) carry out the processive addition of *N*-acetylglucosamine from the activated precursor, UDP-*N*-acetylglucosamine (donor) to the 4-hydroxyl group of the nonreducing end sugar of the growing chitin chain (acceptor) in an SN2 type of reaction in which an aspartate residue acts as the general base and UDP as the leaving group (Merzendorfer 2013). The pKa of this catalytic aspartate is influenced by the presence of other acidic groups near the binding pocket. As expected, these enzymes are integral membrane proteins that secrete the chitin product into the extracellular space, where the chitin chains become incorporated into the cell wall (or to other cell wall polysaccharides in the case of fungi) or into laminae in the cuticular matrix and the lattice-like peritrophic matrices (in arthropods/insects).

5.2.1 A Mechanistic View on Chitin Synthesis, Translocation and Fibrillogenesis

The CHSs belong to GT2 family of processive glycosyltransferases with the characteristic protein folding pattern of GT-A with an open β-sheet surrounded by an α-helix on either side. The insect/arthropod enzymes have two sets of transmembrane segments connected by a large central intracellular loop domain that contains several catalytically critical amino acid residues and motifs. Based on an analogy with bacterial cellulose synthase (BcsA) (Morgan et al. 2016), one set of transmembrane helices next to the catalytic domain lines a channel in the membrane, through which the elongating chitin polymer is thought to be extruded. Energetically, transport and extrusion of the growing chain through this transmembrane channel appears to be driven by the polymerization reaction as long as these processes are coupled (Gohlke et al. 2017).

The intracellular loops are predicted (or shown to) to contain the critical amino acid residues or sequence motifs implicated in catalysis by site-directed mutagenesis in yeast chitin synthases (Yabe et al. 1998) and crystallographic studies of BcsA (Morgan et al. 2016). These are the QRRRW (W residue of this motif sits next to acceptor glucose binding site at the entrance to the transmembrane channel in BcsA), EDR (equivalent to the TED motif at the *N*-terminus of the "finger helix" and the general base for the catalytic process). This finger helix has been found to occupy two different positions ("down" and "up" positions) during pre-translocation and post-translocation, respectively, in cellulose synthases in a process governed by another helix known as "gating helix" or "interfacial helix" that lies at the base of the carbohydrate-conducting channel (Morgan et al. 2016). A similar mechanism has been proposed for yeast CHS3 based on structural homology modeling of the helical regions of this enzyme (Gohlke et al. 2017).

A detailed investigation of another GT2 enzyme, NodC from *Sinorhizobium meliloti*, which synthesizes short chitooligosaccharides (rather than longer chitin chains) was carried out by Dorfmueller et al. (2014). By site-directed mutagenesis and model-building using the cellulose synthase complex associated with cellulose as a template for modeling, they could identify the catalytically critical residues (EDR and QR/QRW motifs) and provide a structural explanation for the inability of NodC to make longer products. The enzyme is only able to form chitooligomers, because the catalytic site is capped by a hydrophilic product-binding pocket and lacks the open transmembrane channel found in processive chitin synthases. They also proposed a biochemical mechanism in which the two terminal sugars of the growing chain rotate during repositioning of the acceptor site for the next round of polymerization. In subsequent steps, the +1 sugar rotates only every other synthetic step to maintain the β-1-4 orientation of the glycosidic bond.

There is evidence that chitin synthases are di/oligomeric complexes that assemble on cell membranes. Using bimolecular fluorescence complementation techniques, Gohlke et al. (2017) demonstrated that yeast CHS3 subunits formed oligomers at the bud neck and lateral plasma membrane. These oligomers presumably form even in the ER indicating oligomerization prior to their final transport to the Golgi, bud neck, or plasma membrane. In *Manduca sexta*, CHS2 in the midgut appears to be present as an oligomer based on gel filtration studies (presumably a trimer) (Maue et al. 2009). Oligomerization of yeast Chs3 appeared to involve interactions involving the *N*-terminal domain and to involve several proteins of the vesicular transport pathway (Sacristan et al. 2013). It is conceivable that like cellulose synthase, larger assemblies of CHS are involved in the coordinated synthesis of chitin fibrils which spontaneously self-assemble to form chitin microfibrils containing 18–24 strands (Neville et al. 1976). While this mechanism would favor the formation of β-chitin, in which the fibrils are in parallel orientation, it does not explain the formation of α-chitin with antiparallel fibers. Additional mechanisms must be involved in post-synthetic regulation of chitin fiber assembly, which will be discussed later in this chapter.

5.2.2 Specialization Among CHS Isoforms

Sequence searches of genomes of chitin-containing organisms indicate a wide range in the number of genes encoding CHSs (Merzendorfer 2011). In fungi, this number can range widely and reach up to 20 or more isoforms in filamentous fungi. But, in most other organisms including nematodes, insects, and arthropods, this number is either one or two even though the chitinous structures formed are highly differentiated and vary widely in their chitin content. The biochemical basis for this variation is not obvious. In insects where *CHS* gene expression has been studied extensively, it is clear that there is specialization in the type of CHS involved in the synthesis of chitin in the cuticle or in the peritrophic matrix. One CHS (called CHS-A) specializes in cuticular chitin synthesis and the other CHS, CHS-B, is utilized exclusively to make PM chitin. The general domain architectures of these two CHS proteins are nearly the same with a couple of notable differences. While both enzymes have a set of transmembrane segments in the *N*-terminal domain the number of TMS are not identical. This TMS domain is followed by a central catalytic domain with a "finger helix" and "gating helix" as in the yeast CHSs, consistent with a processive mechanism. Both types of insect CHS enzymes have a similar number and organization of TMSs in the *C*-terminal domain, but a predicted coiled coil in insect CHS-A enzyme is absent in CHS-B enzymes. Since the CHS-A products are organized into hydrophobic chitin laminae (α-form) but the products of CHS-B are hydrophilic and consist of a criss-crossing matrix of chitin (presumably β-form), some sequence differences such as the coiled-coil region (or some other unidentified difference(s) in structural features between CHS-A and CHS-B) may determine their oligomerization and/or interaction with other proteins and hence the assembly of the chitin products. It is noteworthy that several proteins that are found in the cuticle that are not expressed in the midgut epithelial cells. These include Knickkopf, CPAP3A, CPAP3C and CDA1 and CDA2. In their absence (or on depletion) the chitin fibers in the procuticle become amorphous, less organized, and fibrillar, which could indicate that they may be involved in chitin assembly (Chaudhari et al. 2011; Petkau et al. 2012; Pesch et al. 2015; Noh et al. 2018a, b). See also Sect. 7.

Additional complexity may arise from the use of alternate splicing, which is well studied in the case of CHS-A from insects. In all CHS-A enzymes from insects, there are two alternative forms of an exon that encodes 59 amino acids long segment that includes the first of the two TMS segments near the *C*-terminus of CHS proteins. This allows control of expression of CHS-A proteins with either exon-encoded segment in specific tissues or at developmental stages (Tellam et al. 2000; Arakane et al. 2005; Hogenkamp et al. 2005; Zhang et al. 2010; Wang et al. 2012; Yang et al. 2013). The forms with the alternate exon b appear to be expressed predominantly in tracheal tissue. RNA interference (RNAi) to silence expression of these isoforms produce distinctly different phenotypes compared to RNAi of the other CHS-A isoform with exon a.

Another set of alternate exons involving an upstream exon (exon 2) of CHS-A has been reported in two lepidopteran species, *Ostrinia furnacalis* and *Bombyx mori* (Qu and Yang 2011, 2012). This set of alternate exons appears to be preferentially expressed in epidermal cells. The expression of these alternate transcripts of *CHS*s has been explored. All four combinations of transcripts involving these two pairs of exons were shown to be differentially expressed. Transcripts with exon 2a were the predominant forms at all developmental stages. Transcripts with exon 2b were derived from a promoter downstream of the main promoter and missed all of exon 1. Silencing of this alternatively spliced transcript affected only the formation of the head capsule in the third instar larvae. Transcripts with exon 19a were expressed at all developmental stages except pupal day 3. Transcripts with exon 19b were predominant in day 3 embryos and day 3 pupae indicating a special requirement for this transcript at these stages. RNAi of transcripts with either 2a or 19a exon sequences produced similar molting defects. RNAi of exon 2b containing transcripts resulted in "double head" molting phenotype in third instar larvae. Collectively, these results indicate that each of the four transcripts derived from this *CHS-A* gene is uniquely required at particular developmental stages or in specific tissues.

A comparable study with *B. mori CHS-A*, which also has this alternate exon 2 and an alternative promoter has been carried out. Expression analysis of the two alternatively spliced transcripts of exon 2 of *CHS-A* in *B. mori* has indicated differential regulation of these two transcripts during development and in different sexes, with the males showing higher levels of this transcript in the wing at mid-pupal stage. Depletion of this transcript by RNAi resulted in decreased chitin content of the wing and vein crimpling (Xu et al. 2017). The two promoters differed in their expression response when 20-hydroxyecdysone (20HE) was injected into 3-day old fifth instar larvae or when the epidermis tissue was incubated with 20-HE. These results have led to the suggestion that lepidopteran insects may have utilized an alternative promoter to regulate wing development in the mid-pupal stage in this order of insects.

An alternative splice variant of *CHS-B* has also been described in *Heliothis zea*. But the protein product was predicted to be missing the catalytic domain, while retaining the transmembrane segments. The physiological significance of this protein is not yet established (Shirk et al. 2015).

5.3 Chitinases

Chitinases belong to either family 18 or family 19 glycosyl hydrolases (Henrissat 1991) that share no amino acid sequence similarities. Family 18 chitinases are widely distributed in all kingdoms, whereas family 19 enzymes are restricted to plants. In this chapter, we will restrict our attention to family 18 chitinases. They are exclusively endochitinases cleaving internal β-1-4 bonds in chitin polymers. These hydrolytic enzymes yield β-anomeric products at the reducing ends and hence use a retaining mechanism. The analysis of crystal structures of several family 18 chitinases with bound substrates or products has provided clues concerning their catalytic mechanism. It indicated that in addition to the proton donor glutamate, the C2 *N*-acetyl group of the sugar bound in the -1 position of the substrate-binding site has a role in catalysis. This mechanism has been called substrate-assisted catalysis (anchimeric assistance). It involves the formation of a positively charged oxazolinium ion intermediate, in which a covalent bond is formed between the carbonyl oxygen of the acetamido group and the C1 atom of the sugar, thus stabilizing the transition state intermediate (Brameld et al. 1998). Additional support for this mechanism comes from the inhibition of this class of enzymes by allosamidine, which forms a non-hydrolyzable analog of the transition state oxazolinium intermediate, and the fact that an obvious nucleophile near the substrate-binding site in family 18 chitinases is absent (Koga et al. 1997; Tews et al. 1997). The crystal structures of substrate–enzyme complexes also indicate that the reducing end of the sugar at the -1 subsite is in the boat conformation, which is energetically unfavorable. Additionally, as it is close to the catalytically important aspartate residue, this arrangement may provide an explanation for hydrolysis at this site (Chen et al. 2014).

The catalytic domains of family 18 chitinases assume the typical $\beta_8\alpha_8$ TIM barrel structure with eight β-strands in the center of the barrel and eight α-helices constituting the outer surface of the barrel. Four conserved motifs at characteristic positions in this structure have been identified. The conserved motif I is in the third β-strand and has the consensus sequence KXX(V/L/I)A(V/L)GGW. The second motif, FDG (L/F)DLDWE(Y/F)P in the β4 strand contains the catalytically critical glutamate, which has been shown to be the proton donor in the hydrolytic reaction. The other two conserved motifs, MXYDL(R/H)G and GAM(T/V)WA(I/L)DMDD are in the β6 and β8-strand respectively (Arakane and Muthukrishnan 2010). In the crystal structure of a group I chitinase from *O. furnacalis*, a deep and long substrate-binding cleft was observed, which is lined by several aromatic amino acids interacting with the sugar residues of the substrate that is open at both ends. In addition, the surface contains a flat plane with additional aromatic amino acids that are positioned close to the cleft near where the reducing end of the substrate is located. Mutations of any one of these aromatic amino acids, especially the one closest to the binding cleft affected the enzyme's activity on long substrates (but not the shorter substrates) as well as binding affinity to chitin substrate (Chen et al. 2014), emphasizing the role of stacking interactions between these residues and several sugar residues of long substrates.

Most insects have multiple chitinases with the number of proteins ranging from one to >20. Based on phylogenetic analyses, tissue and/or developmental stage specificity of expression and presence of different protein domains, insect chitinases, and chitinase-like proteins have been divided into as many as 11 groups (Tetreau et al. 2015a, b). These domains include a catalytic domain, chitin-binding domains, a serine–threonine-rich glycosylation domain and other domains. All of them have at least one catalytic domain, though some of them are devoid of chitinase activity (group V proteins, e.g., Zhu et al. 2008b). Almost all of them have signal peptides that get cleaved inside the ER and the rest of the protein packaged in secretory vesicles are transported into the extracellular space via a default secretory pathway. However, there are a couple of exceptions. Members of group III and group VIII chitinases have transmembrane segments and hence are predicted to remain on the plasma membrane with the catalytic domains facing the extracellular space. Overall, the insect chitinases have diverse domain organizations, tissue and developmental patterns of expression, and differences in their catalytic parameters and substrate preferences. This apparently leads to functional specialization among the chitinases, with some of them having roles in chitin degradation during molts, while others may have functions such as cuticle and peritrophic matrix remodeling, digestion and defense or function as growth factors.

5.3.1 Group I Chitinases

The roles of individual groups of chitinases or individual chitinases have been analyzed by RNAi studies in a variety of insects including *T. castaneum*, *D. melanogaster*, *O. nubilalis*, *Nilaparvata lugens*, *S. exigua*, *S. litura*, and others. These studies have indicated that group I chitinases are clearly implicated in the turnover and shedding of the exuvium, the remaining old cuticle, especially at the pupal to the adult stage. Upon RNAi to knockdown specific chitinase genes involved in molting, the insects are trapped in the pupal cuticle and cannot digest away the old cuticle (Zhu et al. 2008a; Pesch et al. 2016a, b; Li et al. 2015; Su et al. 2016). In most insects, there is a single copy of a group I chitinase characterized by the presence of a signal peptide, followed by a catalytic domain, a linker domain, and a chitin-binding domain (CBD) belonging to CBM14 family. However, in mosquito genomes, this gene has duplicated to have four or five copies of the gene (Zhang et al. 2011b). In *Locusta migratoria,* there are two copies of this gene encoding chitinases, with one of them missing the CBD (Li et al. 2015). Even though both genes show similar expression profiles and regulation by 20-hydroxyecdysone, only one is required for molting. A similar situation seems to exist in *Nilaparvata lugens,* where a gene duplication has occurred and only the group I chitinase with CBD appears to be required for molting, whereas the other one devoid of a CBD is not. Curiously, this gene is expressed only in male reproductive tissues and accessory glands (Xi et al. 2015). The transcripts for this group of genes are generally expressed at all stages of development but are elevated during the pupal stages suggesting a critical role in adult eclosion.

5.3.2 Group II Chitinases

Group II chitinases have an *N*-terminal signal peptide, followed by four or more catalytic domains interspersed with multiple CBDs. Typically, one or two catalytic domains have substitutions of the proton donor, glutamate, in conserved domain II indicating that they may not have catalytic activity, while retaining substrate-binding ability. However, these chitinases do have two or more catalytic domains at the *C*-terminal region that are presumed to have catalytic activity, with all four conserved domains intact and multiple CBDs. In fact, individual domains of this chitinase from *O. furnacalis* have been expressed and purified from yeast cells and crystallized. They possess endochitinase activity and act additively rather than synergistically (Chen et al. 2018a, b). RNAi to silence the expression of this chitinase gene results in inhibition of egg hatch as well as molting arrest at all stages without affecting new cuticle deposition in *T. castaneum* (Zhu et al. 2008c). Down-regulation of the transcripts of group II genes also results in molting phenotypes (double cuticle, entrapment in pupal cuticle and lethality) in other insects including *Chilo suppressalis* and *O. furnacalis* (Su et al. 2006; He et al. 2013). Analyses of molting fluids for fragments of this protein by tryptic peptide analyses have indicated that this protein and a group I chitinase are indeed secreted into the molting fluid (Qu et al. 2014). Since molting defects are seen after RNAi of a group II chitinase gene in multiple insect species, it may have a critical role to play in molting that is not fulfilled by group I chitinases alone. It is likely that both group I and group II chitinases are complementary and essential for cuticle degradation. Possibly, the group II enzyme carries out the initial decrystallization of α-chitin followed by an endo-type of attack along with a group I chitinase.

5.3.3 Group III Chitinases

Group III chitinases have an *N*-terminal transmembrane domain followed by two catalytic domains and ending with a *C*-terminal CBD belonging to CBM14 family (ChtBD2). Chitinases with the same domain organization are found in all insects analyzed and several other subphyla of arthropods suggesting an ancient origin for this group of chitinases (Tetreau et al. 2015a; Liu et al. 2018). Each catalytic domain has the characteristic $\beta_8\alpha_8$ TIM barrel structure of GH18 family of chitinases and the two domains are linked by a spacer domain. They are predicted to be membrane-anchored and, in fact, expression in a baculovirus system supports this notion (Noh et al. 2018a, b). However, this protein has also been localized in distal parts of the cuticle including the layers immediately below the epicuticle as revealed by TEM studies suggesting the release of this enzyme after insertion into the plasma membrane (Noh et al. 2018a, b). Both domains act as endochitinases and there is no synergism between the two domains as observed in some bacterial systems (Liu et al. 2018). This class of enzymes prefers single-stranded chitin substrates and is devoid of any activity on insoluble chitin.

RNAi to silence group III chitinase genes has been carried out in different insects. In *T. castaneum*, there was no inhibitory effect on molting and adults did emerge with high frequency. However, there was defective abdominal contraction and forewing expansion during the pupal stage. Pupal maturation was also affected with defective leg folding and pronotum was not fully extended. When RNAi was carried out during the prepupal stage, pupation and adult emergence occurred at the normal time, but the elytra were significantly shorter and the hindwings did not fold properly. The elytra had a rough and wrinkled appearance (Zhu et al. 2008c). The structural abnormalities of the cuticle were investigated in much greater detail by Noh et al. (2018a, b). The adults had significantly shorter wings and legs. Even though the insects moved freely, they toppled over frequently and had difficulties in uprighting their bodies. Similarly, Chen et al. (2017) reported that RNAi to silence expression of a group III chitinase gene in white-backed planthopper, (*Sogatella furcifera*) affected molting and wing development, which resulted in multiple phenotypes including "wasp-waisted" adult insects, elongated distal wing pads and thin thorax–abdominal junctions. Similarly a group III chitinase from rice striped stem borer, *Chilo suppresalis* resulted in wings that were curled and did not spread properly (Su et al. 2016).

5.3.4 TEM Analyses of Cuticle

Analyses of larval body wall or pupal procuticle following RNAi for the group III chitinase in *T. castaneum* indicated the defective organization of the alternating electron-dense and electron-lucent layers resulting in fuzzy boundaries between adjacent cuticle layers as well as irregular pore canals (Noh et al. 2018a, b). Both observations suggest a loss of structural integrity of the cuticle. The ultrastructural defects persisted in the adult cuticle, as well as with the laminae becoming less compact compared to controls. The pore canals were rudimentary and devoid of long pore canal fibers. The defects were also seen in cuticles of other tissues such as legs and ventral abdomen. Immunolocalization studies have indicated that even though this protein is predicted to have a transmembrane segment, it is found in distal layers of the procuticle, reaching even the laminae just below the epicuticle (Noh et al. 2018a, b). This enzyme has also been detected in the mesocuticle and endocuticle of adult elytra which have a different morphology and assume a brick-like or "Balken" structure. After down-regulation of this group III chitinase, the cuticular layers that form after adult eclosion are abnormal and contain amorphous fibers. The requirement of a group III chitinase was demonstrated not only for "hard" cuticles but also for soft cuticles (Noh et al. 2018a, b). Since purified catalytic domains of group III chitinases act only on soluble chitin but not on insoluble chitin (Liu et al. 2018), it has been proposed that they are not involved in chitin degradation during molting but in some step of chitin maturation during cuticular chitin deposition and pore canal formation (Noh et al. 2018a, b; Liu et al. 2018). The presence of a *C*-terminal CBD and two tandem catalytic domains suggests that this enzyme acts on nascent chitin at two places in the same chitin chain, to release chitin fragments of a uniform size and possibly to aid in the formation of antiparallel chitin bundles and/or higher order chitin fiber assemblies.

5.3.5 Group IV Chitinases

Unlike members of the group I, II, and III which generally seem to have single copies (with some notable exceptions; see Sect. 2.1), group IV chitinases have numerous representatives in many insects and presumably arose from several gene duplication events. Often they are clustered in the genome. They are generally expressed in the gut tissue at both larval and adult stages (Zhu et al. 2008c; Khajuria et al. 2010; Su et al. 2016). They have a single catalytic domain and some of them lack a CBD. But all of them do have signal peptides and presumably are secreted into the gut lumen. Representative members of this group of chitinases from *D. melanogaster*, one with a CBD and one without a CBD, have been expressed and purified from the culture medium of insect cells and shown to possess catalytic activity. Their catalytic efficiencies are comparable to those of group I enzymes, even though their affinities for long substrates are less than those of the group I chitinases (Zhu et al. 2008b). They all have the 4 conserved sequence motifs of family 18 chitinases including the conserved motif II (FDGLDWEYP), which contains the catalytically critical proton donor, glutamate (Zhu et al. 2004).

RNAi to silence expression of individual group IV chitinases has yielded mixed results concerning their essentiality for insect survival. Down-regulation of individual group IV chitinases did not result in any noticeable phenotype or molting defects or survival in *T. castaneum* larvae. RNAi of some combinations of these genes also did not result any phenotypes (Zhu et al. 2008c). This observation was attributed to the redundancy of the members of this group and the ability of different members to compensate for the functions of other members of the same group. However, RNAi of Cht2 in *D. melanogaster* did result in larval lethality and a thinning of the procuticle (Pesch et al. 2016a, b). Feeding of dsRNA with a group IV chitinase from *Ostrinia nubilalis* resulted in the elevation of chitin content of the PM and a reduction in body weight (Khajuria et al. 2010). In *Chilo suppresalis,* injection of dsRNA targeting two different group IV chitinases expressed predominantly in larval midgut resulted in the death of more than two-thirds of the larvae. Interestingly, while one of these proteins had the proton donor glutamate in consensus region II, the other enzyme had substitutions of an aspartate as well as the proton donor, glutamate. These data suggest a role for group IV chitinases in

PM chitin turnover, but this needs further examination as the authors also reported an effect on molting (Su et al. 2016).

5.3.6 Group V Chitinases

This family of proteins was initially identified as growth promoting substances in conditioned medium derived from cultures of imaginal discs (Kawamura et al. 1999). A total of six proteins belonging to this family were identified from *D. melanogaster*. All of them have a leader peptide and are secreted into the culture medium. They have a single GH18 domain and do not have a CBD or other domains. A crystal structure of one of these proteins is available for the imaginal disc growth factor 2 (DmIDGF2) (Varela et al. 2002). It has the typical $\beta_8\alpha_8$ TIM barrel structure of GH18 family of hydrolases, but has two prominent insertions, one between the β_4 strand and the α_4 helix and the second insertion is between β_7 and α-7. The first insertion is highly conserved among chitinases of group V and has the consensus sequence, KPRKVGXX(L/I)GSXWKFKKXF(T/S)GDXVVDE. It seems to be devoid of a defined structure as it is not visible in the crystal structure. This solvent exposed structure must interact with some other cellular component because of its conserved sequence. Further, this presumed loop structure undergoes proteolysis at a precise location between the F and T residues in DmIDGF2 (Varela et al. 2002). This cleavage does not seem to affect its function in cell proliferation. There is some evidence for a similar cleavage of *T. castaneum* IDGF4 (Zhu et al. 2008b). Both DmDS47 (a group V chitinase-like protein) and TcIDGF2 do bind to colloidal chitin (Zhu et al. 2008b). The second insertion is of variable length and has been implicated in determining the specificity of bacterial chitinases in acting as an endo- versus exochitinase as well as affecting its processivity (Zees et al. 2009; Li and Greene 2010).

Another notable feature of group V chitinases is that the conserved region II of this type of insect chitinases contains one or more substitutions. Several (but not all) lack the proton donor glutamate. Even some IDGFs with this glutamate residue are devoid of enzymatic activity (Zhu et al. 2008b). Since one of the two insertions found in all group V chitinase-like proteins follows immediately after the consensus sequence II, it is likely that this insertion interferes with the binding of the substrate and/or the catalytic function of this glutamate. The finding that at least some members of this group of proteins do bind to colloidal chitin, but are devoid of catalytic activity, makes the latter possibility more likely.

Group V chitinase-like proteins are expressed at all developmental stages and RNAi studies have indicated that several of them have essential biological functions in insects (Zhu et al. 2008c; Pesch et al. 2016a, b; Xi et al. 2015). In *T. castaneum,* TcIDGF4 did not affect pupation, but did affect adult eclosion. But the precise step affected has not been determined. In *D. melanogaster*, depletion of transcripts for four of the IDGFs resulted in "double cuticle" phenotypes involving the head skeleton and the posterior spiracles in the larval stages indicating molting defects in larval stages. No molting defects were seen at pupal stages (Pesch et al. 2016a, b). Some IDGFs have been suggested to have a role in cuticle maintenance and in epithelial defense). In *D. melanogaster* stage 18 embryos, IDGFs showed expression in cuticle-forming tissues including epidermis, posterior spiracles and tracheal tubules with differences among different IDGFs. In larval and pupal stages, the expression levels were relatively low, except for IDGF5 in larval stage 3 and high levels of IDGF5 in the adult stages. RNAi of *DmIDGF* genes also led to deformed cuticle with reduced thickness (Pesch et al. 2016a, b). In *T. castaneum,* RNAi to silence expression of *TcIDGF4* resulted in arrest of pupal development and death, whereas RNAi for *TcIDGF2* did not produce any phenotype. In *Nilaparvata lugens,* IDGF was highly expressed in fat body and female reproductive organs but RNAi did not result in any observable effects on molting, survival or fecundity (Xi et al. 2015). In several cases, the precise location of expression (e.g., pupal or adult tracheae) of individual IDGFs have not been determined.

5.3.7 Group VI Chitinases

This group was not recognized as a separate group in earlier classifications, but it deserves a separate grouping. First, the chitinases of this group differ from those of the prominent group I (and group IV) in some significant ways while sharing some characteristics. They have a signal peptide at the *N*-terminus immediately preceding a GH18 chitinase domain with all the signature motifs of active chitinases. The GH18 domain is immediately followed by a CBM14 domain. At the *C*-terminus, there is a long domain rich in serine, threonine, proline and glutamic acid (the so-called PEST domain). It resembles the mucin domain and several members of this family have been classified as mucins or as acidic chitinases. In some members of this family in insects, there is a second CBD close to the *C*-terminus. These proteins are presumed to be heavily *O*-glycosylated. In *T. castaneum*, they are expressed at all developmental stages at nearly constant levels and mostly in the carcass with low-level expression in the posterior midgut. Notably, their expression was barely detectable in the anterior and posterior midgut (Zhu et al. 2008c).

5.3.8 Other Chitinase Families

Not enough studies have been done either to identify the phylogenetic relationship among members of these families involving a large number of insect species or their tissue specificity or developmental patterns of expression. There are one or two RNAi studies with some of these families. But further studies are needed to draw general conclusions about the roles of these chitinases in insects in general or in other arthropods.

5.4 *N*-acetylglucosaminidases

Among the several families of *N*-acetylglucosaminidases (NAG), only the members of the family 20 glycosylhydrolases (GH) are implicated in chitin remodeling. Although GH20 enzymes can act on different substrates such as glycans, glycoproteins, and glycolipids with *N*-acetylhexosamines, we will focus exclusively on chitinolytic glucosaminidases in this review. Along with the endo-acting chitinases, NAGs, which are also secreted into molting fluid (Qu et al. 2014; Zhang et al. 2014) are required for complete depolymerization of polymeric chitin to the monosaccharide, *N*-acetylglucosamine, which can be utilized as building block for new chitin upon activation by UDP. They have been purified from different sources including the molting fluid, hemolymph, integument and gut (Dziadik-Turner et al. 1981; Koga et al. 1982; Nagamatsu et al. 1995; Zen et al. 1996; Filho et al. 2002; Tomiya et al. 2006; Leonard et al. 2006; Yang et al. 2008). NAGs typically cleave oligosaccharides from the nonreducing end and release *N*-acetylglucosamine or *N*-acetylgalactosamine. They tend to prefer *N*-acetylglucosamine over *N*-acetylgalactosamine, though this preference can be highly variable.

The crystal structure of *O. furnacalis* HEX1 (homolog of TcNAG1) has been resolved at 2.1Å and has provided us with some structural insights (Liu et al. 2012). It is a homodimer with a side-by-side symmetry and *N*-glycosylated at Asn164 and Asn375. Each monomer has two identifiable domains. The *N*-terminal Domain 1, which follows a signal peptide, is about 200 amino acids long and is stabilized by six disulfide bonds. It has a six-stranded antiparallel β-strand conserved among most members of this family and an α-helix and a β-strand at the *N*-terminus, both of which are implicated in dimerization. The second domain of about 300 amino acids comprises the standard $\beta_8\alpha_8$ TIM barrel structure with some structural features unique to GH20 enzymes. The active site consisting of the conserved catalytic triad of Asp249, His303, and Glu368 is flanked by three tryptophans and an aspartate and tyrosine, which form several H-bonds with the inhibitor-substrate (TMG-chitotriomycin) and contains the catalytic water molecule (Liu et al. 2012). The mode of action of this enzyme is a substrate-assisted catalysis that involves the acetamido group of the substrate and the formation of the oxazolinium intermediate as in GH18 chitinases. This gene is highly expressed in the integument at the pharate pupal stage but remained at a nearly constant lower level in the alimentary tract during the last instar mid-larval stage and pharate pupal stage indicating an important role in molting and chitin turnover, presumably in response to rising ecdysteroid titers. Administration of dsRNA in the middle of the feeding stage of the last instar did not seem to affect pupation but caused pupal lethality at different periods of the pupal stage, with a range of phenotypes that differed in the extent of removal of the old pupal cuticle.

Analyses of genomes of insects indicate the presence of multiple GH20 enzymes. In *T. castaneum*, at least four enzymes (called TcNAG1, TcNAG2, TcNAG3, and TcFDL) have been implicated in molting. RNAi to suppress the expression of any one of these four genes affected molting to varying degrees depending on the timing of injection of the dsRNA for the targeted *NAG* gene (Hogenkamp et al. 2008). Even though all molts are susceptible to RNAi of each of these genes, dsRNA-mediated silencing of *TcNAG1* seems to result in mortality especially at the pupal to adult molt (90%). This gene is most highly expressed in the integument at the pupal stage. Other genes including *TcNAG2* and *TcFDL* had a similar mortality (80%) at the pupal stage, but considerably lower mortality (10-20%) at earlier stages. The morphology of these moribund insects was reminiscent of those caused by RNAi-silencing of the *TcCHT5* gene, encoding a group I chitinase, which is the most prominent chitinolytic enzyme secreted into the molting fluid at the pharate pupal stage (Kramer et al. 1993; Zhu et al. 2008c). The insects failed to get rid of their old cuticle even though the newly formed cuticle was visible under the old cuticle. It is known that NAGs and CHTs are secreted into the molting fluid with similar timings and hormonal control and to synergize the actions of each other (Fukamizo and Kramer 1985). It has been shown that the activity of group I chitinases is inhibited by high concentrations of the product chitooligosaccharides. It is likely that the synergism of the CHT + NAG combination is due to relief from product inhibition. Moreover, it is possible that the substrate specificities and/or kinetic constants of the NAGs for different chitooligosaccharides differ to achieve high-efficiency hydrolysis on crystalline chitin by the binary system consisting of CHTs and NAGs. This would explain why RNAi of each of the *NAGs* can produce some phenotype in at least a percentage of the animals. The other explanation is that the tissue specificity of expression of the different *NAG* genes is different. For example, *TcNAG1* is highly expressed both in the integument and midgut whereas *TcNAG3* is expressed at high levels in the midgut (Hogenkamp et al. 2007).

A phylogenetic analysis of genomes from sixteen insects including the cigarette beetle, *Lasiderma sericorne*, has shown that all insects have the four groups of NAGs (including FDLs) indicating their evolutionary importance (Chen et al. 2018a b). The ortholog of *TcNAG1* in the cigarette beetle, *L. serricorne, LsNAG1,* exhibits periodic peaks of expression at late larval and late pupal stages. RNAi-silencing of this gene by injection of dsRNA, resulted in defective molting and high mortality.

5.5 Molting Fluid-Associated Chitinolytic Enzymes

When insects molt to next stage of development, their old cuticle first separates from the epidermal cells (apolysis) and then a new envelope and epicuticle start to form all over the body plan before deposition of new layers of the chitin–protein-containing procuticle. Molting fluid containing proteolytic and chitinolytic enzymes is secreted in the space between the two cuticular layers for the digestion/dissolution of the innermost layers of the old cuticle, which are presumably not tanned or otherwise cross-linked (Reynolds and Samuels 1996). Since crystalline chitin is not a good substrate for many chitinolytic enzymes and is protected by being embedded in a matrix of proteins, many of which have a chitin-binding domain (ChtBD2 or the Rebers & Riddiford (R & R) domain, or other types of CBDs), the proteases must first unpack the chitin fibrils from the protein matrix in a process that is the reverse of the assembly of chitin–protein laminae. One can anticipate that the actions of proteases first weaken the bonds between these chitin-binding proteins and chitin bundles leading to the disorganization of the stacked cuticular laminae and the liberation of chitin–protein bundles of ~300 nm often seen in extracts of cuticle following treatment with harsh solvents (Fabritius et al. 2009; Kaya et al. 2014). Proteins that presumably join short chitin–protein nanofibers into longer fibers of the stacks of laminae may be targeted first. The 18–24 chains of chitin bundles that constitute the α-chitin crystallites first proposed by Neville et al. (1976) are most likely to fray near their ends (or surface-exposed chains) due to thermodynamic fluctuations (Beckham and Crowley 2011) and or the actions of oxidative enzymes (Vaaje-Kolstad et al. 2010) (as in microbial systems) or equivalent chitin depolymerizing enzymes (lytic polysaccharide monooxygenases, LPMOs) in arthropods. Group II chitinase also are plausibly involved in this process, as they possess several CBDs that bind chitin and inactive as well as active GH18 domains capable of endochitinase activity. The synergist actions of chitinases and NAG1 may efficiently depolymerize crystalline chitin to the monomeric GlcNAc sugars, which are likely to be resorbed by epidermal cells by endocytosis. When the old cuticle is nearly completely detached, other physiological processes such as abdominal contractions triggered by neuropeptides lead to sloughing of the old cuticle (Reynolds and Samuels 1996; Arakane et al. 2008a, 2008b).

With the expectation that the composition of the molting fluid proteins will provide some clues about the enzymes and their targets during molting/apolysis, Qu et al. (2014) and Zhang et al. (2014) analyzed the molting fluid from two developmental stages of the silkworm using tandem mass spectrometry. Surprisingly, they found a total of only four different chitinolytic enzymes in these fluids even though the genome encodes more than two dozen enzymes potentially capable of chitin hydrolysis. One group I chitinase, one group II chitinase, and one bacterial-type chitinase along with an exochitinase belonging to GH20 family accounted for most of the chitinolytic activities at either stage. This finding is consistent with the expression of only these chitinolytic enzyme encoding genes in the integumental tissue in many insects. Other enzymes detected in the molting fluid include chitin deacetylases (CDA1, CDA2, and CDA4). Several chitin-binding proteins including six cuticular proteins analogous to peritrophins (CPAP3s) were identified. Also present were several proteins with R & R domains and two belonging to the CPH family of cuticular proteins. In addition, the molting fluid in the pharate adult stage contains Knickkopf, which is known to be a GPI-anchored protein with a role in protecting and organizing chitin into laminae (Moussian et al. 2005; Chaudhari et al. 2011). Whether these proteins were released due to digestion by molting-associated proteases/phospholipases or merely dissociated from the chitin bundles in the cuticular laminae during apolysis is not established.

5.6 Chitin Deacetylases (CDAs)

CDAs (E.C.3.5.1.41) and chitooligosaccharide deacetylases (3.5.1.105) belong to family 4 carbohydrate esterases (CE4; www.cazy.org). CDAs are widely distributed in microorganisms and in arthropods, but only the microbial enzymes have been well characterized. They can act on nascent chitin (in coordination with chitin synthases), soluble forms of chitin such as glycol chitin, or on chitooligosaccharides, but generally have very little activity on crystalline (α or β) or colloidal chitin. But their activity on crystalline chitin can be enhanced by pre-treatment with lytic polysaccharide monooxygenases (LPMOs) that oxidatively cleave the glycoside linkage in chitin and increase the accessibility of the acetyl groups on the surface of the substrate (Liu, et al. 2017; see also Chap. 6 of this book). The CDAs have a ~150 amino acids long NodB homology domain found in the well-characterized NodB chitooligosaccharide deacetylase from *Rhizobium meliloti* (see also Chap. 3 of this book). This domain is a distorted β8α8 TIM barrel domain that is missing one β/α repeat unit and displays several loops characteristic of particular CDAs. Typically, the CE4 domain contains five conserved motifs. The first motif TFDD contains the aspartate (first asp) that acts as the general base followed by the second aspartate which chelates a Zn^{2+} ion. The two histidine residues in a second motif HS/TXXH are also involved in Zn^{2+} binding forming a loop with the His…His…Asp (Asp from motif 1) triad. The consensus of the third conserved motif is RXPY, which constitutes one side of the active site with motif 4 (DXXDW/Y) representing the other side of the active site. The catalytic acid residue is provided by the His in motif 5 (IV/ILXHD), which forms a hydrophobic pocket binding the methyl group of the C2-acetate (Grifoll-Romero et al. 2018). Recent crystal structures have provided additional insights regarding the roles of small versus large loops in determining the substrate specificity or pattern of deacetylation of chitooligosaccharides. The reaction is a metal-assisted general acid–base reaction that involves a catalytic water molecule bound in the active site pocket. Unfortunately, there is no structural information on insect CDAs. However, it is clear from chemical and spectroscopic analyses of cuticles that only about 10–20% of the total sugars in chitin are deacetylated. But there is no evidence for long stretches of deacetylation in insect chitin chains, because the end products of digestion of insect cuticles with a mixture of endo- and exochitinases are mostly monosaccharides and disaccharides with little evidence for larger chitosan oligosaccharides (Charles Specht, unpublished data).

5.6.1 Arthropod/Insect CDAs

Following the first identification of an insect CDA from *Trichoplusia ni* PM (Guo et al. 2005), a search of the completed genomes of twelve arthropods (10 insects, an arachnid (deer tick), and a crustacean (water flea)) for genes encoding CDAs and related proteins has revealed that all of them have multiple genes encoding CDAs that can be arranged phylogenetically into five groups with distinct compositions of subdomains. While all five groups of CDAs have the CE4 (NodB) domain, some groups have additional domains including a low density lipoprotein receptor a domain (LDLa; PMID:7548065) domain, a CBM14 chitin-binding domain and a long serine–threonine–proline–glutamine-rich linker domain (Dixit et al. 2008; Tetreau et al. 2015a, b). While all five groups have the five conserved sequences implicated in catalysis, metal or substrate/product binding, some of them have mutations of catalytically critical residues leading to the suggestion that they may not all be enzymatically active. The absence of definitive assays for CDA activity and the varying modes of action among CDAs (multichain, sequential, and random deacetylations) have further complicated functional analyses (Grifoll-Romero et al. 2018). Most assays involve the use of soluble chitin substrates and the detection of chitosan produced, which requires sequential deacetylation of the chitin substrate. This hampers detection of activity using enzymes that randomly deacetylate chitin or deacetylate chitin only at specific positions (at or near the ends of chitin chains, for example). However, there are some limited examples of successful detection of CDA activity as well as several examples of inability to detect deacetylase activity with purified enzymes and chitin substrates. Assays involving chitooligosaccharides as substrates and detection of the product acetate formed have proven to be the most promising assays.

Studies using homozygous mutants of *D. melanogaster* have provided evidence concerning the physiological importance of group I CDAs. These mutations in CDA1 and CDA2 (both belonging to group I CDAs) termed *serpentine* and *vermiform* affected the development of dorsal tracheal trunks during embryonic development resulting in tortuous tracheal tubules (Luschnig et al. 2006; Wang et al. 2006). In *T. castaneum*, injection of dsRNA for *CDA1* and *CDA2*, which are mainly expressed in epidermis and tracheal tissues, resulted in the arrest of every molt and in mortality (Arakane et al. 2009). Similarly, injection of dsRNA for *CDA2* on day 2 of the fifth instar nymphs of *Locusta migratoria* resulted in molting failure (Yu et al. 2016). Administration of dsRNA for *Leptinotarsa decemlineata CDA2* resulted in some larvae being trapped in their exuviae, some pupae with abnormal morphology and death at the pharate adult stage, and adults that were small, wrinkled wings, as well as adult mortality. These results suggest that depletion of CDAs leads to defective molting at all developmental stages and cuticular abnormalities in multiple insects (Wu et al. 2018).

The roles of CDA belonging to the other groups are not clear-cut. RNAi-silencing of *TcCDA6*, *7*, *8* and *9* (belonging to group V), which are only expressed in the gut tissue, did not reveal any developmental phenotypic abnormalities irrespective of whether one or all of these CDA transcripts were down-regulated (Arakane et al. 2009). However, overexpression of a group V CDA from *Helicoverpa armigera* (using a recombinant baculovirus containing this gene) resulted in the accelerated killing of the host insects, *Spodoptera frugiperda* and *S. exigua* (Jakubowska et al. 2010), when compared with insects infected with the control virus. This result suggests that the permeability of the PM might have been altered by exposure to high concentrations of this CDA.

Detailed studies on the structure of the insect cuticle using TEM analyses of cuticles from control insects and those subjected to RNAi-silencing of *CDA1* and *CDA2* from *Locusta migratoria* and *T. castaneum* have been carried out. In *L. migratoria*, pore canals in the nymphal cuticle appear to follow a crescent path in the cuticular laminae. Such a helicoidal path is consistent with the cuticle containing a stack of chitin fibril planes that are rotated with respect to one another by a small angle. However, after depletion of transcripts for *LmCDA2*, the newly formed pore canals appear to lose this helicoidal appearance and assume a straight appearance consistent with a parallel stack of chitin fibrils. Further, immunolocalization of LmCDA2 placed this protein at the apical surface of the procuticle just below the epicuticle, leading to the hypothesis that as the cuticular layers are being assembled, LmCDA2 is required to organize them in a helicoidal fashion.

Different results were obtained after similar analyses of elytral cuticles of *T. castaneum* following RNAi-silencing of *TcCDA1* or *TcCDA2*. At the larval stage, the body wall cuticle of this beetle is indeed helicoidal, but at the pupal stage when CDAs continue to be expressed at high levels, the pore canals follow a nearly straight path all the way to the epicuticle suggesting a parallel stacking of chitin fibers. In these insects, after RNAi-silencing of either *TcCDA*, the cuticle organization became disorganized, with disruption of the laminae and loss of clear-cut boundaries between the dark and light regions of the cuticular laminae and loss of pore canal integrity (Noh et al. 2018a, b). When both CDAs were down-regulated, there was no evidence of laminae. The pore canals were also distorted with loss of the long electron-lucent (presumably chitin) fibers found in the middle of the pore canals. The chitin fibrils were much smaller and did not form thick and long bundles leading to the hypothesis that CDAs are needed for higher order organization of chitin bundles in the cuticle. Based on gel electrophoresis, some of the TcCDA1 and TcCDA2 proteins appeared to form a dimer. Further, they were both localized in the assembly zone immediately above the plasma membrane, when gold-labeled antibodies were used to detect them in TEM. This observation suggests that CDAs may have a role in assembly of nascent chitin crystallites.

5.7 Chitin-Remodeling Proteins

Besides the enzymes that are known to remodel chitin by acting on the polymer directly, there are other proteins that bind to chitin and participate in organizing chitin fibrils into higher order structures. Examples of this class of proteins are Knickkopf (KNK) and several cuticular proteins belonging to the CPAP families, PMP families, and the R& R families. They have one or more of the chitin-binding motifs belonging to the CBM14 family or they have one of three consensus sequences known as Rebers & Riddiford consensus sequences. It is likely that there are other uncharacterized chitin-binding proteins with distinct biochemical functions with other motifs or domains. In this chapter, we will focus only on those proteins known to affect chitin-containing structures.

5.7.1 Knickkopf Family of Proteins

Knickkopf (German word meaning literally kinked head) was initially identified as a gene that affected cuticle integrity in mutational screens of *D. melanogaster* (Weischaus et al. 1984) and was cloned by Ostrowski et al. (2002). Since then orthologs and paralogs of this gene have been identified in a variety of insects and nematodes. There are three members in this family in insect genomes. All three of them have two DM13 domain, a dopamine monooxygenase *N*-terminal domain (DOMON) and a long *C*-terminal domain whose function remains undetermined. The DM13 and DOMON (Pfam 10517 and 03351 domains, respectively) are associated with electron transport proteins involved in hydroxylation and oxidative cross-linking of proteins. The DOMON domain proteins also contain *C*-terminal domains that are predicted to provide thiol groups as binding sites for a heme prosthetic group associated with cytochrome (Iyer et al. 2007). Other ligands such as dopamine and carbohydrates may also be potentially linked to this *C*-terminal domain. The DM13 domain has a fold rich in β-strands.

In *D. melanogaster*, mutations of the KNK gene result in embryonic death and loss of chitin organization in tracheal tubes which take on a cystic appearance (Moussian et al. 2005). They seem to be deficient in the fibrillar chitin fibers and take on an amorphous appearance. In *T. castaneum*, RNAi-silencing of this major *KNK* that is expressed in the integument (but not in the gut) results in a clear reduction in cuticular chitin, loss of laminar organization of the cuticle, severe molting defects and lethality at all developmental stages (Chaudhari et al. 2011). The other two *KNK* paralogs (*KNK2* and *KNK3*) are essential for adult molting and cause lethality but seem to affect only body wall denticles and tracheal taenidia, indicating their unique functions in specialized cuticles (Chaudhari et al. 2014).

The TcKNK protein has been expressed in a baculovirus system and shown to be a GPI-anchored protein that can be released from insect cells by treatment with phosphoinositide-specific phospholipase C (Chaudhari et al. 2011). In vivo, this protein is, in fact, distributed throughout the procuticle and seems to be co-localized with chitin throughout the procuticle. The purified protein does bind to colloidal chitin and seems to protect it from chitinases in vivo, because the reduction in chitin content of the pharate adult brought about by RNAi of *TcKNK* is ameliorated by the simultaneous down-regulation of the two major chitinase(s) in the molting fluid by RNAi (Chaudhari et al. 2011). But the chitin that accumulates in these insects lacking both KNK and chitinases is not organized in the form of alternating light and dark laminae or long pore canal fibers. Instead smaller, thinner filaments accumulate in the pharate adult elytral cuticle indicating that KNK not only protects chitin fibers but also organizes them into thicker and longer bundles that further assemble into orderly horizontal laminae and long vertical pore canal fibers. The precise mechanism of how KNK (most likely in cooperation with several other proteins) brings about this process has not been elucidated. However, the timing and level of expression of KNK relative to the period (and amount) of chitin synthesis appear to be important in maintaining the structural integrity of the wing cuticle, as well as other cuticular structures such as bumps in *D. melanogaster* epidermal tissue (Li et al. 2017).

5.7.2 CPAP Proteins

The insect cuticle contains numerous proteins that exhibit substantial variations in number and composition depending on the type of cuticle and the developmental stage. There are compositional differences between larval and adult cuticle and between soft and hard cuticle at the same developmental stage (Dittmer et al. 2012; Zhou et al. 2016). These proteins interact with chitin and alter its physical properties such as viscoelastic properties and permeability characteristics. The CPAP proteins are characterized by the presence of one (CPAP1s) or three (CPAP3s) CBD domains. This domain is characterized by the presence of a 6-cysteine containing chitin-binding motif with a characteristic spacing between successive cysteines. These cysteines are most likely involved in disulfide bond formation and assume well-defined three-dimensional structures. CPAP1 family and CPAP3 families are further subdivided into 16 and 7 subgroups, respectively, based on the sequence conservation between the cysteines (Tetreau et al. 2015b; Jasrapuria et al. 2010). The sequence conservation in the CBD region within each subgroup is quite high, at least for some groups suggesting a biological need for conservation of these sequences. The CBDs have a strong affinity for chitin and the affinity of proteins containing these domains seems to increase with the number of CBDs (Arakane et al. 2003). The evolutionary conservation of these subgroups in several insect orders indicates their essential nature and functional specialization over a long period of insect/arthropod evolution.

A detailed study of the chitin- and chitosan-binding properties of six CPAP3 proteins from *B. mori* has been carried out using purified preparations following expression in *E. coli* (Qu et al. 2017). All of these proteins bound strongly to crystalline chitin and colloidal chitin. However, they differed in their affinity to partially deacetylated chitin. BmCPAP3-D1 had the highest affinity for chitosan (70 or 100% deacetylation). Curiously, this protein was up-regulated during pupal–adult transition suggesting a physiological role for this protein at this developmental stage.

In CPAP1 family, the CBD domain is located generally near the *N*-terminus but may be located in the middle or near the *C*-terminus in others. The CPAP1 proteins show substantial variation in their total length and are expressed mostly in cuticle-forming tissues but not in the gut. In contrast, the CPAP3 proteins have a narrow size variation and the three CBD domains account for almost all of their total length except for the two short spacers between them. They all have a signal peptide, consistent with their being extracellular proteins, where they can interact with chitin. CPAP proteins may have a structural function or an enzymatic function in which the CBD may serve to anchor the protein on chitin.

The roles of CPAP1 and CPAP3 have been investigated by mutational and/or RNAi studies in several insects (Jasrapuria et al. 2012; Petkau et al. 2012; Pesch et al. 2015). These studies have indicated that at least some of these proteins are essential for survival, molting, cuticle integrity, and fecundity and that their unique functions cannot be substituted by other CPAP proteins. Among the CPAP1 family members, only three of them have been shown to have essential functions based on RNAi studies in *T. castaneum*. These include *TcCPAP1-C, TcCPAP1-H*, and *TcCPAP1-J*. Interestingly, these are the subgroups that show the highest levels of sequence conservation in the CBDs among insects (68–85%). Depletion of these transcripts individually leads to lethality at the pharate adult stage, loss of chitin and/or structural integrity of the elytral cuticle or embryonic arrest (Jasrapuria et al. 2012). The functions of the other CPAP1 proteins have not been investigated in detail.

The *CPAP3* genes have been studied in greater detail in several orders of insects. Knocking down *CPAP3* genes of *T. castaneum* individually caused morphological effects that were varied and led to a wide range of phenotypes. The observed phenotypes included molting defects, mortality, depletion of fat body, underdeveloped ovaries, loss of fecundity, joint defects, and rough elytra (Jasrapuria et al. 2012). The two exceptions were *TcCPAP3A2* and *TcCPAP3E*, which did not yield any observable phenotypes after injection of the corresponding dsRNAs.

In *D. melanogaster*, homozygous mutants of the *CPAP3-A* gene (called *Obst-A*) exhibited growth reduction, molting defects, and defective wound healing (Petkau et al. 2012). The corresponding protein was localized (or present) in the assembly zone where it may participate in the assembly of chitin along with other proteins that also localize to (or present) this region including CDA1 and CDA2 and KNK (Noh et al. 2018a, b; Pesch et al. 2015). It is noteworthy that reduction in the amount of any of these proteins results in loss of laminar organization of the epidermal procuticle. The precise roles of each of these chitin-remodeling proteins remain to be established.

5.7.3 Peritrophic Matrix Proteins (PMPs)

Besides being the supporting matrix of the cuticle, chitin is also found in the peritrophic matrix, which is also an extracellular matrix elaborated by the epithelial cells lining the midgut (Hegedus et al. 2009; Merzendorfer et al. 2016). The enzyme responsible for the synthesis of chitin associated with PM is chitin synthase B (CHS-B), which is a paralog of chitin synthase A (CHS-A) which makes cuticular chitin, as discussed in Sect. 5.2.2. CHS-B resides in the apical tips of brush border microvilli of the midgut (Zimoch and Merzendorfer 2002). When the gene encoding CHS-B in *T. castaneum* is silenced by RNAi, the structural integrity and barrier function of the peritrophic matrix is lost (Kelkenberg et al. 2015). In addition, several other proteins that accumulate in the assembly zone (Knickkopf, CDAs, and CPAPs) and are implicated in the higher order assembly of chitin fibers are not expressed in the midgut-lining cells. In addition, the assortment of proteins expressed in these cells do not include many of the R & R proteins, but consist of another family of proteins, named PMPs. This class of proteins contains CBM14 domains closely related to, but slightly different from, the CBDs found in CPAP1 and CPAP3 proteins of insect cuticles (Jasrapuria et al. 2010; Tetreau et al. 2015b). These dissimilarities probably account for the fundamental differences in the properties of the chitin polymers found in these two locations. While the cuticular chitin is rigid and hydrophobic, the PM-associated chitin is a gel that is highly hydrated, flexible and permeable to small solutes and water.

Unlike cuticular chitin, which exists in close association with the epidermal cells as a multilaminar hydrophobic structure, the PM is a sieve-like structure with criss-crossing chitin fibers that delaminate away from the finger-like protrusions of microvilli of midgut-lining cells (Harper and Hopkins 1997; Hopkins and Harper 2001). Multiple laminations of chitin networks are often found especially in the middle and posterior parts of the midgut. Chitin fibers in PM are mostly arranged as orthogonal or hexagonal lattices or as random felts. There are large pores in the PM that allow for the passage of digestive enzymes secreted by the epithelial cells lining the midgut and for digestion products to pass through the PM in opposite directions (Bolognesi et al. 2008). In *T. castaneum*, the size exclusion decreases along the length of the midgut, allowing larger molecules to cross the PM in the anterior parts (size >40 nm), intermediate in the middle midgut (size 8–9 nm), but much smaller in the posterior parts (size 1–2 nm) (Agrawal et al. 2014). There is evidence suggesting that the permeability may be influenced by particular PMPs that are associated with PM at specific regions of the midgut. The PMPs vary substantially in their sizes depending on the number of CBD domains in each protein. This number varies from 1 to as many as 19, which may allow binding/cross-linking of different chitin chains by a single protein forming a three-dimensional network (Jasrapuria et al. 2010; Dinglassen et al. 2009; Venancio et al. 2009; Toprak et al. 2015; Shen and Jakobs-Lorena 1998, Agrawal et al. 2014). It is likely that the ratio of protein to chitin is not uniform along the length of the PM from the anterior side to the posterior. The posterior parts of the PM are mechanically stronger than the anterior part, which sometimes exists as a thin gel-like substance (also termed peritrophic gel; Terra 2001). The expression of the individual *PMP* genes varies along the midgut in *T. castaneum*, where the expression of the genes encoding smaller PMPs is predominantly in the anterior and middle parts and the expression of genes encoding the largest PMPs is confined to the posterior midgut. The proteins with the most CBDs were also confined to the posterior PM (Jasrapuria et al. 2012; Agrawal et al 2014).

Another function of the PM-associated proteins may also be to influence the permeability characteristics of the PM. Some of the PMPs have large mucin domains that are rich in serine and threonine residues and are indeed glycosylated. (Schorderet et al. 1998; Vuocolo et al. 2001; Agrawal et al. 2014; Toprak et al. 2010, 2015). In *T. castaneum*, RNAi of the two PMPs demonstrated to be glycosylated (TcPMP3 and TcPMP5B) resulted in a substantial increase in exclusion size compared to the other TcPMPs that do not have mucin domains (Agrawal et al.

2014). Whether this is a direct effect on pore size or due to binding of other proteins to the sugar moieties is unclear. In fact, feeding of lectins has been shown to cause lethality of *L. cuprina* larvae (Eisemann et al. 1994). An additional way in which PMPs can affect PM permeability is due to charge effects. It is known that cations pass through the PM more readily than anions in the presence of glycosylated proteins (Miller and Lehane 1993; Barbehenn 2001).

5.7.4 CPR Family Proteins

Besides CPAP proteins, there is a large assortment of cuticular proteins (CPRs) with the Rebers & Riddiford (R & R) chitin-binding sequences. Homology modeling of this consensus sequence of 66 amino acids has revealed the structure of an antiparallel β-sheet half barrel, into which a chitin chain could be fitted, and in which the aromatic amino acid residues are stacked on the sugar planes (Hamodrakas et al. 2005). Some of the CPR proteins have been shown to bind to chitin (Willis et al. 2012). The RR1 proteins are generally localized in the soft procuticle and intersegmental membranes and the RR2 proteins are consistently found in the hard cuticle (Zhou et al. 2016). Depletion of two major cuticular proteins belonging to the RR2 family that is found in the laminae and pore canals of hard cuticles of *T. castaneum* (TcCPR27 and TcCPR18) resulted in wrinkled elytra, unfolded hindwings and death. On the other hand, there were alterations in the dynamic mechanical properties of the elytra that were suggestive of a greater degree of cross-linking of proteins. The horizontal laminar organization was less organized, and the pore canals were distorted; they lacked the thick pore canal fibers (Noh et al. 2014). Another *T.castaneum* RR1 protein, TcCPR4, was found exclusively in the pore canals. RNAi-silencing of this gene results in abnormally shaped pore canals filled with loose fibers rather than well-organized bundles of chitin around the periphery with a central column of a thick chitin bundle (Noh et al. 2015). Taken together, these results indicate that CPR cuticular proteins also contribute to the overall organization of chitin in all cuticles.

5.7.5 Other Chitin Modifying Enzymes

Besides the enzymes and binding proteins that interact with chitin as described above, there are other proteins that modify chitin and participate in chitin remodeling. Most prominent among them are lytic polysaccharide monooxygenases (LPMO) which is the subject of another chapter in this book (Aachman and Eijsink, Chap. 6). Other enzymes including transglutaminases or laccases may be involved in cross-linking of cuticular proteins and possibly chitin and they are also not covered in this chapter.

5.7.6 Overview of Chitin Metabolism and Enzymes Involved

Figure 5.1 summarizes the roles of enzymes involved in chitin metabolism. Starting from the activated precursor UDP-*N*-acetylglucosamine and a growing chitin chain as a primer, chitin is synthesized by either CHS-A or CHS-B. Chitin deacetylases

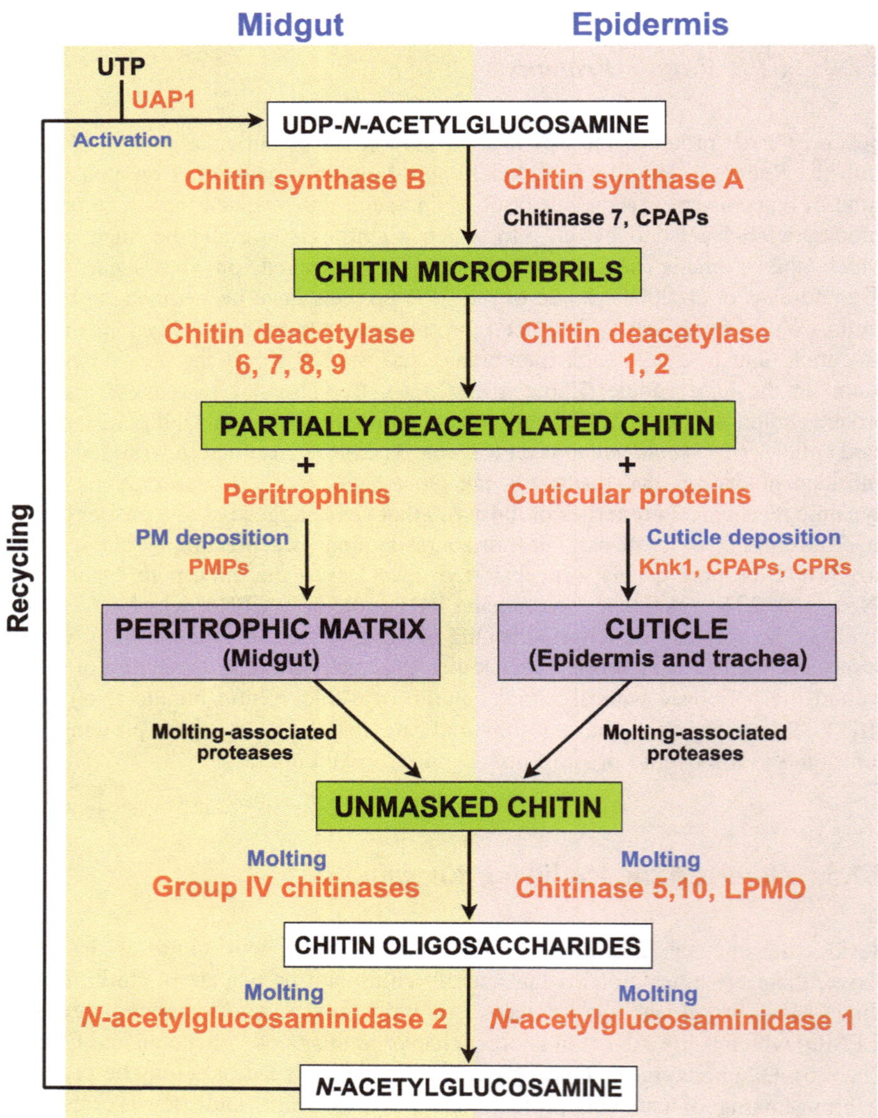

Fig. 5.1 Overview of chitin metabolism in insects

partially deacetylate the nascent chitin in the PM or the cuticle. The partially deacetylated chitin fibers associate with either cuticular proteins or PMPs to form the matrices containing chitin and proteins. At the time of molting, appropriate proteases unmask the chitin chains, which are then acted upon by a mixture of chitinases and *N*-acetylglucosaminidases to yield monomeric *N*-acetylglucosamines, which are then recycled after activation by UTP-*N*-acetylglucosamine pyrophosphorylase (UAP).

5.8 Conclusions

From the time the chitin polymeric chains emerge from the catalytic center of oligomeric assemblies of CHS enzymes, they interact with other nascent chitin chains and a whole assortment of chitin-remodeling enzymes and proteins that determine its final properties such as crystallinity, size, rigidity, elasticity, state of hydration, higher order organization, and cross-linking with other components including phenolics and minerals. These allow the cells that make chitin a variety of choices leading to the production of extracellular matrices with a wide range of physiochemical and biological properties appropriate for the anatomical region and the developmental stage of the organism. Likewise, during the process of disassembly of these matrices, a whole assortment of depolymerizing enzymes and proteins are involved that work in concert to allow remodeling and reutilization of the components of the old matrix. These processes are under a variety of hormonal and developmental controls that offer numerous possibilities for the performance of these organisms in a hostile environment. Further, a better understanding of these complex processes will allow the production of biomimetic agents with desirable properties for industrial and biomedical applications.

Acknowledgements This work was supported by the Basic Science Research Program through the National Research Foundation of Korea (NRF) funded by the Korea government (MSIT) (2018R1A2B6005106) to YA.

References

Agrawal S, Kelkenberg M, Begum K, Steinfeld L, Williams CE, Kramer KJ, Beeman RW, Park Y, Muthukrishnan S, Merzendorfer H (2014) Two essential peritrophic matrix proteins mediate matrix barrier functions in the insect midgut. Insect Biochem Mol Biol 49:24–34

Arakane Y, Zhu Q, Matsumiya M, Muthukrishnan S, Kramer KJ (2003) Properties of catalytic, linker and chitin-binding domains of insect chitinase. Insect Biochem Mol Biol 33:631–648

Arakane Y, Hogenkamp DG, Zhu YC, Kramer KJ, Specht CA et al (2004) Characterization of two chitin synthase genes of the red flour beetle, *Tribolium castaneum,* and alternate exon usage in one of the genes during development. Insect Biochem Mol Biol 34:291–304

Arakane Y, Muthukrishnan S, Kramer KJ, Specht CA, Tomoyasu Y et al (2005) The Tribolium chitin synthase genes TcCHS1 and TcCHS2 are specialized for synthesis of epidermal cuticle and midgut peritrophic matrix. Insect Mol Biol 14:453–463

Arakane Y, Li B, Muthukrishnan S, Beeman RW, Kramer KJ, Park Y (2008a) Functional analysis of four neuropeptides, EH, ETH, CCAP and bursicon and their receptors in adult ecdysis behavior of the red flour beetle, *Tribolium castaneum*. Mech Dev 125:984–995

Arakane Y, Specht CA, Kramer KJ, Muthukrishnan S, Beeman RW (2008b) Chitin synthases are required for survival, fecundity and egg hatch in the red flour beetle, *Tribolium castaneum*. Insect Biochem Mol Biol 38:959–962

Arakane Y, Dixit R, Begum K, Park Y, Specht CA et al (2009) Analysis of functions of the chitin deacetylase gene family in *Tribolium castaneum*. Insect Biochem Mol Biol 39:355–365

Arakane Y, Muthukrishnan S (2010) Insect chitinase and chitinase-like proteins. Cell Mol Life Sci 67:201–216

Barbehenn RV (2001) Roles of peritrophic membranes in protecting herbivorous insects from ingested plant allelochemicals. Arch Insect Biochem Physiol 47:86–99

Beckham GT, Crowley MF (2011) Examination of the α-chitin structure and decrystallization thermodynamics at the nanoscale. J Phys Chem B 115:4516–4522

Bolognesi R, Terra WR, Ferreira C (2008) Peritrophic membrane role in enhancing digestive efficiency. Theoretical and experimental models. J Insect Physiol 54:1413–1422

Brameld KA, Shrader WD, Imperiali B, Goddard WAIII (1998) Substrate assistance in the mechanism of family 18 chitinases: theoretical studies of potential intermediates and inhibitors. J Mol Biol 280:913–923

Chaudhari SS, Arakane Y, Specht CA, Moussian B, Boyle DL et al (2011) Knickkopf protein protects and organizes chitin in the newly synthesized insect exoskeleton. Proc Natl Acad Sci USA 108:17028–17033

Chaudhari SS, Moussian B, Specht CA, Arakane Y, Kramer KJ, Beeman RW, Muthukrishnan S (2014) Functional specialization among members of Knickkopf family of proteins in insect cuticle organization. PLoS Genet 2014(10):e1004537

Chen L, Liu T, Zhou Y, Chen Q, Shen X, Yang Q (2014) Structural characteristics of an insect group I chitinase, an enzyme indispensable to moulting. Acta Crystallographica D Biol Crystallogr 70:932–942

Chen C, Yang H, Tang B, Yang WJ, Jin DC (2017) Identification and functional analysis of chitinase 7 gene in white-backed planthopper, *Sogatella furcifera*. Comput Biochem Physiol B: Biochem Mol Biol 208–209:19–28

Chen X, Xu K, Yan X, Chen C, Cao Y, Wang Y, Li C, Yang W (2018a) Characterization of a β-*N*-acetylglucosaminidase gene and its involvement in the development of *Lasioderma serricorne* (Fabricius). Stored Product Res 77:156–165

Chen W, Qu M, Zhou Y (2018b) Qing Y (2018) Structural analysis of group II chitinase (ChtII) catalysis completes the puzzle of chitin hydrolysis in insects. J Biol Chem 293:2652–2660

Cohen E (2010) Chitin biochemistry: synthesis, hydrolysis and inhibition. In: Jérôme C, Stephen JS (eds) Advances in insect physiology, vol 38. Academic Press, New York, pp 5–74

Dinglasan RR, Devenport M, Florens L, Johnson JR, McHugh CA, Donnelly-Doman M, Carucci DJ, Yates JR, 3r, Jacobs-Lorena, M (2009) The *Anopheles gambiae* adult midgut peritrophic matrix proteome. Insect Biochem Mol Biol 39:125–134

Dittmer NT, Hiromasa Y, Tomich JM, Lu N, Beeman RW, Kramer KJ, Kanost MR (2012) Proteomic and transcriptomic analyses of rigid and membranous cuticles from the elytra and hindwings of the red flour beetle. Tribolium castaneum. J. Proteome Research 11:269–278

Dixit R, Arakane Y, Specht CA, Richard C, Kramer KJ et al (2008) Domain organization and phylogenetic analysis of proteins from the chitin deacetylase gene family of *Tribolium castaneum* and three other species of insects. Insect Biochem Mol Biol 38:440–451

Dorfmueller HC, Ferenbach AT, Borodkin VS, van Aalten DM (2014) A structural and biochemical model of processive chitin synthesis. J Biol Chem 289:23020–23028

Dziadik-Turner C, Koga D, Mai MS, Kramer KJ (1981) Purification and characterization of two β-*N*-acetylhexosaminidases from the tobacco hornworm, Manduca sexta (L.) (Lepidoptera: Sphingidae). Arch Biochem Biophys 21:546–560

Eisemann CH, Donaldson RA, Pearson RD, Cadogan LC, Vuocolo T, Tellam RL (1994) Larvicidal activity of lectins on *Lucilia cuprina*: Mechanism of action. Ent Exp Appl 72:1–10

Fabritius HO, Sachs C, Triguero PR, Raabe D (2009) Influence of structural principles on the mechanics of a biological fiber-based composite material with hierarchical organization: The Exoskeleton of the Lobster *Homarus americanus*. Adv Mater Sci 21:391–400

Filho BP, Lemos FJ, Secundino NF, Pascoa V, Pereira ST, Pimenta PF (2002) Presence of chitinase and beta-*N*-acetylglucosaminidase in the *Aedes aegypti*: A chitinolytic system involving peritrophic matrix formation and degradation. Insect Biochem Mol Biol 32:1723–1729

Fukamizo T, Kramer KJ (1985) Mechanism of chitin hydrolysis by the binary chitinase system in insect moulting fluid. Insect Biochem 15:141–145

Gohlke S, Muthukrishnan S, Merzendorfer H (2017) *In Vitro* and *In Vivo* studies on the structural organization of Chs3 from *Saccharomyces cerevisiae*. Int J Mol Sci 18:E702

Grifoll-Romero L, Pascual S, Aragunde H, Biarnés X, Planas A (2018) Chitin deacetylases: structures. Specific Biotech Appl Polymers 10:352

Guo W, Li G, Pang Y, Wang P (2005) A novel chitin-binding protein identified from the peritrophic membrane of the cabbage looper, *Trichoplusia ni*. Insect Biochem Mol Biol 35:1224–1234

Harper MS, Hopkins TL (1997) Peritrophic membrane structure and secretion in European corn borer larvae (*Ostrinia nubilalis*). Tissue Cell 29:463–475

He B, Chu Y, Yin M, Müllen K, An C, Shen J (2013) Fluorescent nanoparticle delivered dsRNA toward genetic control of insect pests. Adv Mater 25:4580–4584

Hegedus D, Erlandson M, Gillott C, Toprak U (2009) New insights into peritrophic matrix synthesis, architecture, and function. Annu Rev Entomol 54:285–302

Henrissat B (1991) A classification of glycosylhydrolases based on amino acid sequence similarities. Bochem J. 280:309–316

Hogenkamp DG, Arakane Y, Zimoch L, Merzendorfer H, Kramer KJ et al (2005) Chitin synthase genes in *Manduca sexta*: characterization of a gut-specific transcript and differential tissue expression of alternately spliced mRNAs during development. Insect Biochem Mol Biol 35:529–540

Hogenkamp DG, Arakane Y, Kramer KJ, Muthukrishnan S, Beeman RW (2008) Characterization and expression of the & #x03B2;-N-acetylhexosaminidase gene family of Tribolium castaneum. Insect Biochem Mol Biol 38:478–489

Hopkins TL, Harper MS (2001) Lepidopteran peritrophic membranes and effects of dietary wheat germ agglutinin on their formation and structure. Arch Insect Biochem Physiol 47:100–109

Hamodrakas SJ, Willis JH, Iconomidou VA (2005) A structural model of the chitin-binding domain of cuticle proteins. Insect Biochem Molec Biol 32:1577–1583

Intra J, Pavesi G, Horner DS (2008) Phylogenetic analyses suggest multiple changes of substrate specificity within the glycosyl hydrolase 20 family. BMC Evol Biol 8:214

Iyer LM, Anantharaman V, Aravind L (2007) The Domon domains are involved in heme and sugar recognition. Bioinformatics 23:2660–2664

Jakubowska AK, Caccia S, Gordon KH, Ferre J, Herrero S (2010) Downregulation of a chitin deacetylase-like protein in response to baculovirus infection and its application for improving baculovirus infectivity. J Virol 84:2547–2555

Jasrapuria S, Arakane Y, Osman G, Kramer KJ, Beeman RW, Muthukrishnan S (2010) Genes encoding proteins with peritrophin A-type chitin-binding domains in *Tribolium castaneum* are grouped into three distinct families based on phylogeny, expression and function. Insect Biochem Mol Biol 40:214–227

Jasrapuria S, Specht CA, Kramer KJ, Beeman RW, Muthukrishnan S (2012) Gene families of cuticular proteins analogous to peritrophins (CPAPs) in *Tribolium castaneum* have diverse functions. PLoS ONE 7:e49844

Kawamura K, Shibata T, Saget O, Peel D, Bryant PJ (1999) A new family of growth factors produced by the fat body and active on *Drosophila* imaginal disc cells. Development 126:211–219

Kaya M, Baran T, Erdoğan S, Menteş A, Özüsağlam MA, Çakmak YS (2014) Physicochemical comparison of chitin and chitosan obtained from larvae and adult Colorado potato beetle (*Leptinotarsa decemlineata*). Mater Sci Eng 45:72–81

Kelkenberg M, Odman-Naresh J, Muthukrishnan S, Merzendorfer H (2015) Chitin is a necessary component to maintain the barrier function of the peritrophic matrix in the insect midgut. Insect Biochem Mol Biol 56:21–28

Khajuria C, Buschman LL, Chen MS, Muthukrishnan S, Zhu KY (2010) A gut-specific chitinase gene essential for regulation of chitin content of peritrophic matrix and growth of *Ostrinia nubilalis* larvae. Insect Biochem Mol Biol 40:621–629

Kimura S (1976) Insect haemolymph exo-β -*N*-acetylglucosaminidase from *Bombyx mori*. Purification and properties. Biochim Biophys Acta 446:399–406

Koga D, Mai MS, Dziadik-Turner C, Kramer KJ (1982) Kinetics and mechanism of exochitinase and ß-*N*-acetylhexosaminidase from the tobacco hornworm, *Manduca sexta* L. (Lepidoptera: Sphingidae). Insect Biochem 12:493–499

Koga D, Sasaki Y, Uchiumi Y, Hirai N, Arakane Y, Nagamatsu Y (1997) Purification and characterization of *Bombyx mori* chitinases. Insect Biochem Mol Biol 27:757–767

Kramer KJ, Corpuz L, Choi HK, Muthukrishnan S (1993) Sequence of a cDNA and expression of the gene encoding epidermal and gut chitinases of *Manduca sexta*. Insect Biochem Mol Biol 23:691–701

Leonard R, Rendic D, Rabouille C, Wilson IB, Preat T, Altmann F (2006) The *Drosophila* fused lobes gene encodes an *N*-acetylglucosaminidase involved in *N*-glycan processing. J Biol Chem 281:4867–4875

Li H, Greene LH (2010) Sequence and structural analysis of the chitinase insertion domain reveals two conserved motifs involved in chitin binding. PLoS ONE 5:e8654

Li D, Zhang J, Wang Y, Liu X, Ma E et al (2015) Two chitinase 5 genes from *Locusta migratoria*: Molecular characteristics and functional differentiation. Insect Biochem Mol Biol 58:46–54

Li K, Zhang X, Lou W, Zhang J, Moussian B (2017) Timed Knickkopf function is essential for wing cuticle formationin Drosophila melanogaster. Insect Biochem Mol Biol 89:1–10

Liu T, Zhang H, Liu F, Wu Q, Shen X, Yang Q (2011) Structural determinants of an insect & #x03B2;-N-acetyl-D-hexosaminidase specialized as a chitinolytic enzyme. J Biol Chem 286:4049–4058

Liu F, Liu T, Qu M, Yang Q (2012) Molecular and biochemical characterization of a novel β-*N*-acetyl-D-hexosaminidase with broad substrate-spectrum from the Asian corn borer. O. furnacalis. Int J Biol Sci 8:1085–1096

Liu XJ, Li F, Li DQ, Ma EB, Zhang WQ, Zhu KY, Zhang JZ (2013) Molecular and functional analysis of UDP-*N*-acetylglucosamine pyrophosphorylases from the migratory locust. Locusta migratoria. PLoS ONE 8:e71970

Liu T, Zhu W, Wang J, Zhou Y, Duan Y, Qu M, Yang Q (2018) The deduced role of a chitinase containing two non-synergistic catalytic domains. Acta Cryst D74:21–29

Liu Z, Gay LM, Tuveng TR, Agger JW, Westereng B, Mathiesen G, Horn SJ, Vaaje-Kolstad G, van Aalten DMF, Eijsink VGH (2017) Structure and function of a broad-specificity chitin deacetylase from *Aspergillus nidulans* FGSC A4. Sci Rep 7:1746

Lu YM, Zen KC, Muthukrishnan S, Kramer KJ (2002) Site-directed mutagenesis and functional analysis of active site acidic amino acid residues D142, D144 and E146 in *Manduca sexta* (tobacco hornworm) chitinase. Insect Biochem Mol Biol 32:1369–1382

Luschnig S, Batz T, Armbruster K, Krasnow MA (2006) *Serpentine* and *vermiform* encode matrix proteins with chitin binding and deacetylation domains that limit tracheal tube length in *Drosophila*. Curr Biol 16:186–194

Maue L, Meissner D, Merzendorfer H (2009) Purification of an active, oligomeric chitin synthase complex from the midgut of the tobacco hornworm. Insect Biochem Mol Biol 39:654–659

Merzendorfer H (2013) Chitin synthesis inhibitors: old molecules and new developments. Insect Sci 20:121–138

Merzendorfer H, Zimoch L (2003) Chitin metabolism in insects: structure, function and regulation of chitin synthases and chitinases. J Exp Biol 206:4393–4412

Merzendorfer H, Kelkenberg M, Muthukrishnan S (2016) Peritrophic matrices. In: Moussian B, Cohen E (eds) Extracellular composite matrices in arthropods, Chap. 8. Springer, Switzerland, pp 255–324

Miller N, Lehane MJ (1993) Ionic environment and the permeability properties of the peritrophic membrane of *Glossina morsitans morsitans*. J Insect Physiol 631:139–144

Morgan JL, McNamara JT, Fischer M, Rich J, Chen HM, Withers SG, Zimmer J (2016) Observing cellulose biosynthesis and membrane translocation in crystallo. Nature 531:329–334

Moussian B, Tang E, Tonning A, Helms S, Schwarz H et al (2005) *Drosophila* Knickkopf and Retroactive are needed for epithelial tube growth and cuticle differentiation through their specific requirement for chitin filament organization. Development 133:163–171

Muthukrishnan S, Merzendorfer H, Arakane Y, Kramer KJ (2012) Chitin metabolism in insects. In: Gilbert LI (ed) Insect biochemistry and molecular biology. Elsevier, San Diego, pp.193–235

Muthukrishnan S, Merzendorfer H, Arakane Y, Yang Q (2016) Chitin metabolic pathways in insects and their regulation. In: Moussian B, Cohen E (eds) Extracellular composite matrices in arthropods, Chap. 2. Springer, Switzerland, pp 31–65

Muzzarelli RAA (1973) Chitin. In: Muzzarelli RAA (ed) Natural chelating polymers: alginic acid, chitin, and chitosan. Pergamon Press, New York, pp 83–252

Nagamatsu Y, Yanagisawa I, Kimoto M, Okamoto E, Koga D (1995) Purification of a chitooligosaccharidolytic beta-*N*-acetylglucosaminidase from *Bombyx mori* larvae during metamorphosis and the nucleotide sequence of its cDNA. Biosci Biotechnol Biochem 59:219–225

Neville AC, Parry DA, Woodhead-Galloway J. (1976) The chitin crystallite in arthropod cuticle. J Cell Sci 21:73–82

Noh MY, Kramer KJ, Muthukrishnan S, Arakane Y (2014) Two major cuticular proteins are required for assembly of horizontal laminae and vertical pore canals in rigid cuticle of *Tribolium castaneum*. Insect Biochem Mol Biol. 53:22–29

Noh M, Muthukrishnan S, Kramer KJ, Arakane Y (2018a) A chitinase with two catalytic domains is required for organization of the cuticular extracellular matrix of a beetle. PLOS Genetics| https://doi.org/10.1371/journal.pgen.1007307

Noh M, Muthukrishnan S, Kramer KJ, Arakane Y (2018b) Group I chitin deacetylases are essential for higher order organization of chitin fibers in beetle cuticle. J Biol Chem 293:6985–6995

Ostrowski S, Dierick HA, Bejsovec A (2002) Genetic control of cuticle formation during embryonic development of *Drosophila melanogaster*. Genetics 161:171–182

Pesch YY, Riedel D, Behr M (2015) Obstructor A organizes matrix assembly at the apical cell surface to promote enzymatic cuticle maturation in *Drosophila*. J Biol Chem 290:10071–10082

Pesch Y, Riedel D, Patil KR, Loch G, Behr M (2016a) Chitinases and Imaginal disc growth factors organize the extracellular matrix formation at barrier tissues in insects. Sci Rep 6, Article number:18340

Pesch Y, Riedel D, Loch G, Behr M (2016b) Drosophila chitinase 2 is expressed in chitin producing organs for cuticle formation. Arthropod Struct Dev 46:4–12

Petkau G, Wingen C, Jussen LC, Radtke T, Behr M (2012) Obstructor-A is required for epithelial extracellular matrix dynamics, exoskeleton function, and tubulogenesis. J Biol Chem 287:21396–21405

Qu M, Yang Q (2011) A novel alternative splicing site of class A chitin synthase from the insect *O. furnacalis*- Gene organization, expression pattern and physiological significance. Insect Biochem Mol Biol 41:923–931

Qu M, Yang Q (2012) Physiological significance of alternatively spliced exon combinations of the single-copy gene class A chitin synthase in the insect *O. furnacalis* (Lepidoptera). Insect Mol Biol 21:395–404

Qu M, Ma L, Chen P, Yang Q (2014) Proteomic analysis of insect molting fluid with a focus on enzymes involved in chitin degradation. J Proteome Res 13:2931–2940

Qu M, Ren Y, Liu Y, Yang Q (2017) Studies on the chitin/chitosan binding properties of six cuticular proteins analogous to peritrophin 3 from *Bombyx mori*. Insect Molec Biol 26432–439

Reynolds SE, Samuels RI (1996) Physiology and biochemistry of insect molting fluid. Adv Insect Physiol 26:157–232

Sacristan C, Manzano-Lopez J, Reyes A, Spang A, Muniz M, Roncero C (2013) Oligomerization of the chitin synthase Chs3 is monitored at the Golgi and affects its endocytic recycling. Mol Microbiol 90:252–266

Schorderet S, Pearson RD, Vuocolo T, Eisemann C, Riding GA, Tellam RL (1998) cDNA and deduced amino acid sequences of a peritrophic membrane glycoprotein, 'peritrophin-48', from the larvae of *Lucilia cuprina*. Insect Biochem Mol Biol 28:99–111

Shen Z, Jacobs-Lorena M (1998) A type I peritrophic matrix protein from the malaria vector *Anopheles gambiae* binds to chitin. Cloning, expression, and characterization. J Biol Chem 273:17665–17670

Shirk PD, Perera OP, Shelby KS, Furlong RB, LoVullo ED, Pophamc HJR (2015) Unique synteny and alternate splicing of the chitin synthases in closely related heliothine moths. Gene 574:121–139

Su C, Tu G, Huang S, Yang Y, Shahad MF, Li F (2016) Genome-wide analysis of chitinase genes and theirvaried functions in larval moult, pupation and eclosion in the rice striped stem borer, *Chilo suppressalis*. Insect Mol Biol 25:401–412

Tellam RL (1996) The peritrophic matrix. In: Lehane MJ, Billingsley PF (eds) Biology of the insect midgut. Chapman-Hall, Cambridge, pp. 86–114

Tellam RL, Vuocolo T, Johnson SE, Jarmey J, Pearson RD (2000) Insect chitin synthase cDNA sequence, gene organization and expression. Eur J Biochem 267:6025–6043

Terra WR (2001) The origin and functions of the insect peritrophic membrane and peritrophic gel. Arch Insect Biochem Physiol 47:47–61

Tetreau G, Cao XL, Chen YR, Muthukrishnan S, Jiang HB et al (2015a) Overview of chitin metabolism enzymes in *Manduca sexta*: Identification, domain organization, phylogenetic analysis and gene expression. Insect Biochem Mol Biol 62:114–126

Tetreau G, Dittmer NT, Caoc X, Agrawal S, Chen Y-R et al (2015b) Analysis of chitin-1 binding proteins from *Manduca sexta* provides new insights into evolution of peritrophin A type chitin-binding domains in insects. Insect Biochem Mol Biol 62:27–41

Tews I, van Scheltinga T, Perrakis A, Wilson KS, Dijkstra BW (1997) Substrate-assisted catalysis unifies two families of chitinolytic enzymes. J Am Chem Soc 119:7954–7959

Tomiya N, Narang S, Park J, Abdul-Rahman B, Choi O, Singh S, Hiratake J, Sakata K, Betenbaugh MJ, Palter KB, Lee YC (2006) Purification, characterization, and cloning of a *Spodoptera frugiperda* Sf9 beta-*N*-acetylhexosaminidase that hydrolyzes terminal *N*-acetylglucosamine on the *N*-glycan core. J Biol Chem 281:19545–19560

Toprak U, Baldwin D, Erlandson M, Gillott C, Hegedus DD (2010) Insect intestinal mucins and serine proteases associated with the peritrophic matrix from feeding, starved and moulting *Mamestra configurata* larvae. Insect Mol Biol 19:163–175

Toprak U, Erlandson M, Baldwin D, Karcz S, Wan L, Coutu C, Gillott C, Hegedus DD (2015) Identification of the *Mamestra configurata* (Lepidoptera: Noctuidae) peritrophic matrix proteins and enzymes involved in peritrophic matrix chitin metabolism. Insect Sci. https://doi.org/10.1111/1744-7917.12225

Vaaje-Kolstad G, Westereng B, Horn SJ, Liu Z, Zhai H, Sorlie M, Eijsink GH (2010) An oxidative enzyme boosing the enzymatic conversion of recalcitrant polysaccharides. Science 330:219–222

Varela PF, Llera AS, Mariuzza RA, Tormo J (2002) Crystal structure of imaginal disc growth factor-2. J Biol Chem 277:13229–13236

Venancio TM, Cristofoletti PT, Ferreira C, Verjovski-Almeida S, Terra WR (2009) The *Aedes aegypti* larval transcriptome: a comparative perspective with emphasis on trypsins and the domain structure of peritrophins. Insect Mol Biol 18:33–44

Vuocolo T, Eisemann CH, Pearson RD, Willadsen P, Tellam RL (2001) Identification and molecular characterisation of a peritrophin gene, peritrophin-48, from the myiasis fly *Chrysomya bezziana*. Insect Biochem Mol Biol 31:919–932

Wang S, Jayaram SA, Hemphala J, Senti KA, Tsarouhas V, Jin H, Samakovlis C (2006) Septate-junction-dependent luminal deposition of chitin deacetylases restricts tube elongation in the *Drosophila* trachea. Curr Biol 16:180–185

Wang Y, Fan HW, Huang HJ et al (2012) Chitin synthase 1 gene and its two alternative splicing variants from two sap-sucking insects, *Nilaparvata lugens* and *Laodelphax striatellus* (Hemiptera: Delphacidae). Insect Biochem Mol Biol 42:637–646

Watanabe T, Kobori K, Miyashita K, Fujii T, Sakai H, Uchida M, Tanaka H (1993) Identification of glutamic acid 204 and aspartic acid 200 in chitinase A1 of *Bacillus circulans* WL-12 as essential residues for chitinase activity. J Biol Chem 268:18567–18572

Wieschaus E, Nüsslein-Volhard C, Jurgens G (1984) Mutations affecting the pattern of the larval cuticle in Drosophila melanogaster: zygotic loci on the X-chromosome and the fourth chromosome. Wilhelm Roux's Arch Dev Biol 193:296–307

Willis JH, Papandreou NC, Iconomidou VA, Hamodrakas SJ (2012) Cuticular proteins. In: Gilbert LI (ed) Insect molecular biology and biochemistry. Chapel Hill (NC): Academic, New Jersey, pp 134–166

Wu JJ, Chen Z-C, Wang Y-W, Fu K-Y, Guo W-C, Li G-Q (2018) Silencing *chitin deacetylase 2* impairs larval-pupal and pupal-adult molts in *Leptinotarsa decemlineata*. Insect Mol Biol In press

Xi Y, Pan PL, Ye YX, Yu B, Xu HJ et al (2015) Chitinase-like gene family in the brown planthopper, *Nilaparvata lugens*. Insect Biochem Mol Biol 24:29–40

Xi Y, Pan PL, Ye YX, Yu B, Zhang CX (2014) Chitin deacetylase family genes in the brown planthopper, *Nilaparvata lugens* (Hemiptera: Delphacidae). Insect Mol Biol 23:695–705

Xu G, Zhang J, Lyu H, Liu J, Ding Y, Feng Q, Zhong Q, Zheng S (2017) BmCHSA-2b, a Lepidoptera specific alternative splicing variant of epidermal chitin synthase, is required for pupal wing development in *Bombyx mori*. Insect Biochem Mol Biol 87:117–126

Yabe T, ToshikoY-O Nakajima T, Masayuki S, Mikio A, Hisafumi Y-O (1998) Mutational analysis of chitin synthase 2 of *Saccharomyces cerevisiae:* Identification of additional amino acid residues involved in its catalytic activity. Eur J Biochem 258:941–947

Yang Q, Liu T, Liu FY, Qu MB Qian XH. (2008) A novel β-*N*-acetyl-D-hexosaminidase from the insect *O. furnacalis* (Guene′e). FEBS J 275:5690–5702

Yang WJ, Xu KK, Cong L, Wang JJ (2013) Identification, mRNA expression, and functional analysis of chitin synthase 1 gene and its two alternative splicing variants in oriental fruit fly, *Bactrocera dorsalis*. Int J Biol Sci 9:331–342

Yu R, Liu W, Li D, Zhao X, Ding G, Zhang M, Ma E, Zhu K-Y, Li S, Mousian B, Zhang J (2016) Helicoidal organization of chitin in the cuticle of the migratory locust requires the function of the chitin deacetylase enzyme (LmCDA2). J Biol Chem 291:24352–24363

Zees AC, Pyrpassopoulos S, Vorgias CE (2009) Insights into the role of the (alpha + beta) insertion in the TIM-barrel catalytic domain, regarding the stability and the enzymatic activity of chitinase A from *Serratia marcescens*. Biochim Biophys Acta 1794:23–31

Zen KC, Choi HK, Nandigama K, Muthukrishnan S, Kramer KJ (1996) Cloning, expression and hormonal regulation of an insect β-*N*-acetylglucosaminidase gene. Insect Biochem Mol Biol 26:435–444

Zhang D, Chen J, Yao Q, Pan Z, Chen J et al (2012) Functional analysis of two chitinase genes during the pupation and eclosion stages of the beet armyworm *Spodoptera exigua* by RNA interference. Arch Insect Biochem Physiol 79:220–234

Zhang J, Lu A, Zhang Q, Ling E (2014) Functional analysis of insect molting fluid proteins on the protection and egulation of ecdysis. J Biol Chem 289:35891–35906

Zhang J, Liu X, Li D, Sun Y, Guo Y et al (2010) Silencing of two alternative splicing-derived mRNA variants of chitin synthase 1 gene by RNAi is lethal to the oriental migratory locust, *Locusta migratoria manilensis* (Meyen). Insect Biochem Mol Biol 40:824–833

Zhang J, Zhang X, Arakane Y, Muthukrishnan S, Kramer KJ, Ma E, Zhu KY (2011a) Comparative genomic analysis of chitinase and chitinase-like genes in the African malaria mosquito (*Anopheles gambiae*). PLoS ONE 6(5):e19899

Zhang J, Zhang X, Arakane Y, Muthukrishnan S, Kramer KJ, Ma E, Zhu KY (2011b) Identification and characterization of a novel chitinase-like gene cluster (AgCht5) possibly derived from tandem duplications in the African malaria mosquito, *Anopheles gambiae*. Insect Biochem Mol Biol 41:521–528

Zhong XW, Wang XH, Tan X, Xia QY, Xiang ZH, Zhao P (2014) Identification and molecular characterization of a chitin deacetylase from *Bombyx mori* peritrophic membrane. Int J Mol Sci 15:1946–1961

Zhou Y, Badgett MJ, Bowen JH, Vannini L, Orlando R, Willis JH (2016) Distribution of cuticular proteins in different structures of adult *Anopheles gambiae*. Insect Biochem Mol Biol 75:45–57

Zhu Q, Arakane Y, Banerjee D, Beeman RW, Kramer KJ, Muthukrishnan S (2008a) Domain organization and phylogenetic analysis of the chitinase-like family of proteins in three species of insects. Insect Biochem Mol Biol 38:452–466

Zhu Q, Arakane Y, Beeman RW, Kramer KJ, Muthukrishnan S (2008b) Characterization of recombinant chitinase-like proteins of *Drosophila melanogaster* and *Tribolium castaneum*. Insect Biochem Mol Biol 38:467–477

Zhu Q, Arakane Y, Beeman RW, Kramer KJ, Muthukrishnan S (2008c) Functional specialization among insect chitinase family genes revealed by RNA interference. Proc Nat Acad Sci USA 105:6650–6655

Zhu Q, Deng Y, Vanka P, Brown SJ, Muthukrishnan S et al (2004) Computational identification of novel chitinase-like proteins in the *Drosophila melanogaster* genome. Bioinformatics 20:161–169

Zimoch L, Merzendorfer H (2002) Immunolocalization of chitin synthase in the tobacco hornworm. Cell Tissue Res 308:287–297

Chapter 6
Chitin-Active Lytic Polysaccharide Monooxygenases

Gaston Courtade and Finn L. Aachmann

Abstract Lytic polysaccharide monooxygenases (LPMOs) are copper-dependent enzymes that catalyze the cleavage of 1,4-glycosidic bonds various plant cell wall polysaccharides and chitin. In contrast to glycoside hydrolases, LPMOs are active on the crystalline regions of polysaccharides and thus synergize with hydrolytic enzymes. This synergism leads to an overall increase in the biomass-degradation activity of enzyme mixtures. Chitin-active LPMOs were discovered in 2010 and are currently classified in families AA10, AA11, and AA15 of the Carbohydrate-Active enZYmes database, which include LPMOs from bacteria, fungi, insects, and viruses. LPMOs have become important enzymes both industrially and scientifically and, in this chapter, we provide a brief introduction to chitin-active LPMOs including a summary of the 20+ chitin-active LPMOs that have been characterized so far. Then, we describe their structural features, catalytic mechanism, and appended carbohydrate modules. Finally, we show how chitin-active LPMOs can be used to perform chemo-enzymatic modification of chitin substrates.

Keywords Lytic polysaccharide monooxygenase (LPMO) · Copper-dependent enzymes · Crystalline polysaccharides

6.1 Introduction

Most Carbohydrate-Active enZYmes (CAZymes) are glycoside hydrolases (GH) that hydrolyze glycosidic bonds by acid/base catalysis. In fact, *Serratia marcescens*, an Enterobacterium regarded as one of the most efficient chitin degrading organisms (Monreal and Reese 1969), contains four GHs involved in chitin degradation. ChiA and ChiB are family 18 GHs exochitinases that processively depolymerize chitin starting from chain ends, producing chitobiose

G. Courtade · F. L. Aachmann (✉)
Department of Biotechnology and Food Science, NOBIPOL, NTNU Norwegian University of Science and Technology, Sem Sælands vei 6/8, N-7491 Trondheim, Norway
e-mail: finn.l.aachmann@ntnu.no

Q. Yang and T. Fukamizo (eds.), *Targeting Chitin-containing Organisms*, Advances in Experimental Medicine and Biology 1142,
https://doi.org/10.1007/978-981-13-7318-3_6

(diGlcNAc) (Igarashi et al. 2014; Hult et al. 2005). ChiC is a family18 GH endoglucanase with two additional domains that hydrolyze glycosidic bonds at random in amorphous regions on a chitin chain. Chitobiase is a family 20 GH that hydrolyzes diGlcNAc into GlcNAc monomers (Vaaje-Kolstad et al. 2013). However, the activity of these GHs is significantly slower on crystalline substrates than on soluble carbohydrate chains (Vermaas et al. 2015). An explanation for this is that hydrolysis is an inherently inefficient method of chitin depolymerization. This is because water molecules are precluded from penetrating the tightly packed hydrogen-bonded network in crystalline chitin fibrils. Furthermore, open chain ends and amorphous regions seldom occur on a crystalline chitin substrate. These features of insoluble chitin are obstacles to efficient enzymatic degradation of chitin by GHs.

Lytic polysaccharide monooxygenases (LPMOs), discovered in 2010, circumvent the challenges faced by GHs and have brought about a paradigm shift regarding the degradation of crystalline polysaccharides (Vaaje-Kolstad et al. 2005, 2010). LPMOs are copper-dependent enzymes that oxidize glycosidic bonds in 1, 4-linked polysaccharides. Originally their oxidative activity was observed on chitin (Vaaje-Kolstad et al. 2010) and cellulose (Forsberg et al. 2011; Quinlan et al. 2011; Phillips et al. 2011; Langston et al. 2011; Eibinger et al. 2014). Interestingly, mixtures of GHs and LPMOs were shown to have much higher polysaccharide-degradation activity than any of the enzymes by themselves. This synergism can be explained by LPMOs binding and cleaving glycosidic bonds in crystalline regions of polysaccharides. New chain ends generated by LPMO activity lead to decreased crystallinity in the area near to the oxidation site, which creates new starting points for GHs, overall boosting the activity of the enzyme mixtures (Vaaje-Kolstad et al. 2005; Harris et al. 2010; Vaaje-Kolstad et al. 2010; Eibinger et al. 2014; Vermaas et al. 2015; Horn et al. 2012) (see Fig. 6.1).

Today, LPMOs have been found in bacteria, fungi, and insects. LPMO activity has been demonstrated for a variety of 1,4-linked substrates (cellulose (Forsberg et al. 2011; Quinlan et al. 2011; Phillips et al. 2011; Langston et al. 2011; Eibinger et al. 2014), chitin (Vaaje-Kolstad et al. 2010), cellooligosacharides (Isaksen et al. 2014; Frandsen et al. 2016), xyloglucan (Agger et al. 2014; Kojima et al. 2016; Bennati-Granier et al. 2015), xylan (Couturier et al. 2018; Frommhagen et al. 2015), starch (Lo Leggio et al. 2015; Vu et al. 2014). LPMOs are classified in the Carbohydrate-Active enZYme database (CAZy) as auxiliary activity (AA) families AA9 (Levasseur et al. 2013) (cellulose-oxidizing fungal LPMOs; formerly classified as glycoside hydrolases GH61), AA10 (Levasseur et al. 2013) (cellulose and chitin-oxidizing bacterial LPMOs; formerly classified as carbohydrate-binding modules CBM33), AA11 (Hemsworth et al. 2014) (chitin-oxidizing fungal LPMOs), AA13 (Lo Leggio et al. 2015) (starch-oxidizing fungal LPMOs), AA14 (Couturier et al. 2018) (xylan-oxidizing fungal LPMOs), and AA15 (Sabbadin et al. 2018) (cellulose and chitin-oxidizing insectile LPMOs). Several chitin-active LPMOs have been characterized, spanning families AA10, AA11, and AA15, as summarized in Table 6.1.

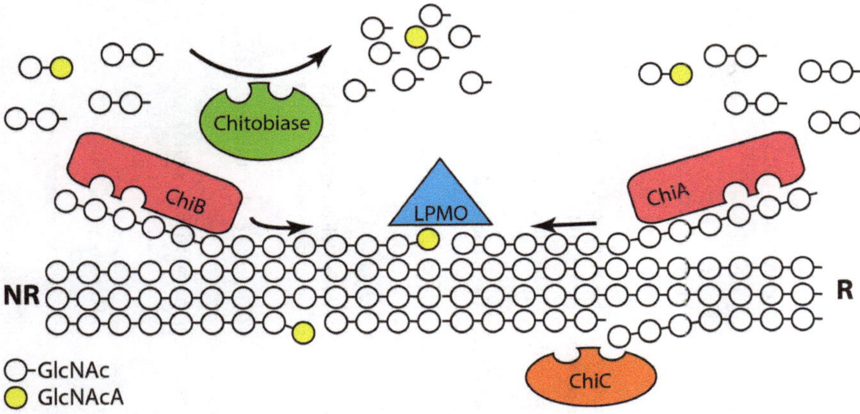

Fig. 6.1 Overview of chitinolytic enzymes in *S. marcescens*. Chitin is shown as chains of GlcNAc (open circles). ChiB (pink) is a GH18 that degrades the chitin chains from their non-reducing ends (NR) and ChiA (pink) is a GH18 that degrades chitin from the reducing ends (R). Both enzymes are processive and produce primarily chitobiose (diGlcNAc) (Igarashi et al. 2014; Hult et al. 2005). ChiC (orange) is a GH18 that randomly hydrolyzes glycosidic bonds in the amorphous regions of chitin, creating starting points for ChiA and ChiB. The LPMO (*Sm*LPMO10A, also known as CBP21) causes oxidative cleavage of glycosidic bonds in the crystalline regions of chitin yielding aldonic acids (GlcNAcA; yellow circles). In this way, *Sm*LPMO10A creates new chain ends for ChiA and ChiB. Chitobiase is a GH20 that converts diGlcNAc and other short chitooligosaccharides to GlcNAc monomers. The figure is adapted from (Vaaje-Kolstad et al. 2013)

6.2 Structural Features of LPMOs

The first LPMO structures were those of the chitin-active *Sm*LPMO10A (Vaaje-Kolstad et al. 2005) (also known as CBP21) and the cellulose-active *Hj*LPMO9A (Karkehabadi et al. 2008) (also known as Cel61B). These two LPMOs belong to families AA10 and AA9, respectively, and there is only 16.5% sequence identity between them. From these and subsequent structures, it became clear that LPMOs from different families have overall similar structures, as well as a conserved "histidine brace" active site (Quinlan et al. 2011) and conserved surface groups involved in substrate interactions. These similarities have been confirmed by a comprehensive number of structural studies, which have been reviewed elsewhere (Frandsen and Lo Leggio 2016).

Overall, the catalytic domains of LPMOs are approximately 150–250 residues in length. During translation, LPMOs are produced with an N-terminal signal sequence of approximately 15–30 amino acids, which is proteolytically cleaved when the protein is translocated to the periplasm (Vaaje-Kolstad et al. 2010; Moser et al. 2008). Mature LPMOs have an N-terminal His residue. Amino acid numbering in LPMOs, depends on the length of the signal sequence; for example, *Sm*LPMO10A has a 27-residue long signal peptide and thus the N-terminal amino

Table 6.1 Overview of characterized chitin-active LPMOs. The PDB codes correspond to structures of the catalytic domains

Family	Organism	Protein name(s)	Associated CBMs	PDB code	Substrates	References
AA10	*Bacillus amyloliquefaciens*	ChbB, *Ba*AA10A, *Ba*CBM33, Rbam17540, BAMF_1859	–	2YOW, 2YOX, 2YOY, 5IUJU	α and β chitin	Hemsworth et al. (2013) and Gregory et al. 2016)
AA10	*Bacillus cereus*	*Bc*LPMO10A	CBM5, Two fibronectin-type III-like domains	–	α and β chitin	Mutahir et al. (2018)
AA10	*Bacillus licheniformis*	*Bl*LPMO10A	–	5LW4	α and β chitin, preference for β chitin	Courtade et al. (2015) and Forsberg et al. (2014)
AA10	*Bacillus thuringiensis*	*Bt*LPMO10A	–	5WSZ	Likely active on chitin	To be published
AA10	*Burkholderia pseudomallei*	*Bp*AA10A	–	3UAM	Likely active on chitin	To be published
AA10	*Cellvibrio japonicus*	*Cj*LPMO10A, CJA_2191	CBM5, CBM73		α and β chitin	Forsberg et al. (2016)
AA10	*Cellvibrio mixtus*	*Cm*AA10	–	–	α chitin, Avicel	Wang et al. (2018)
AA10	*Enterococcus faecalis*	*Ef*AA10A, EF0362, *Ef*CBM33A, *Efa*CBM33	–	4A02, 4ALC, 4ALE, 4ALQ, 4ALR, 4ALS, 4ALT	α and β chitin	Vaaje-Kolstad et al. (2012) and Gudmundsson et al. (2014)
AA10	*Jonesia denitrificans*	Jden_1381, *Jd*LPMO10A	CBM5, GH18	5AA7, 5VG0, 5VG1	α and β chitin, preference for β chitin	Mekasha et al. (2016) and Bacik et al. (2017)
AA10	*Listeria monocytogenes*	*Lm*LPMO10, Lmo2467	Fibronectin-type III-like domain, two CBM5/12	5L2 V	Active on chitin	Paspaliari et al. (2015)
AA10	*Micromonospora aurantiaca*	*Ma*LPMO10B, Micau_1630	CBM2	5OPF	β chitin, PASC	Forsberg et al. (2018)
AA10	*Serratia marcescens*	CBP21, *Sm*AA10A, *Sm*LPMO10A	–	2BEM, 2BEN, 2LHS	α and β chitin	

(continued)

Table 6.1 (continued)

Family	Organism	Protein name(s)	Associated CBMs	PDB code	Substrates	References
						Vaaje-Kolstad et al. (2005), Aachmann et al. (2012) and Vaaje-Kolstad et al. (2010)
AA10	*Streptomyces ambofaciens*	SamLPMO10B	–	–	β-chitin	Valenzuela et al. (2017)
AA10	*Streptomyces coelicolor*	*Sc*AA10B, *Sc*LPMO10B, SCO0643, SCF91.03c	–	4OY6, 4OY8	β chitin, PASC, Avicel	Forsberg et al. (2014)
AA10	*Streptomyces griseus*	*Sg*LPMO10F, SGR_6855	–	–	α and β chitin	Nakagawa et al. (2015)
AA10	*Streptomyces lividans*	*Sl*AA10E, SliLPMO10E, SLI_3182	–	5FTZ	β chitin	Chaplin et al. (2016)
AA10	*Thermofibida fusca*	*Tf*AA10A, TfLPMO10A, E7, Tfu_1268	–	4GBO	β chitin, PASC, Avicel	Kruer-Zerhusen et al. (2017) and Forsberg et al. (2014)
AA10	*Entomopoxviruses*	Fusolin	–	4YN1, 4YN2, 4OW5, 4X27, 4X29	Binds to crab shell chitin	Chiu et al. (2015)
AA10	*Vibrio cholera*	*Vc*AA10B, VCA0811, *Vc*GbpAD1, GbpAD1	GbpAD2, GbpAD3, GbpAD4, CBM73	2XWX	Binds to chitooligosaccharides, active on chitin	Wong et al. (2012) and Loose et al. (2014)
AA11	*Aspergillus oryzae*	*Ao*AA11, *Ao*LPMO11, AO090102000501	X278	4MAH, 4MAI	β chitin	Hemsworth et al. (2014)
AA11	*Fusarium fujikuroi*	*Ff*AA11	X278	–	α and β chitin, lobster shell	Wang et al. (2018)
AA15	*Thermobia domestica*	*Td*AA15A, *Td*LPMO15A	–	5MSZ	β chitin, Avicel	Sabbadin et al. (2018)
AA15	*Thermobia domestica*	*Td*AA15B, *Td*LPMO15B	CBM14	–	α and β chitin	Sabbadin et al. (2018)

acid in mature *Sm*LPMO10A should be His28. The N-terminus together with the side chain of the N-terminal His and the side-chain of a more distal His form the "histidine brace" active site of LPMOs (Quinlan et al. 2011). The active site is further shaped by a Tyr (for family AA9, AA10, AA11, AA13, and AA14) or Phe (for families AA10 and AA15; see Fig. 6.2b) and coordinates one copper ion, which classifies it as a type II copper site (see Fig. 6.2b) (Peisach and Blumberg 1974). The active site can bind other metal ions such as Zn(II) instead of copper, albeit at lower affinities and producing an inactive form of the enzyme (Aachmann et al. 2012). The N-terminal histidine of LPMOs produced by filamentous fungi carries a τ-nitrogen methylation (Quinlan et al. 2011; Petrović et al. 2018).

LPMOs are mostly composed of β-strands, and their core is composed of a β-sandwich fold encompassing two β-sheets, including 8–9 β-strands altogether. The structure is stabilized by hydrophobic residues, as well as by one or two disulfide bridges. The β-sandwich core is decorated with loops and helices, particularly, in the first 60–70 amino acids, yielding, and overall pyramidal shape. The base of the pyramid corresponds to the substrate binding surface (Fig. 6.2c) and has polar residues that mediate substrate interactions and key aromatic residues (e.g., Tyr54 in *Sm*LPMO10A) to orient the LPMO with the crystalline polysaccharide substrates. The copper site is located on the center of this substrate binding surface (see Fig. 6.2a) and when an LPMO binds its substrate, the active site is positioned near the glycosidic bond (Frandsen and Lo Leggio 2016).

Important residues for chitin-binding have been identified on the basis of hydrogen/deuterium exchange monitored by NMR spectroscopy (Aachmann et al. 2012), by site-directed mutagenesis (Vaaje-Kolstad et al. 2005; Loose et al. 2018), and by molecular dynamics (MD) simulations (Bissaro et al. 2018). These studies indicate that polar residues located on the surface around the copper site (Fig. 6.2c) contribute to binding, while an aromatic residue (Tyr54 in *Sm*LPMO10A) appears to be important for orienting the LPMO on the substrate, possibly through π-interactions with the C-H bonds in the GlcNAc pyranose ring.

6.3 Catalytic Mechanism of LPMOs

Even though LPMOs were first thought to be non-catalytic (Vaaje-Kolstad et al. 2005), it was shown in 2010 that the products from LPMO-treated chitin were oxidized on the C1 carbon in the scissile glycosidic bond. In cellulose, C1, C4 or both may be oxidized, but only C1 oxidation has thus far been detected for LPMO activity on chitin (Beeson et al. 2012; Vaaje-Kolstad et al. 2010; Borisova et al. 2015; Forsberg et al. 2014). While, there is currently no consensus on specific details of the reaction mechanism, such as the nature of the reactive oxygen species (Walton and Davies 2016; Bissaro et al. 2017; Hangasky et al. 2018), there is agreement regarding the overall nature of the reaction (summarized in Fig. 6.3). The LPMO reaction (Fig. 6.3) involves the reduction of Cu(II) to Cu(I) in the active site by an electron donor that can either be another oxidoreductase, such as cellobiose

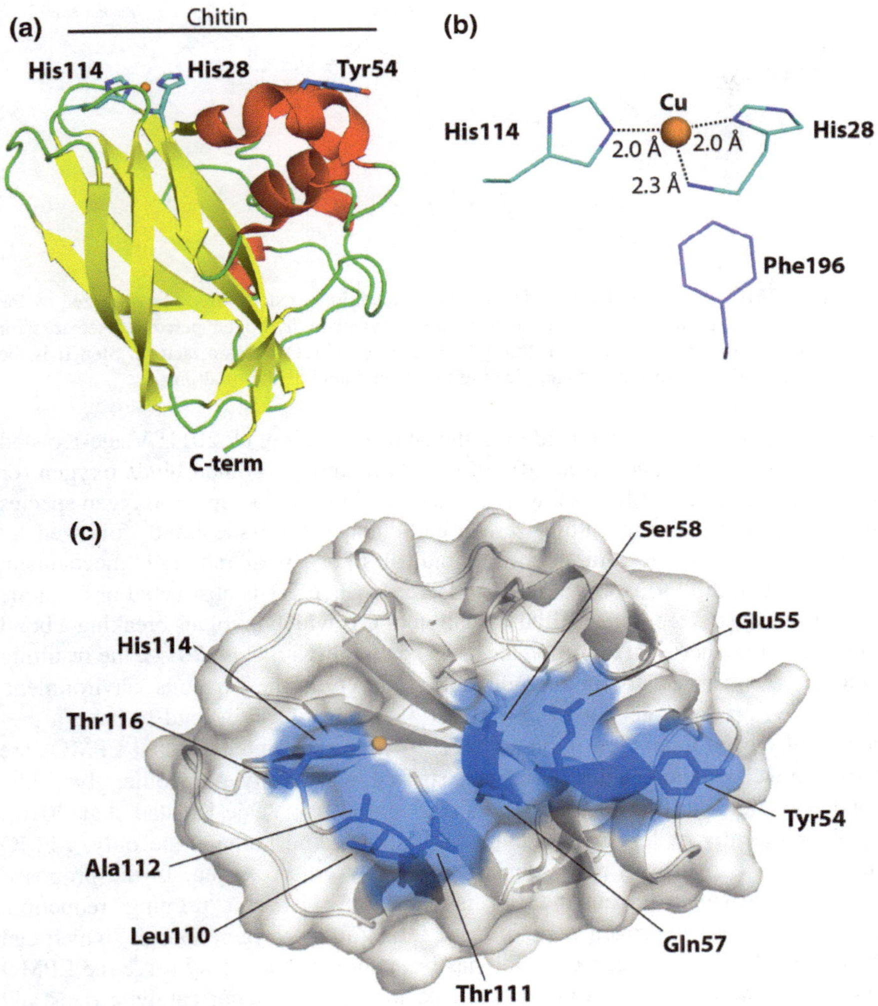

Fig. 6.2 Structural features of LPMOs. a An Overview of the fold of *Sm*LPMO10A, showing β-strands (yellow), helices (red), and loops (green), as well as the copper site (cyan) with a bound copper ion (orange), and Tyr54, which mediates substrate binding. **b** Histidine brace copper site of *Sm*LPMO10A, showing the coordinating histidines (His28 and His114), as well as Phe196 in the axial position. **c** Surface view of *Sm*LPMO10A, with a bound copper ion (orange). Important residues for chitin binding, as determined by H/D exchange monitored by NMR spectroscopy (Aachmann et al. 2012) are colored blue, they are Gln53, Tyr54, Glu55, Gln57, Ser58,Leu110, Thr111, Ala112, His114, and Thr116. The structures were made using PyMol and the PDB code 2BEM. The copper coordinates were extracted from the PDB code 2YOX

dehydrogenase (CDH) (Phillips et al. 2011; Tan et al. 2015; Langston et al. 2011; Loose et al. 2016), lignin-derived phenols (Westereng et al. 2015), glucose-methanol-choline oxidases (GMCs) (Kracher et al. 2016), or an organic

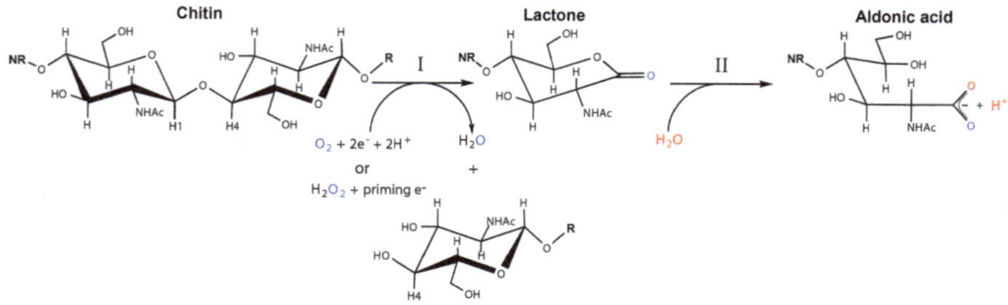

Fig. 6.3 LPMO oxidation of chitin. The reaction occurs in two steps. Step I is catalyzed by the LPMO and requires the addition of electrons and oxygen or hydrogen peroxide (see text for details). Oxidation of the C1-carbon (through abstraction of H1) forms a lactone. Step II is the spontaneous hydration of the lactone, yielding an aldonic acid as the final product

compound, such as ascorbic acid or gallic acid (Quinlan et al. 2011; Vaaje-Kolstad et al. 2010; Frommhagen et al. 2016). LPMO-bound Cu(I) then binds oxygen (or hydrogen peroxide, see below), which is activated to form a copper–oxygen species that abstracts a hydrogen atom from a carbon in the scissile bond, followed by hydroxylation of the resulting substrate radical, likely by a "rebound" mechanism. This results in the formation of either a lactone, when H1 is abstracted or a ketone when H4 is abstracted. After initial oxidation of C1, which involves breaking a bond with a binding energy of approximately 95 kcal/mol (Kari et al. 2014), the resulting lactone is converted to an aldonic acid (GlcA) in the aqueous environment. Solubilized products generated by LPMO oxidation of chitin tend to have a predominantly even numbered polymerization degree. This indicates that LPMOs are active on well-ordered chains in a crystal surface, where only every other glycosidic bond is available to LPMO attack (Forsberg et al. 2011; Vaaje-Kolstad et al. 2010).

More recently, it has been shown that another, and perhaps the only, LPMO co-substrate is hydrogen peroxide instead of molecular oxygen. In the proposed hydrogen peroxide mechanism, Cu(II) is reduced to Cu(I) in a "priming" reduction. Then Cu(I) binds hydrogen peroxide and forms a Cu–oxygen species, which can then abstract H1. Hydrogen peroxide has been shown to not only drive the LPMO reaction in the absence of oxygen but also to promote higher catalytic rates and yields of oxidized products. Moreover, product yields have been shown to be in stoichiometric ratio with hydrogen peroxide and in super-stochiometric ratio with respect to the reducing agent (Bissaro et al. 2017). In the O_2-based mechanism, each reaction cycle requires the delivery of two electrons, which implies stoichiometric consumption of reductant (Fig. 6.3).

6.3.1 Detecting LPMO Activity

Traditionally, LPMO activity has been measured by setting up a batch reaction, taking out samples at different timepoints and measuring the products of the reaction chromatographically (Vaaje-Kolstad et al. 2010; Loose et al. 2014). Such an approach is time consuming if the goal is to simply detect the presence of LPMO activity. Recently, Breslmayr and colleagues have developed an LPMO assay based on the peroxidase activity of the enzyme (Breslmayr et al. 2018). In this assay, an LPMO-containing sample (cell extract or culture medium) is used to oxidize a chromogenic substrate (2,6-DMP: 2,6-dimethoxyphenol or hydrocoeruligone). The activity can then be measured in a straightforward manner by monitoring absorbance at 469 nm. Of note, the efficiency of this assay may vary between LPMOs.

6.4 Multimodular LPMOs

While many LPMOs exist solely as a catalytic domain, others are found tethered to carbohydrate-binding modules (CBMs) through peptide linkers of various lengths and compositions. Similar to the catalytic domains of LPMOs and other carbohydrate-active enzymes, CBMs are divided into families according to the similarity of their amino acid sequences. Currently, there are 83 distinct CBM families (www.cazy.org) and these CBMs display significant variation in terms of substrate specificity. CBMs are thought to have two main functions that arise as a consequence of the binding event. They target the "correct" substrate for their attached catalytic domain (McLean et al. 2002), and they create proximity between the catalytic domain and the substrate. While extensive reviews of CBMs have been published elsewhere (Boraston et al. 2004), it is of interest to focus on CBMs that are associated with chitin-active LPMOs, particularly CBM2, CBM5 and CBM73 in AA10s, and CBM14 in AA15s, (see Table 6.2).

While CBM14 may bind both insoluble chitin and chitin oligomers (Crasson et al. 2017), CBM2, CBM5 (Boraston et al. 2004) and CBM73 (Forsberg et al. 2016) are specific for binding insoluble, crystalline substrates like chitin. A comparison of the substrate-binding face of the CBMs (Fig. 6.4) reveals that their binding surfaces are flat and that they have a high abundance of aromatic residues (Type A, Boraston et al. 2004). Both of these features are hallmarks of interactions with a crystalline substrate, as aromatic side chains facilitate π-interactions with C–H bonds in carbohydrates (Hudson et al. 2015).

For LPMOs, the presence of a CBM significantly enhances LPMO binding to its substrate (Forsberg et al. 2014) and results in higher yields of oxidized products,

Table 6.2 Overview of CBMs commonly associated with chitin-active LPMOs. Adapted from www.cazy.org

CBM family	Fold type	Approx. sequence length (# of a.a.)	Example of associated LPMO	LPMO substrate
2	β-sandwich	100	*Sg*LPMO10F	α and β chitin
5	Ski boot	60	*Cj*LPMO10A	
14	Hevein-like	50	*Td*LPMO15B	
73	Unknown	65	*Cj*LPMO10A	

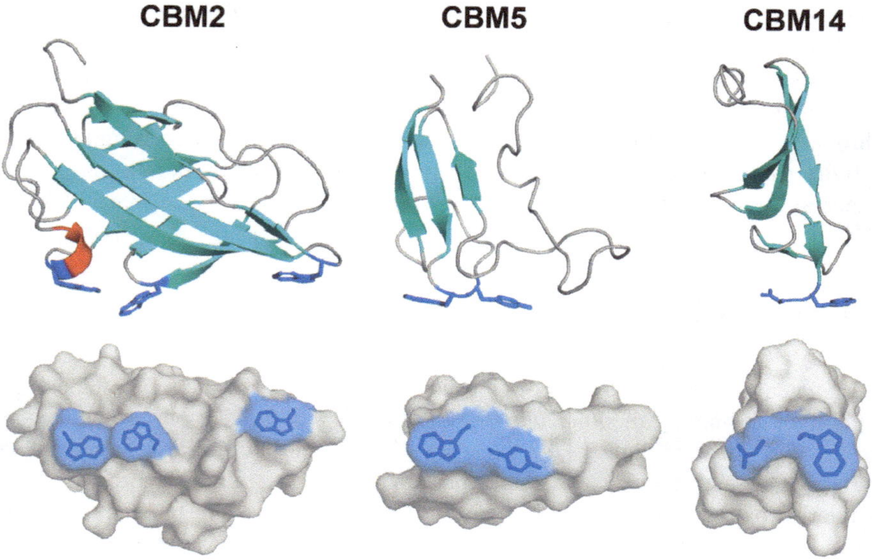

Fig. 6.4 Structure comparison of the overall fold and substrate-binding surface of CBMs. The figure shows representative structures of CBM2 (PDB ID: 1EXG (Xu et al. 1995)), CBM5 (PDB ID: 1AIW (Brun et al. 1997)), and CBM14 (PDB ID: 5HBF (Fadel et al. 2016)). The top panel shows the cartoon representation of the structures (β-strands: cyan, α-helices: red, loops: gray), and the putative substrate binding residues are shown as blue sticks. The bottom panel shows the surface representation of the proteins, viewed from the substrate binding surface, where the putative substrate binding residues are highlighted. These CBMs are associated with other CAZymes (not LPMOs) but are shown here for illustrative purposes

compared to versions of the enzyme comprising only the catalytic LPMO domain (Forsberg et al. 2014). It is, however, not well understood why some LPMOs have CBMs (in some cases more than one, e.g., *Cj*LPMO10A and *Bc*LPMO10A) (Forsberg et al. 2016; Mutahir et al. 2018), while others exist only as a single catalytic domain. Even with the availability of studies on LPMO-associated CBMs, where the presence of CBMs correlate with higher yields (Crouch et al. 2016; Forsberg et al. 2016, 2018) and the mechanism through which CBMs enhance activity and the interplay between CBMs, linker regions, and catalytic domains are not well understood. One likely option is that by keeping the LPMO closely associated with the substrate, the CBM prevents nonproductive reactions between a reduced LPMO and its co-substrate that could lead to enzyme inactivation (Bissaro et al. 2017; Forsberg et al. 2018; Mutahir et al. 2018). A recent study has shown that the "anchoring" of a CBM2 to the substrate may lead to multiple localized oxidation of the substrate by the LPMO (Courtade et al. 2018).

6.5 Chemo-Enzymatic Modification of LPMO Products

LPMO activity is usually determined by measuring the release of oxidized, soluble chitooligosaccharides by mass spectrometry (MS) and high-performance chromatography (HPLC) (Vaaje-Kolstad et al. 2010; Loose et al. 2014). Such an approach leads to underestimation of the total performance of LPMOs since oxidation events that do not result in oligosaccharide release are not detected. Vuong and colleagues (2017) developed a method to detect C1-oxidizing activity of cellulose-active LPMOs by covalently linking a water-soluble fluorophore to oxidized positions within the cellulose fiber. The coupling reaction (Fig. 6.5) can be divided into two steps. In step 1, the aldonic acid [1] formed at C1 from the LPMO reaction is activated by a carbodiimide coupling reagent (EDC: 1-ethyl-3-[3-(dimethylamino)propyl]carbodiimide) to create an active-ester intermediate ([2]: O-acylisourea intermediate). In step 2, reductive amination is carried out using a compound with a primary amine (ANDA: fluorophore primary amine, 7-amino-1,3-naphthalenedisulfonic acid). This results in a nucleophilic substitution reaction where an amide bond is formed between the carboxyl group at C1 and the compound [3].

The possibility of chemo-enzymatic modification on a chitin surface using aldonic acids created by LPMOs was first recognized by Wang and coworkers (Wang et al. 2018). LPMO activity in chitin is selective for C1. This leads to oxidation of the reducing end, which can be used for further modifications, similar to what has been done for TEMPO-oxidized polysaccharides (Wang et al. 2018). Similar to the EDC activation in step 1, an active-ester intermediate could also be formed using an Oxyma-derived uranium salt (COMU: (1-cyano-2-ethoxy-2-oxoethylidenaminooxy) dimethyl aminomorpholinocarbenium hexafluoro phosphate). This activation creates a good leaving group for a subsequent substitution reaction with compounds with a primary amine group, leading to the formation of a stable amide bond. Using this approach, the possibility of using a chitin-active LPMO for grafting a fluorescent probe, a peptide and gold nanoparticles onto a

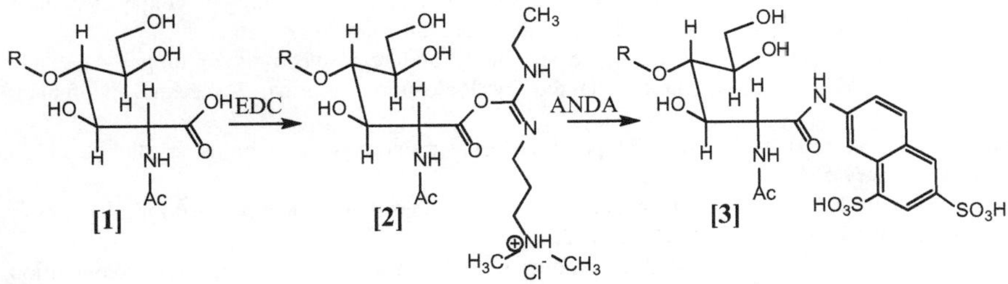

Fig. 6.5 Labeling scheme of LPMO products. The aldonic acid [1] formed at C1 reacts with water-soluble EDC (1-ethyl-3-[3-(dimethylamino)propyl]carbodiimide) yielding an active-ester intermediate (O-acylisourea intermediate) [2], allowing the primary amine of the fluorophore ANDA (7-amino-1,3-naphthalenedisulfonic acid) to form an amide group, creating a fluorescence-labeled product [3]

chitin surface was demonstrated (Wang et al. 2018). Such chemo-enzymatic approach represents a simple and environmentally friendly method for the greener production of functionalized chitin-based biomaterials.

6.6 Further Perspectives

LPMOs are important enzymes both for industry and from a scientific point of view. LPMOs are still in the spotlight of current enzyme research and several intriguing aspects of these powerful enzymes are the target of ongoing intense investigations. These aspects include but are not limited to, the diversity of their roles in nature, their structure, their substrate specificities, their interplay with appended carbohydrate-binding modules, and their catalytic mechanism.

Owing to their unique oxidative mechanism, LPMOs play an important role in the saccharification of polysaccharides, including chitin. This role may be exploited both for the production of chitooligosaccharides and for chemo-enzymatic modification of chitin substrates. In this context, chitin-active LPMOs will likely also play a role in the production of novel, functionalized nanomaterials.

References

Aachmann FL et al (2012) NMR structure of a lytic polysaccharide monooxygenase provides insight into copper binding, protein dynamics, and substrate interactions. Proc Natl Acad Sci USA 109(46):18779–18784

Agger JW et al (2014) Discovery of LPMO activity on hemicelluloses shows the importance of oxidative processes in plant cell wall degradation. Proc Natl Acad Sci USA 111(17):6287–6292

Bacik J et al (2017) Neutron and atomic resolution X-ray structures of a lytic polysac- charide monooxygenase reveal copper-mediated dioxygen binding and evidence for N-terminal deprotonation. Biochemistry 56:2529–2532

Beeson WT et al (2012) Oxidative cleavage of cellulose by fungal copper-dependent polysaccharide monooxygenases. J Am Chem Soc 134(2):890–892

Bennati-Granier C et al (2015) Substrate specificity and regioselectivity of fungal AA9 lytic polysaccharide monooxygenases secreted by Podospora anserina. Biotechnol Biofuels 8 (90):90–103

Bissaro B et al (2018) How a lytic polysaccharide monooxygenase binds crystalline chitin. Biochemistry 57(12):1893–1906

Bissaro B et al (2017) Oxidative cleavage of polysaccharides by monocopper enzymes depends on H_2O_2. Nat Chem Biol 13(10):1123–1128

Boraston AB et al (2004) Carbohydrate-binding modules: fine-tuning polysaccharide recognition. Biochem J 382:769–781

Borisova AS et al (2015) Structural and functional characterization of a lytic polysaccharide monooxygenase with broad substrate specificity. J Biol Chem 290(38):22955–22969

Breslmayr E et al (2018) A fast and sensitive activity assay for lytic polysaccharide monooxygenase. Biotechnol Biofuels 11(1)

Brun E et al (1997) Solution structure of the cellulose-binding domain of the endoglucanase Z secreted by *Erwinia chrysanthemi*. Biochemistry 36(51):16074–16086

Chaplin AK et al (2016) Heterogeneity in the Histidine-brace Copper Coordination Sphere in Auxiliary Activity Family 10 (AA10) Lytic Polysaccharide Monooxygenases. J Biol Chem 291 (24):12838–50

Chiu E et al (2015) Structural basis for the enhancement of virulence by viral spindles and their in vivo crystallization. Proc Natl Acad Sci 112(13):201418798

Courtade G et al (2015) ^{1}H, ^{13}C, ^{15}N resonance assignment of the chitin-active lytic polysaccharide monooxygenase *Bl*LPMO10A from *Bacillus licheniformis*. Biomol NMR Assign 9(1):207–210

Courtade G et al (2018) The carbohydrate-binding module and linker of a modular lytic polysaccharide monooxygenase promote localized cellulose oxidation. J Biol Chem 293 (34):13006–13015

Couturier M et al (2018) Lytic xylan oxidases from wood-decay fungi unlock biomass degradation. Nat Chem Biol 14:306–310

Crasson O et al (2017) Human chitotriosidase: catalytic domain or carbohydrate binding module, who's leading HCHT's biological function. Scientif Rep 7:2768–2777

Crouch LI et al (2016) The contribution of non-catalytic carbohydrate binding modules to the activity lytic polysaccharide monooxygenases. J Biol Chem 291(14):7439–7449

Eibinger M et al (2014) Cellulose surface degradation by a lytic polysaccharide monooxygenase and its effect on cellulase hydrolytic efficiency. J Biol Chem 289(52):35929–35938

Fadel F et al (2016) X-Ray crystal structure of the full length human chitotriosidase (CHIT1) reveals features of its chitin binding domain. PLoS One 11(4):1–15

Forsberg Z et al (2011) Cleavage of cellulose by a CBM33 protein. Protein Sci Publ Protein Soc 20(9):1479–1483

Forsberg Z, Røhr AK et al (2014a) Comparative study of two chitin-active and two cellulose-active AA10-type lytic polysaccharide monooxygenases. Biochemistry 53 (10):1647–1656

Forsberg Z et al (2016) Structural and functional analysis of a lytic polysaccharide monooxygenase important for efficient utilization of chitin in *Cellvibrio japonicus*. J Biol Chem 291(14):7300–7312

Forsberg Z, Mackenzie AK et al (2014b) Structural and functional characterization of a conserved pair of bacterial cellulose-oxidizing lytic polysaccharide monooxygenases. Proc Natl Acad Sci USA 111(23):8446–8451

Forsberg Z et al (2018) Structural determinants of bacterial lytic polysaccharide monooxygenase functionality. J Biol Chem 293(4):1397–1412

Frandsen KEH et al (2016) The molecular basis of polysaccharide cleavage by lytic polysaccharide monooxygenases. Nat Chem Biol 12:298–303

Frandsen KEH, Lo Leggio L (2016) Lytic polysaccharide monooxygenases: a crystallographer's view on a new class of biomass-degrading enzymes. IUCrJ 3:448–467

Frommhagen M et al (2015) Discovery of the combined oxidative cleavage of plant xylan and cellulose by a new fungal polysaccharide monooxygenase. Biotechnol Biofuels 8(101):101–113

Frommhagen M et al (2016) Lytic polysaccharide monooxygenases from *Myceliophthora thermophila* C1 differ in substrate preference and reducing agent specificity. Biotechnol Biofuels 9(1):186

Gregory RC et al (2016) Activity, stability and 3-D structure of the Cu(II) form of a chitin-active lytic polysaccharide monooxygenase from *Bacillus amyloliquefaciens*. Dalton Trans 45:16904–16912

Gudmundsson M et al (2014) Structural and electronic snapshots during the transition from a Cu (II) to Cu(I) metal center of a lytic polysaccharide monooxygenase by X-ray photoreduction. J Biol Chem 289(27):18782–92

Hangasky JA, Iavarone AT, Marletta MA (2018) Reactivity of O_2 versus H_2O_2 with polysaccharide monooxygenases. Proc Nat Acad Sci

Harris PV et al (2010) Stimulation of lignocellulosic biomass hydrolysis by proteins of glycoside hydrolase family 61: structure and function of a large, enigmatic family. Biochemistry 49 (15):3305–3316

Hemsworth GR et al (2014) Discovery and characterization of a new family of lytic polysaccharide monooxygenases. Nat Chem Biol 10(2):122–126

Hemsworth GR et al (2013) The copper active site of CBM33 polysaccharide oxygenases. J Am Chem Soc 135(16):6069–6077

Horn SJ et al (2012) Novel enzymes for the degradation of cellulose. Biotechnol Biofuels 5:45–56

Hudson KL et al (2015) Carbohydrate-aromatic interactions in proteins. J Am Chem Soc 137:15152–15160

Hult E et al (2005) Molecular directionality in crystalline β-chitin: hydrolysis by chitinases A and B from *Serratia marcescens* 2170. Biochem J 388:851–856

Igarashi K et al (2014) Two-way traffic of glycoside hydrolase family 18 processive chitinases on crystalline chitin. Nat Commun 5:1–7

Isaksen T et al (2014) A C4-oxidizing lytic polysaccharide monooxygenase cleaving both cellulose and cello-oligosaccharides. J Biol Chem 289(5):2632–2642

Kari J et al (2014) Kinetics of cellobiohydrolase (Cel7A) variants with lowered substrate affinity. J Biol Chem 289(47):32459–32468

Karkehabadi S et al (2008) The first structure of a glycoside hydrolase family 61 member, Cel61B from *Hypocrea jecorina*, at 1.6 Å resolution. J Mol Biol 383(1):144–154

Kojima Y et al (2016) A lytic polysaccharide monooxygenase with broad xyloglucan specificity from the brown-rot fungus *Gloeophyllum trabeum* and its action on cellulose-xyloglucan complexes. Appl Environ Microbiol 82(22):6557–6572

Kracher D et al (2016) Extracellular electron transfer systems fuel cellulose oxidative degradation. Science (New York, N.Y.), 352(6289):1098–1101

Kruer-Zerhusen N et al (2017) Structure of a *Thermobifida fusca* lytic polysaccharide monooxygenase and mutagenesis of key residues. Biotechnol Biofuels 10(1):1–12

Langston JA et al (2011) Oxidoreductive cellulose depolymerization by the enzymes cellobiose dehydrogenase and glycoside hydrolase 61. Appl Environ Microbiol 77(19):7007–7015

Levasseur A et al (2013) Expansion of the enzymatic repertoire of the CAZy database to integrate auxiliary redox enzymes. Biotechnol Biofuels 6:41–64

Loose JSM et al (2014) A rapid quantitative activity assay shows that the Vibrio cholerae colonization factor GbpA is an active lytic polysaccharide monooxygenase. FEBS Lett 588 (18):3435–3440

Loose JSM et al (2016) Activation of bacterial lytic polysaccharide monooxygenases with cellobiose dehydrogenase. Protein Sci

Loose JSM et al (2018) Multipoint precision binding of substrate protects lytic polysaccharide monooxygenases from self-destructive off-pathway processes. Biochemistry 57(28):4114–4124

Lo Leggio L et al (2015) Structure and boosting activity of a starch-degrading lytic polysaccharide monooxygenase. Nat Commun 6:5961–5969

McLean BW et al (2002) Carbohydrate-binding modules recognize fine substructures of cellulose. J Biol Chem 277(52):50245–50254

Mekasha S et al (2016) Structural and functional characterization of a small chitin-active lytic polysaccharide monooxygenase domain of a multi-modular chitinase from Jonesia denitrificans S. Ferguson. FEBS Lett 590(1):34–42

Monreal J, Reese ET (1969) The chitinase of Serratia marcescens. Can J Microbiol 15(7):689–696

Moser F et al (2008) Regulation and characterization of *Thermobifida fusca* carbohydrate-binding module proteins E7 and E8. Biotechnol Bioeng 100(6):1066–1077

Mutahir Z et al (2018) Characterization and synergistic action of a tetra-modular lytic polysaccharide monooxygenase from Bacillus cereus. FEBS Lett 592(15):2562–2571

Nakagawa YS et al (2015) A small lytic polysaccharide monooxygenase from Streptomyces griseus targeting α-and β-chitin. FEBS J

Paspaliari DK et al (2015) *Listeria monocytogenes* has a functional chitinolytic system and an active lytic polysaccharide monooxygenase. FEBS J 282(5):921–936

Peisach J, Blumberg WE (1974) Structural implications derived from the analysis of electron paramagnetic resonance spectra of natural and artificial copper proteins. Arch Biochem Biophys 165(2):691–708

Petrović DM et al (2018) Methylation of the N-terminal histidine protects a lytic polysaccharide monooxygenase from auto-oxidative inactivation. Protein Sci 27:1635–1650

Phillips CM et al (2011) Cellobiose dehydrogenase and a copper-dependent polysaccharide monooxygenase potentiate cellulose degradation by *Neurospora crassa.* ACS Chem Biol 6 (12):1399–1406

Quinlan RJ et al (2011) Insights into the oxidative degradation of cellulose by a copper metalloenzyme that exploits biomass components. Proc Natl Acad Sci USA 108(37):15079–15084

Sabbadin F et al (2018) An ancient family of lytic polysaccharide monooxygenases with roles in arthropod development and biomass digestion. Nat Commun 9(756):756–767

Tan T-C et al (2015) Structural basis for cellobiose dehydrogenase action during oxidative cellulose degradation. Nat Commun 6(May):7542–7552

Vaaje-Kolstad G et al (2010) An oxidative enzyme boosting the enzymatic conversion of recalcitrant polysaccharides. Science (New York, NY), 330(6001):219–222

Vaaje-Kolstad G et al (2012) Characterization of the chitinolytic machinery of Enterococcus faecalis V583 and high-resolution structure of its oxidative CBM33 enzyme. J Mol Biol 416 (2):239–254

Vaaje-Kolstad G, Houston DR et al (2005a) Crystal structure and binding properties of the *Serratia marcescens* chitin-binding protein CBP21. J Biol Chem 280(12):11313–11319

Vaaje-Kolstad G et al (2013) The chitinolytic machinery of *Serratia marcescens*-a model system for enzymatic degradation of recalcitrant polysaccharides. FEBS J 280(13):3028–3049

Vaaje-Kolstad G, Horn SJ et al (2005b) The non-catalytic chitin-binding protein CBP21 from *Serratia marcescens* is essential for chitin degradation. J Biol Chem 280(31):28492–28497

Valenzuela SV et al (2017) Fast purification method of functional LPMOs from Streptomyces ambofaciens by affinity adsorption. Carbohyd Res 448:205–211

Vermaas JV et al (2015) Effects of lytic polysaccharide monooxygenase oxidation on cellulose structure and binding of oxidized cellulose oligomers to cellulases. J Phys Chem B 119 (20):6129–6143

Vu VV et al (2014) A family of starch-active polysaccharide monooxygenases. Proc Natl Acad Sci USA 111(38):13822–13827

Vuong TV et al (2017) Microplate-based detection of lytic polysaccharide monooxygenase activity by fluorescence-labeling of insoluble oxidized products. Biomacromol 18(2):610–616

Walton PH, Davies GJ (2016) On the catalytic mechanisms of lytic polysaccharide monooxygenases. Curr Opin Chem Biol 31:195–207

Wang D, Li J, Wong ACY et al (2018a) A colorimetric assay to rapidly determine the activities of lytic polysaccharide monooxygenases. Biotechnol Biofuels 11(1):1–11

Wang D, Li J, Salazar-Alvarez G et al (2018b) Production of functionalised chitins assisted by fungal lytic polysaccharide monooxygenase. Green Chem 20(9):2091–2100

Westereng B et al (2015) Enzymatic cellulose oxidation is linked to lignin by long-range electron transfer. Scientif Rep 5:18561–18577

Wong E et al (2012) The Vibrio cholerae colonization factor GbpA possesses a modular structure that governs binding to different host surfaces. PLoS Pathog 8(1):1–12

Xu GY et al (1995) Solution structure of a cellulose-binding domain from *Cellulomonas fimi* by nuclear magnetic resonance spectroscopy. Biochemistry 34(21):6993–7009

Chapter 7
Bacterial Chitinase System as a Model of Chitin Biodegradation

Takafumi Itoh and Hisashi Kimoto

Abstract Chitin, a structural polysaccharide of β-1,4-linked *N*-acetyl-D-glucosamine residues, is the second most abundant natural biopolymer after cellulose. The metabolism of chitin affects the global carbon and nitrogen cycles, which are maintained by marine and soil-dwelling bacteria. The degradation products of chitin metabolism serve as important nutrient sources for the chitinolytic bacteria. Chitinolytic bacteria have elaborate enzymatic systems for the degradation of the recalcitrant chitin biopolymer. This chapter introduces chitin degradation and utilization systems of the chitinolytic bacteria. These bacteria secrete many chitin-degrading enzymes, including processive chitinases, endo-acting non-processive chitinases, lytic polysaccharide monooxygenases, and *N*-acetyl-hexosaminidases. Bacterial chitinases play a fundamental role in the degradation of chitin. Enzymatic properties, catalytic mechanisms, and three-dimensional structures of chitinases have been extensively studied by many scientists. These enzymes can be exploited to produce a range of chitin-derived products, e.g., biocontrol agents against many plant pathogenic fungi and insects. We introduce bacterial chitinases in terms of their reaction modes and structural features.

Keywords Bacterial chitinase · Bacterial chitinolysis · Endo-acting non-processive chitinase · Processive chitinase

7.1 Introduction

Chitin, a linear homopolymer of β-1,4-linked *N*-acetyl-D-glucosamine (GlcNAc) residues, is widespread in nature and serves as a structural component in the cell walls of fungi, and the exoskeletons of insects, arthropods, sponges, and squid pens

T. Itoh (✉) · H. Kimoto
Faculty of Bioscience and Biotechnology, Fukui Prefectural University,
4-1-1 Matsuokakenjyoujima, Eiheiji-cho, Yoshida-gun, Fukui 910-1195, Japan
e-mail: ito-t@fpu.ac.jp

H. Kimoto
e-mail: kimoto@fpu.ac.jp

Q. Yang and T. Fukamizo (eds.), *Targeting Chitin-containing Organisms*, Advances in Experimental Medicine and Biology 1142,
https://doi.org/10.1007/978-981-13-7318-3_7

(Gooday 1990a, b). Chitin is classified by its crystalline forms, such as α-, β-, and γ-chitin, differing in the orientation of the microfibrils. To form complex structures in the cell walls, the crystalline chitin often associates with some proteins, or with other polysaccharides, such as glucans and mannans (Attwood and Zola 1967; Austin et al. 1981 Schaefer et al. 1987; Merzendorfer and Zimoch 2003). In accordance with the abundance and ubiquity of chitin, chitin-degrading enzymes (or chitin-related proteins) have ubiquitous distribution across bacteria, fungi (Gooday 1990a), archaea (Andronopoulou and Vorgias 2004), algae (Shirota et al. 2008; Kitaoku et al. 2017), plants (Grover 2012), and animals (Gooday 1990a). Although annual chitin production in natural ecosystems is estimated to be 10^{10}–10^{12} tons, most of the chitin is recycled by bacteria and fungi as their carbon and nitrogen sources (Tharanathan and Kittur 2003).

Bacterial responses to chitin include chemotaxis and chemotropism, when the bacteria either migrate toward the chitin source or grow toward it; adhesion to chitin; secretion of extracellular chitin-degrading enzymes; and uptake of chitin oligosaccharides (Keyhani and Roseman 1999; Li and Roseman 2004; Meibom et al. 2005). Bacterial chitinolysis, the main step in chitin utilization, typically consists of three enzymatic steps, namely breaking down of crystalline chitin, hydrolyzing chitin chains into dimers, and dividing the dimers into monomers. In the extracellular environments, the first two steps are usually catalyzed by three types of enzymes, lytic polysaccharide monooxygenases (LPMO), endo-acting non-processive chitinases, and processive chitinases. LPMOs, classified as auxiliary activity (AA) families 10 and 11 in the CAZy database (Lombard et al. 2014), cleave glycosidic bonds on the surface of crystalline chitin by an oxidative reaction and introduce oxidized ends to promote further degradation by other secreted chitinases. Chitinases are typically grouped into glycoside hydrolase families 18 and 19 (GH18 and GH19). GH19 chitinases are found mostly in plants, nematodes, and some members of Streptomycetaceae (Ohno et al. 1996; Shimosaka et al. 2001; Lacombe-Harvey et al. 2018). The amino acid sequences of the two GH families share little or no similarities and their catalytic mechanisms are entirely different. It has been hypothesized that GH18 and GH19 chitinases have evolved separately (Fukamizo 2000). These chitinases are often regulated and induced by *N*-acetylglucosamine (GlcNAc monomer), a product of chitin hydrolysis (Bassler et al. 1991), or chitin oligomers $(GlcNAc)_n$, where n = 2 to 6 (Uchiyama et al. 2003). GlcNAc has also been reported to act as a suppressor of chitinase expression in some strains of *Streptomyces* (Miyashita et al. 2000) and *Paenibacillus* (Itoh et al. 2013). After the extracellular hydrolysis of chitin polysaccharides into monomers and oligomers, the products are imported into the periplasm or cytoplasm of the bacteria by efficient uptake systems. In some bacterial strains, phosphoenolpyruvate-dependent phosphotransferase systems (PTS) and ATP-binding cassette (ABC) transporters are responsible for the respective uptakes of GlcNAc monomers and oligomers (Berg et al. 2007; Colson *et al.* 2008; Świątek et al. 2012a). In the cytoplasm or the periplasmic space, β-*N*-acetyl-hexosaminidases, usually classified as GH3 or GH20, cleave the imported chitin oligomers into GlcNAc (Ito et al. 2013; Macdonald et al. 2015).

Chitinolytic bacteria, such as *Serratia marcescens* (Vaaje-Kolstad et al. 2013), *Bacillus circulans* WL-12 (Watanabe et al. 1990), *Streptomyces coelicolor* A3(2) (Saito et al. 2007), *Vibrio cholerae* (Li and Roseman 2004), and *Paenibacillus sp.* (Itoh et al. 2013; Kusaoke et al. 2017), produce many chitinases for efficient degradation of chitin. Nucleotide sequences of these chitinases are often almost completely conserved. This genetic conservation could be explained by the multiple gene duplication in a single cell. However, there are some bacteria that possess gene clusters of chitinases with unique sequences, thought to be acquired from other organisms via lateral gene transfer (Hunt et al. 2008).

7.2 Chitin Biodegradation System of *Serratia marcescens*

Serratia marcescens is one of the most efficient bacteria for the degradation of chitin (Monreal and Reese 1969) and has been extensively studied as the model for bacterial chitinolysis. *Serratia* is a genus of rod-shaped Gram-negative bacteria in the family Enterobacteriaceae. Production of chitin-degrading enzymes in *S. marcescens* can be induced by the presence of chitin in the culture medium (Monreal and Reese 1969). *S. marcescens* strain QMB1466 produces five different chitinolytic enzymes (Fuchs et al. 1986). These enzymes are named as ChiA, ChiB, ChiC1, ChiC2, and CBP21; ChiA and ChiB, two processive chitinases, produce disaccharides from chitin chain by sliding along the chain in opposite directions, i.e., ChiA from the reducing end and ChiB from the nonreducing end (Horn et al. 2006); ChiC1 and ChiC2, endo-acting non-processive chitinases, hydrolyze the chitin polymers randomly; and CBP21 is an LPMO that introduces chain breaks by oxidative cleavages (Vaaje-Kolstad et al. 2013) (Fig. 7.1). ChiC2 results from a posttranslational cleavage of ChiC1; the hydrolytic activity of ChiC2 is lower than that of ChiC1 on crystalline chitin (Suzuki et al. 1999). The biological function of

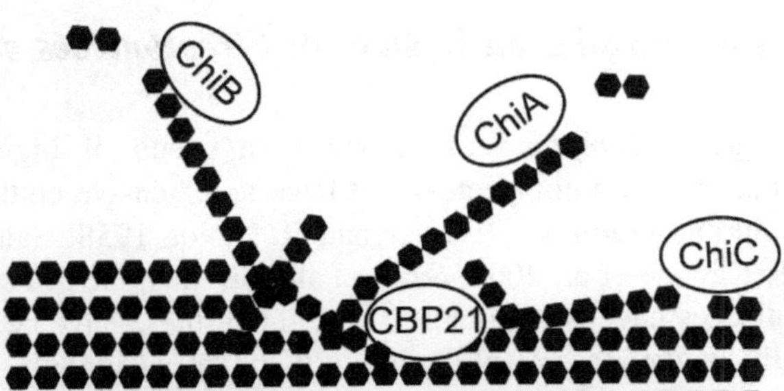

Fig. 7.1 Chitinolytic enzymes of *Serratia marcescens*. The catalytic domains of *S. marcescens* chitinases (ChiA, ChiB, and ChiC) belong to GH18. ChiA and ChiB are processive chitinases that bind closely to the ends of the detached chitin chain and release $(GlcNAc)_2$. ChiC works as an endo-acting non-processive chitinase. CBP21 is an AA10 LPMO and breaks chitin chain by oxidative cleavages

the posttranslational cleavage of ChiC1 remains unclear. These chitin-degrading enzymes work synergistically on chitin degradation (Suzuki et al. 2002). In the cytoplasm or the periplasmic space, *N*-acetyl-hexosaminidase (chitobiase) hydrolyzes oligosaccharides to GlcNAc. According to the CAZy database, in *S. marcescens*, all chitinases (ChiA, B, C1, and C2) belong to GH18, chitobiases belong to GH20, and CBP21 belongs to AA10.

7.3 Chitin Biodegradation System of *Bacillus circulans* WL-12

Bacillus circulans WL-12, a Gram-positive and rod-shaped cell, has been identified as a chitinolytic bacterium that degrades chitin in the cell walls of yeasts and fungi and is known to secrete multiple polysaccharide-degrading enzymes into the culture media (Rombouts and Phaff 1976; Watanabe et al. 1990). When grown in the presence of chitin, over 10 chitinases have been detected in the supernatant of the bacterial culture (Watanabe et al. 1990; Alam et al. 1996). These chitinases are derived from three genes, *chiA*, *chiC*, and *chiD*. Their gene products, ChiA, ChiC, and ChiD, are multi-modular chitinases and their posttranslational cleavages produce multiple chitinases in the culture supernatant. ChiA1, one of the cleavage products, the highest colloidal chitinase activity among others; it yields predominantly a dimer $(GlcNAc)_2$ by binding to insoluble chitin. The structure of ChiA1 contains an N-terminal GH18 catalytic domain, two fibronectin type III (FnIII) domains, and a C-terminal carbohydrate-binding module belonging to CBM12. The catalytic domain of ChiA1 has a deep substrate-binding cleft. On the cleft surface, aromatic residues are linearly located and are important for crystalline chitin hydrolysis (Watanabe et al. 2003).

7.4 Chitin Biodegradation System of *Streptomyces species*

Bacteria in the genus *Streptomyces* belong to a group of high GC content, Gram-positive bacteria (Actinomycetes) that have an extensive chitinolytic system supported by multiple chitinases (Berger and Reynolds 1958; Saito et al. 1999; Schrempf 2001; Kawase et al. 2006; Saito et al. 2007). In the *S. coelicolor* A3(2) genome, 13 chitinases have been found to date; 11 of these have been classified in the GH18 family of chitinases: subfamily A (Chi18aC, Chi18aD, Chi18aE, and Chi18aJ), subfamily B (Chi18bA, Chi18bB, and Chi18bI), and subfamily C (Chi18cH, Chi18cK, Chi18cL, and Chi18 cM), and two in the GH19 family of chitinases (Chi19F and Chi19G) (Kawase et al. 2006). Besides chitinases, *S. coelicolor* A3(2) secretes chitosan-related enzymes, such as GH46 chitobiase (Ghinet et al. 2010), GH20 chitobiase (Saito et al. 2013), and CE4 chitin deacetylase (Świątek et al. 2012b). The metabolism of chitin oligosaccharides in

S. coelicolor A3(2) is well understood (Świątek et al. 2012a, b) (Fig. 7.2). The degradation monomer product, GlcNAc, is imported and phosphorylated to *N*-acetyl-D-glucosamine-6-phosphate (GlcNAc-6P) by the PTS through the cell membrane. In the PTS reactions, a phosphoryl group is transferred from phosphoenolpyruvate to phosphotransferase enzyme I (EI) and then the group is transferred from EI to histidine protein (HPr). The phosphoryl group of HPr is further transferred to the enzyme complex of enzyme IIA (EIIA), enzyme IIB (EIIB), and enzyme IIC (EIIC). GlcNAc, transported by enzyme IIC (EIIC), is phosphorylated to GlcNAc-6P by EIIA. Some ABC transporters, like NgcEFG-MsiK and DasABC-MsiK, can also transport $(GlcNAc)_2$ dimers. Transported $(GlcNAc)_2$ dimer is divided into GlcNAc monomers by β-*N*-acetyl-D-glucosaminidase (DasD) in the cytoplasm. An *N*-acetyl-D-glucosamine kinase (NagK) phosphorylates GlcNAc to GlcNAc-6P, which is then deacetylated to glucosamine-6-phosphate (GlcN-6P) by *N*-acetyl-D-glucosamine-6-phosphate deacetylase (NagA). Thereafter, glucosamine-6-phosphate deaminase (NagB) deaminates GlcN-6P to fructose-6-phosphate (Frc-6P) that would enter the glycolytic pathway.

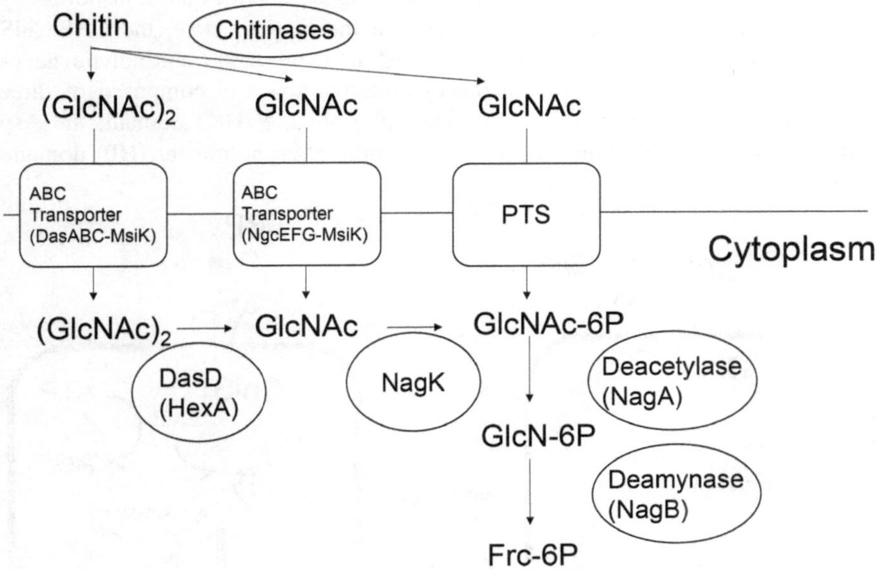

Fig. 7.2 Chitin uptake by *Streptomyces* species. Chitin is hydrolyzed into GlcNAc and $(GlcNAc)_2$ by GH18 and GH19 chitinases in the extracellular space. The degradation product, GlcNAc, is transported through PTS across the cell membrane and is phosphorylated to GlcNAc-6P by the PTS. An ABC transporter, NgcEFG-MsiK can import GlcNAc and $(GlcNAc)_2$. Another ABC transporter, DasABC-MsiK, can also import $(GlcNAc)_2$. Imported $(GlcNAc)_2$ is hydrolyzed into GlcNAc by chitobiase (DasD). GlcNAc is phosphorylated to GlcNAc-6P by GlcNAc kinase (NagK). GlcNAc-6P deacetylase (NagA) converts GlcNAc-6P to GlcN-6P, which is then deaminated to Frc-6P by GlcN-6P deaminase (NagB). Frc-6P would enter the glycolytic pathway

7.5 Chitin Biodegradation System of Marine Bacteria *Vibrio species*

Despite being a ubiquitous biopolymer, chitin is not known to accumulate in ocean sediments; marine chitinolytic bacteria, mainly from the family Vibrionaceae, contribute to the rapid recycling of chitin (Hirono et al. 1998). Chitinolytic and signal transduction systems of *Vibrio* species have been widely studied. The system constitutes five processes: finding chitin (chemotaxis); adhering to chitin via the bacterial cell surface; degrading chitin to oligosaccharides; transporting the degraded oligosaccharides to the cytoplasm by a sugar-specific porin (chitoporin) on the outer membrane and by a sugar-specific ABC transporter on the inner membrane; and converting the degradation products to Frc-6P in the cytosol for the glycolytic pathway (Keyhani and Roseman 1999; Li and Roseman 2004; Meibom et al. 2004; Hunt et al. 2008). Inducing chitinolytic enzymes of *Vibrio* species is intricately regulated by histidine kinases and two-component systems. In the resting state, a binding protein for chitin oligosaccharides (CBP) binds to the periplasmic domain of the membrane protein ChiS. When secreted chitinases degrade chitin to oligosaccharides in the extracellular space, the oligosaccharides are transported by the chitoporin and are bound by the CBP. In the binding state, the CBP/ChiS complex dissociates and transports the signal to express the chitinolytic genes (Fig. 7.3). The domain of ChiS in the cytoplasmic space is composed of three subdomains: the ATP-dependent His kinase/phosphatase (HK) domain; the Asp response regulator (RR) domain; and the histidine phosphotransfer (HP) domain.

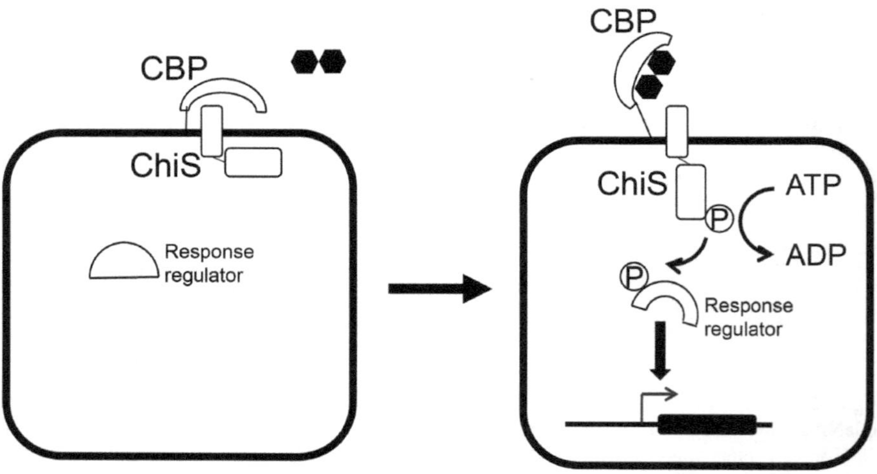

Fig. 7.3 Induction of chitinolytic enzymes of *Vibrio* species. The induction is regulated by a histidine kinase and a two-component system (ChiS). In the resting state, chitin-binding protein (CBP) binds to the periplasmic domain of the membrane protein ChiS. When the oligosaccharides bind to CBP, the CBP/ChiS complex dissociates and transports the signal for the expression of chitin-degrading enzymes

The phosphoryl group is transferred sequentially from ATP to HK, to RR, to HP, and finally to Asp on a response regulator that interacts with the genome to induce the expression of chitinolytic enzymes (Li and Roseman 2004).

7.6 Chitin Biodegradation Systems of *Paenibacillus* sp. str. IK-5 and FPU-7

Similar to the *Streptomyces* species, *Paenibacillus* sp. str. IK-5 produces a chitosanase (ChiE) and a GH19 chitinase (ChiD) in addition to two GH18 chitinases (ChiA and ChiB) and an AA10 LPMO (ChiC) (Kusaoke et al. 2017; Fig. 7.4). The two chitinases (ChiA and ChiB) contain one GH18 catalytic domain and two CBM5 chitin-binding modules. The AA10 LPMO (ChiC) also has a CBM5 module in addition to the catalytic domain. The chitinase ChiD has a GH19 catalytic domain. Chitosanase (ChiE) contains a GH8 catalytic domain at the N-terminus and two discoidin domains (CBM32) at the C-terminus. The CBM32 domains of ChiE can bind specifically and tightly to chitosan. These five enzymes (ChiA, ChiB, ChiC, ChiD, and ChiE) form a huge protein complex called a "chitinasome". When the culture medium contains only chitin, the complex consists of four enzymes (ChiA, ChiB, ChiC, and ChiD); when the culture medium contains both chitin and chitosan, the complex comprises all five enzymes. *Paenibacillus* sp. str. IK-5 cells also produce multi-modular chitobiase composed of the GH20 catalytic domain and S-layer homology domains (SLH) on their surface.

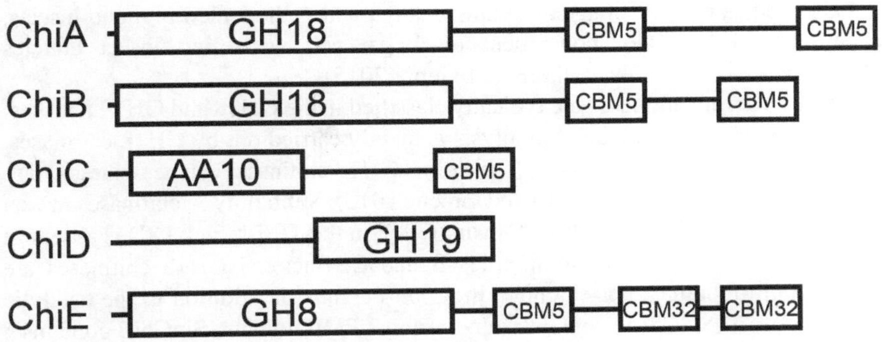

Fig. 7.4 Schematic representations of the *P*. str. IK-5 chitin and chitosan degrading enzymes. The enzymes (ChiA, ChiB, ChiC, ChiD, and ChiE) are produced by *P*. str. IK-5 and have signal peptides for secretion at the N-terminus. ChiA and ChiB are GH18 chitinases with catalytic domains at the N-terminus and two CBM 5 domains at the C-terminus. ChiC has an AA10 catalytic domain as LPMO and a CBM5 at the C-terminus. The catalytic domain of ChiD belongs to GH19. Chitosanase ChiE contains a GH8 catalytic domain at the N-terminus and two CBM32 at the C-terminus. These chitin and chitosan-related enzymes (ChiA, ChiB, ChiC, ChiD, and ChiE) are found as a huge protein complex (chitinasome). When the culture medium contains chitin and chitosan, ChiE is assembled into four enzyme complexes (ChiA, ChiB, ChiC, and ChiD)

Paenibacillus sp. str. FPU-7 (*P.* str. FPU-7), a bacterium with high chitinolytic activity, was recently isolated from soil using a medium containing solid chitin flakes (Itoh et al. 2013). Subject to carbon catabolite repression by GlcNAc, the bacterium constitutively secretes several chitinases into the culture medium. Genes for at least six chitinases (ChiA, ChiB, ChiC, ChiD, ChiE, and ChiF) are found in the genome, each containing a GH18 catalytic domain and one or more auxiliary domains, such as CBM5, CBM12, and FnIII. On a synthetic substrate, pNP-$(GlcNAc)_2$, ChiE shows high activity; ChiA, ChiD, and ChiF are less active; ChiB and ChiC are moderately active. All the chitinases, except ChiD, are highly active on insoluble colloidal chitin. Efficient degradation of chitin flakes by *P.* str. FPU-7 requires not only the chitinase secretions in the culture medium, but also the living cells, suggesting an involvement of cell-bound chitinase(s). Apart from the six chitinases extensively studied in its genome, chitinase ChiW is expressed on the cell surface of *P.* str. FPU-7 in the presence of $(GlcNAc)_2$. ChiW produces $(GlcNAc)_2$ as the final reaction product on the cell surface. The product $(GlcNAc)_2$ would then be transported into the cytosol by some transporters and induce the chitin-related gene expression such as ChiW.

7.7 Overview of Bacterial GH18 Chitinases

Chitinolytic bacteria secrete multiple chitinases as described above. Some bacterial chitinases can be active over a range of temperatures and pH (Meena et al. 2014; Rathore and Gupta 2015) and, thus, these enzymes are used in many industrial processes. Some of these enzymes also have antifungal and antibacterial properties and are used as biocontrol agents against many plant pathogenic fungi and insects, since chitin is a major constituent of fungal cell walls and insect cuticles (Bhattacharya et al. 2007; Rathore and Gupta 2015).

These bacterial chitinases are typically classified in the GH18 and GH19 families; as described above, bacterial chitinolysis is mainly carried out by GH18 chitinases. Based on their amino acid sequences, bacterial GH18 chitinases can be separated into three subfamilies, A, B, and C (Li and Greene 2010). Subfamily A chitinases have a small subdomain, chitin insertion domain (CID), in the TIM-barrel. CID is absent in the chitinases belonging to subfamilies B and C. Bacterial GH18 chitinases are usually multi-modular. For example, in *S. marcescens*, in addition to the catalytic domains, ChiA (SmChiA) contains an N-terminal FnIII domain; SmChiB contains a CBM5 at the C-terminus; SmChiC contains a FnIII domain and CBM12 at the C-terminus (Vaaje-Kolstad et al. 2013). Similarly, *B. circulans* WL-12 ChiA1 (BcChiA1) contains two FnIII and CBM12 domains at the C-terminus (Watanabe et al. 2003). Considering the enzymatic reaction modes, GH18 chitinases are classified into two types: processive chitinases and endo-acting non-processive chitinases. These chitin-degrading enzymes synergistically operate on the polymers. First, non-processive chitinases and LPMO break the glycosidic bond of surface-bound crystalline chitin. Then, the newly produced terminuses of the detached chains from the solid surface become the sites for the attachment of processive chitinases.

Many crystal structures of bacterial GH18 chitinases are available on the Protein Data Bank. The typical catalytic domain of GH18 chitinase folds a TIM-barrel consisting of a substrate-binding cleft (Perrakis et al. 1994). GH18 chitinases hydrolyze chitin through a substrate-assisted mechanism that retains the stereo-chemistry of the anomeric carbon of the newly synthesized reducing end (Tews et al. 1997; Brameld and Goddard 1998; van Aalten et al. 2001) (Fig. 7.5). The catalytic residues are in a conserved DxDxE motif located on fourth β-strand of the core barrel. The catalysis is initiated with a distortion into a boat (or skewed boat) conformation of the GlcNAc residue at the -1 subsite. Substrate binding leads the rotation of a second aspartic acid residue of the DxDxE motif from the first aspartic acid to the catalytic glutamic acid of the DxDxE motif. The second aspartic acid residue then forms hydrogen bonds with the glutamic acid and the *N*-acetyl group of GlcNAc at the -1 subsite. The catalytic glutamic acid protonates the glycosidic bond followed by a nucleophilic attack on the anomeric carbon by the oxygen atom of the *N*-acetyl group. After the scission of the glycosidic bond, an oxazolinium ion intermediate is formed and subsequently hydrolyzed to complete the reaction. Other

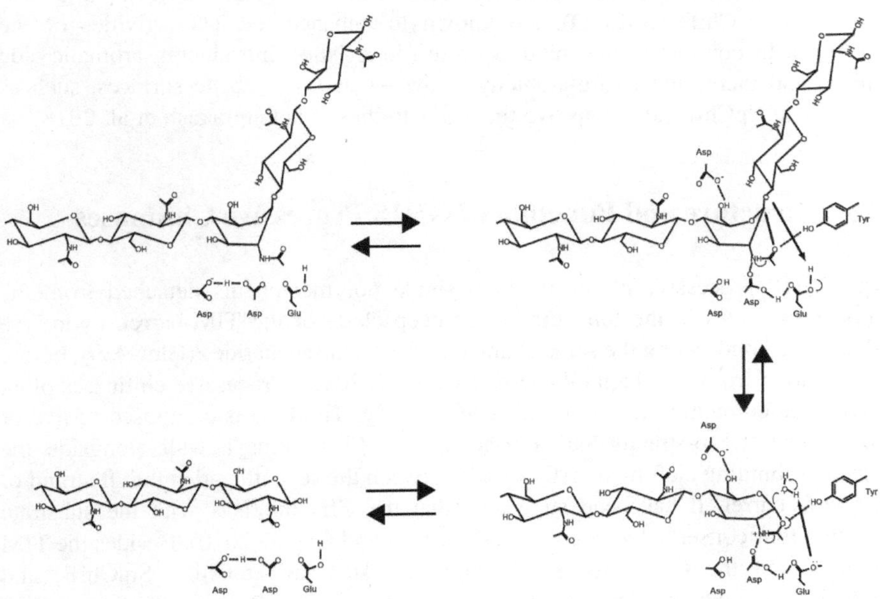

Fig. 7.5 Substrate-assisted catalytic mechanism of GH18 chitinase. The catalysis is initiated when the GlcNAc residue at the -1 subsite is distorted into a boat conformation. The catalytic glutamic acid protonates the glycosidic bond. This is followed by a nucleophilic attack on the anomeric carbon by the oxygen atom of *N*-acetyl group of the substrate GlcNAc at the -1 subsite. After the scission of the glycosidic bond, an oxazolinium ion intermediate is formed and is subsequently hydrolyzed

conserved residues, tyrosine, and aspartic acid (or asparagine), interact with GlcNAc residues at the subsite −1 and thus, serve as important residues for catalysis.

Chitooligosaccharides (CHOS) produced by the hydrolysis of chitin have potential biological applications in the food, medicine, and agriculture sectors. Depending on the length and pattern of acetylation, CHOS exhibit antimicrobial, antitumor, and immuno-enhancing properties (Mallakuntla et al. 2017). Since the chemical synthesis of CHOS involves nonspecific random hydrolysis, controlling the length of the product is very difficult. Besides hydrolytic activities, some GH18 chitinases exhibit transglycosylation (TG) activities. The enzymes can condensate other saccharide molecules without a water molecule as an acceptor, with the oxazolinium ion intermediate as a donor (Umemoto et al. 2015). TG activities, focusing on the enzymatic production of longer CHOS from shorter oligosaccharides, have received considerable attention. TG activities of GH18 are improved by mutagenesis. Mutations on specific sites can reduce the hydrolytic activities and optimize the subsite affinities and/or stability of the oxazolium ion intermediate. Mutations of aspartates in the DXDXE sequence motif affect the stability of the oxazolium ion intermediate and reduce hydrolytic activities. As a result, these mutant enzymes exhibit improved TG activities. Mutations of aromatic residues in the minus subsite (donor site), such as W167A in the −3 subsite of SmChiA; Y163A and Y390F in the -1 subsite of SmChiA; and Y28A and Y222A in the −1 subsite of *Serratia proteamaculans* ChiD (SpChiD), are known to enhance the TG activities of the enzymes. In contrast to the minus subsite (donor site), introducing aromatic side chains and increasing hydrophobicity at the +1 and +2 subsite surfaces, such as Y226W of SpChiD, also improve the TG activities (Madhuprakash et al. 2018).

7.8 Structure and Function of GH18 Processive Chitinases

The GH18 processive chitinases bind single polymer chains detached from the crystalline chitin in the long tunnels or deep clefts of the TIM-barrel, hydrolyze glycosidic bonds along the same chain, and release disaccharides $(GlcNAc)_2$, before dissociation from the chain (Vaaje-Kolstad et al. 2013). Processive chitinases often have a subdomain, CID, as described previously. The CID is composed of five or six antiparallel β-strands and a α-helix. The CID forms a wall alongside the substrate-binding cleft by inserting itself between the seventh and eighth β-strand of the core barrel. It has been suggested that the CID interacts with the substrate through four conserved amino acid residues (Li and Greene 2010). Besides the TIM domain and the CID, processive chitinases, such as SmChiA, SmChiB, and BcChiA1, contain one or two additional domains. (Fig. 7.6a, b, d). Removal of these extra domains reduces the biological activities of the crystalline forms of chitin. Aromatic residues such as tryptophan that are localized in the substrate-binding clefts function as flexible hydrophobic sheaths for binding of the polymer chain and thereby, improve the processivity of the chitin. The polymer chain can slide during the processive action in the sheath. The W97A mutation at

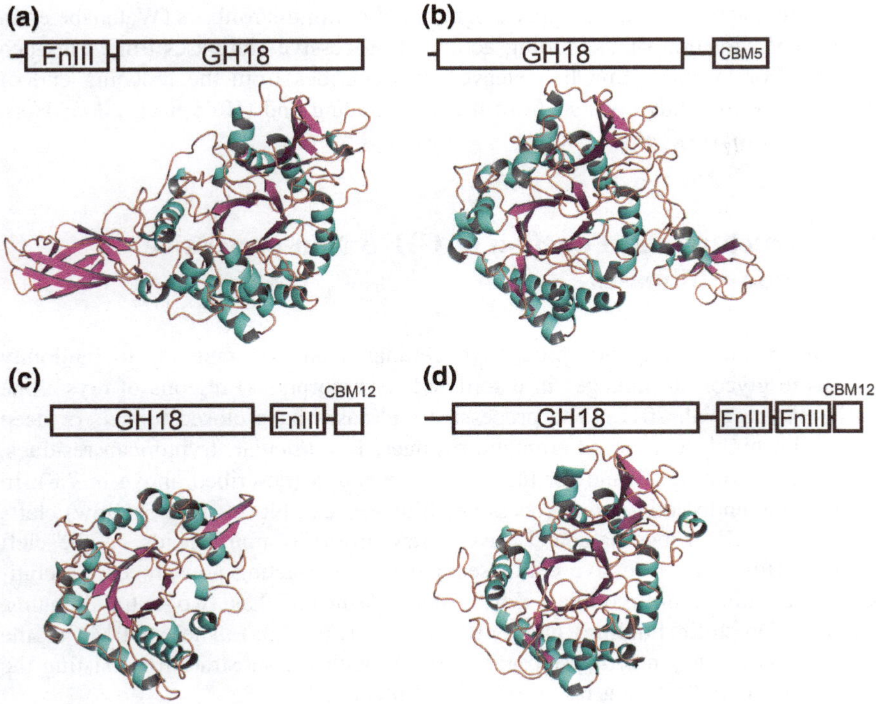

Fig. 7.6 Three-dimensional structures of bacterial GH18 chitinases. **a** SmChiA has a FnIII domain (β-sandwich) at the N-terminus, except for the GH18 catalytic domain (TIM-barrel). GH18 of SmChiA has one additional domain, CID, protruding from the core barrel. **b** SmChiB has a CBM5 domain at the C-terminus, except for the GH18 catalytic domain. GH18 of SmChiB also has a CID. **c** The GH18 catalytic domain of SmChiC. Although SmChiC has FnIII and CBM12 domains at the C-terminus, the crystal structure was determined without these auxiliary domains. **d** The GH18 catalytic domain of BcChiA1. Auxiliary domains of BcChiA1, such as two FnIII and CBM12 domains, are absent in the crystal structure. GH18 of BcChiA1 also has a CID

the subsite +1 of SmChiB reduces the processivity and the degradation activity for crystalline chitin. However, W97A mutation also causes a 29-fold increase in the degradation rate of single chains (Hamre et al. 2014). The processivity seems to be counterbalanced by the catalytic efficiency of the reaction. On the other hand, Watanabe et al. reported that the W433 and Y279 mutants of BcChiA1 exhibited reduced hydrolytic activity against crystalline chitin, colloidal chitin, and $(GlcNAc)_5$ (Watanabe et al. 2001, 2003). The W433 residue interacts with GlcNAc at subsite −1 through a hydrophobic stacking interaction and holds the residue at this position during the catalytic reaction. The Y279 residue assists the formation of an oxazolinium ion intermediate. In addition to the amino acid residues in the substrate-binding cleft, there are two extra tryptophan residues, W122 and W134, on the surface of BcChiA1. These tryptophan residues are located at the extension of the nonreducing end side of the substrate. Mutations at both these sites reduce the hydrolyzing activity against highly crystalline β-chitin microfibrils (Watanabe et al. 2003). The direction of the sliding action of processive chitinases differs in each enzyme. For instance, SmChiA cleaves disaccharides from the reducing end of chitin, while SmChiB does so from the nonreducing end (Hult et al. 2005; Horn et al. 2006; Sikorski et al. 2006).

7.9 Structure and Function of GH18 Non-processive Endo-Chitinases

The GH18 endo-acting non-processive chitinases have a tendency to randomly hydrolyze glycosidic linkages in disordered (or amorphous) regions of crystalline chitin (Payne et al. 2012). The processive chitinases have closed tunnels or deep clefts with highly conserved aromatic residues, in particular, tryptophan residues, that interact with the ligand for the processive action (described above in 7.8). In contrast, the endo-acting non-processive chitinases exhibit open (or shallow) clefts lacking the CID subdomain and have a few aromatic amino acids on the cleft surfaces. Both the processive chitinases and the endo-acting non-processive chitinases are often multi-modular. For instance, SmChiC has two extra domains, CBM12 (for chitin binding) and FnIII domain (Fig. 7.6c). These CBMs of the non-processive enzymes may loosely associate with the substrate for assisting the activity of the endo-acting non-processive enzymes.

7.10 Structure and Function of the Carbohydrate-Binding Modules of GH18 Chitinases

In addition to the catalytic domain, bacterial GH18 chitinases often contain single or multiple CBMs or FnIII domains, as described above. These extra domains can be found either in the N-terminal or in the C-terminal of the enzyme (Fig. 7.6). The structures and functions of the CBMs responsible for binding crystalline polysaccharides, such as chitin and cellulose, are well studied (Boraston 2005; Georgelis, et al. 2012). Although their contribution to the efficiency of catalysis is not yet completely understood, CBMs associate with the crystalline polysaccharide substrates and position them correctly to the catalytic domains. The four types of roles that CBMs have in chitinolysis include: (a) targeting effect, where CBMs target the enzymes to appropriate regions on their substrates (reducing end, nonreducing end, or internal of polysaccharide chains); (b) proximity effect, where CBMs increase the concentrations of the enzymes in close proximity to their substrates and lead the catalytic domains to efficiently act on their substrates; (c) disruptive effect, where some CBMs bind to polysaccharides and disrupt the surface of tightly packed chains, thereby, loosening and exposing the packed substrates to be attacked by their catalytic domains; and (d) adhesion, where some

CBMs have functions to adhere enzymes onto the surfaces of bacterial cell walls. The catalytic domains break the neighboring substrate polymer chains. Cell walls are complex structures; thus, the binding targets of CBMs are not always the intrinsic substrate polysaccharides for their catalytic domains. Some CBMs have broad specificity for various polysaccharides in the cell walls.

The CBM5 and CBM12 modules consist of 40–60 residues and are often found in GH18 chitinases (Fig. 7.6). Both the families have similar amino acid sequences and have conserved aromatic tryptophan residues that interact with the substrate on the protein surface. The presence of either of these modules increases the substrate affinity and the efficiency of chitin hydrolysis, especially for crystalline chitin (Watanabe et al. 1994; Hashimoto et al. 2000; Uni et al. 2012).

All the FnIIIs domains consist of 80–100 residues and share amino acid sequence similarities. Each FnIII domain folds into a typical immunoglobulin-like (β-sandwich) fold with three and four strands (Fig. 7.6a). The FnIII of SmChiA (also referred to as chitinase A N-terminal domain) contains exposed aromatic residues responsible for substrate binding, thereby, improving the catalytic efficiency of the enzyme (Uchiyama et al. 2001). In contrast, the FnIII domain of BcChiA1 does not have surface-exposed aromatic residues and is not directly involved in the chitin binding (Jee et al. 2002). Studies suggest that the latter types of FnIII domains serve as linker domains for the adequate stabilization of other domains or of the overall structure of the enzyme for a competent degradation.

7.11 Cell Surface-Expressed Multi-modular Chitinase ChiW from *Paenibacillus* sp. FPU-7

The Gram-positive bacterium *P.* str. FPU-7 effectively hydrolyzes chitin with several secreted chitinases, as described above. This bacterium produces a unique chitinase, ChiW that has two catalytic domains. This multi-modular enzyme is expressed on the cell surface and has a high activity toward various chitins, including crystalline chitins (Itoh et al. 2013, 2014, 2016). The cell surface-expressed enzyme, ChiW, enhances chitin degradation when it acts in combination with other secreted chitinases of *P.* FPU-7 (Itoh et al. 2013). ChiW contains 1,418 amino acids with a secretory signal peptide (150 kDa). The unique multi-modular architecture of ChiW allows it to function efficiently on the cell surface via three surface layer homology domains (SLH), a right-handed β-helix domain (CBM54), a Gly-Ser-rich loop, two immunoglobulin-like fold domains (Ig-1 and Ig-2), and two GH18 catalytic domains (Figs. 7.7 and 7.8).

The SLH domains are absent in the ChiW crystal structure. Typical SLH domains are composed of three repeats of highly conserved amino acid sequences ($\sim$18 kDa) and bind noncovalently to glycan backbones of the peptidoglycan of Gram-positive bacteria. Thus, the cell wall is surrounded by proteins congregated with SLH domains as a cell envelope or surface layer (Schneewind and Missiakas 2012).

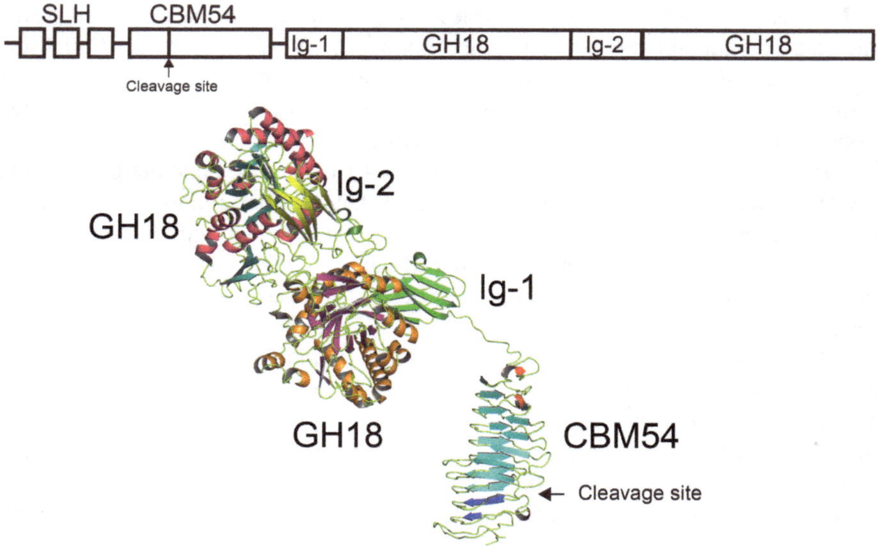

Fig. 7.7 Overall structure of *P.* str. FPU-7 ChiW. The structure is multi-modular with a CBM54 domain and a catalytic region (Ig-1, Ig-2, and two GH18 domains). ChiW is specifically cleaved between Asn282 and Ser283 at the CBM54 domain. The position is pointed out as the cleavage site. SLH domains are absent in the crystal structure

The CBM54 flexibly links to the catalytic region (two Ig-like domains and two GH18 domains) of ChiW via the Gly-Ser-rich loop. The structure of CBM54 consists of a right-handed parallel β-helix fold with 12 coils (Fig. 7.8a). There are 34 β-strands that form 3 parallel β-sheets, named SB1 (made of 12 β-strands), SB2 (made of 12 β-strands), and SB3 (made of 10 β-strands), making 3 distorted faces. Although this fold is often found in enzymes such as carbohydrate lyases, CBM54 of ChiW displays no detectable carbohydrate degradation activities. The domain possesses diverse substrate specificities and can therefore, bind to several cell wall polysaccharides, including chitin, chitosan, β-1,3-glucan, xylan, and cellulose. This domain putatively helps in an efficient decomposition of the cell wall chitin through the contact surface. However, the molecular surface of CBM54 has no distinct cleft or patch surrounded by aromatic residues.

The two GH18 catalytic domains of ChiW have similar structures, consistent with their high degree of amino acid sequence similarity (56% identity) (Fig. 7.8d and e). The structures of the catalytic clefts of the two domains are also similar. Besides the core TIM-barrels, both have two additional subdomains, CID and insertion domain 2; the two subdomains protrude from the barrel and form the walls of a deep active cleft of approximately 42 Å in length and 26 Å in depth. The amino acid residues at the center of the two active sites are almost identical and well conserved in other processive GH18 chitinases. Important residues of SmChiA for saccharide binding, Trp167 at the −3 subsite, Trp539 at the −1 subsite, Trp275 at

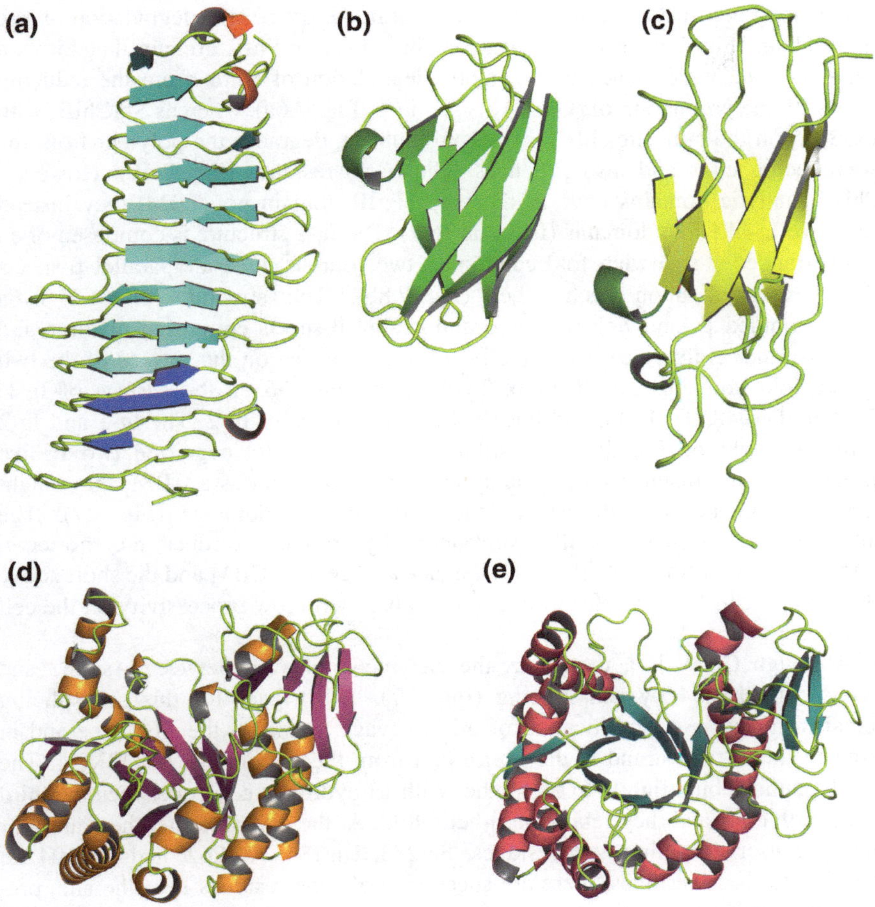

Fig. 7.8 Close-up views of the individual domain of ChiW. **a** The CBM54 domain has a right-handed parallel β-helix structure with 12 coils. **b** The Ig-1 structure is an eight-stranded β-sandwich with two four-stranded antiparallel β-sheets. **c** The Ig-2 structure is a seven-stranded β-sandwich with two antiparallel β-sheets composed of three and four β-strands. **d**, **e** The structures of the two GH18 catalytic domains. The two GH18 catalytic domains have similar structures; besides the core TIM-barrel, they have additional two subdomains, a CID and an insertion domain 2. The two subdomains protrude from the barrel and form the walls of a deep active cleft

the +1 subsite, and Phe396 at the +2 subsite, correspond to the ChiW residues Trp568/Trp1055, Trp905/Trp1396, Trp652/Trp1138, and Trp772/Trp1258, respectively. The catalytic residues of SmChiA, Tyr390, Asp311, Asp313, and Glu315 correspond to Tyr766/Tyr1252, Asp687/Asp1173, Asp689/Asp1175, and Glu691/Glu1177, respectively. These conserved residues indicate that ChiW possesses a catalytic mechanism that is similar to that of SmChiA and other typical processive chitinases. In the processive chitinases, CBMs or FnIII domains often locate along their catalytic domains and assist in the processive degradation of one chitin chain (Fig. 7.6). For example, SmChiA has one FnIII domain that forms a minus subsite, which leads to enzymatic degradation of chitin from the reducing ends with the production of $(GlcNAc)_2$ residues (Fig. 7.6a), whereas SmChiB, with a CBM5 on the opposite side forms a plus subsite, degrades the polymer from the nonreducing ends and also produces $(GlcNAc)_2$ residues (Fig. 7.6b). However, ChiW catalytic domains have neither such FnIII domain nor CBM, they instead have two Ig-like fold domains (Ig-1 and Ig-2). The Ig-1 structure is composed of an eight-stranded β-sandwich fold containing two four-stranded antiparallel β-sheets closely stacked upon each other (Fig. 7.8b). The structure of Ig-2 is a seven-stranded β-sandwich with two antiparallel β-sheets composed of three and four β-strands (Fig. 7.8c). Aromatic residues are located on the surface of the two Ig-like fold domains, i.e., Tyr486, Tyr537, and Phe556 on the surface of Ig-1; Tyr939, Tyr948, Tyr1000, and Phe1044 on the surface of Ig-2. The Ig-1 and Ig-2 domains might be functional substitutions of CBM. However, the two Ig-like domains are too distal from the catalytic clefts to function as a CBM; they might serve as linkers or scaffolds for the two catalytic domains (Fig. 7.7). The substrate-binding sites of ChiW, surrounded by aromatic residues, are shorter in length to those of SmChiA. Possibly, the lack of a general CBM and the short active clefts allow ChiW to transfer from chain to chain with low processivity on the cell surface.

Although ChiW is a monomer, the enzyme is cleaved between Asn282 and Ser283 at CBM54 by self-splicing (Fig. 7.7). The trigger for this self-splicing remains unresolved. The location of this cleavage site is on the SB2 face and in front of the 11th β-strand at the fourth coil from the N-terminus of CBM54. The 2 polypeptides bind tightly to each other with 13 hydrogen bonds between the third and fourth coils and they retain the β-helix fold. At the cleavage site, there are four highly conserved amino acid residues, Ser283, His285, Asp262, and Arg304. In addition to these residues, there are successive glycine residues near the site, presumably providing conformational flexibility to the cleavage site. This limited proteolysis occurs using the hydroxyl group of Ser283 as a nucleophile. The amino acid residues of this cleavage site, Asn-Ser, have been found in various self-splicing proteins (Clarke 1994; Hall et al. 1997).

On the other hand, chitinases with two GH18 catalytic domains have been found in viruses, archaea, bacteria, and insects (Hiramatsu et al. 2000; Tanaka et al. 2001; Howard et al. 2004; Arakane and Muthukrishnan 2010). Chitinase, Tk-ChiA, from the hyperthermophilic archaeon *Thermococcus kodakaraensis* KOD1 (Tanaka et al. 2001) and chitinase B from *Microbulbifer degradans* 2–40 (Howard et al. 2004) have an exo-chitinase at the N-terminus and an endo-chitinase at the C-terminus. In some chitinases, the two catalytic domains work synergistically, because their combined activities exhibit a significantly higher chitinolysis than the sum of their individual activities. The two clefts of ChiW are adequately positioned by the two Ig-like domains and cross each other at approximately right angles. This unique spatial arrangement may be important for the efficient degradation of chitin on the cell surface.

References

Alam MM, Mizutani T, Isono M, Nikaidou N, Watanabe T (1996) Three chitinase genes (chiA, chiC, and chiD) comprise the chitinase system of *Bacillus circulans* WL-12. J Ferment Bioeng 82:28–36

Andronopoulou E, Vorgias CE (2004) Multiple components and induction mechanism of the chitinolytic system of the hyperthermophilic archaeon *Thermococcus chitonophagus*. Appl Microbiol Biotechnol 65:694–702

Arakane Y, Muthukrishnan S (2010) Insect chitinase and chitinase-like proteins. Cell Mol Life Sci 67:201–216

Attwood MM, Zola H (1967) The association between chitin and protein in some chitinous tissues. Comp Biochem Physiol 20:993–998

Austin PR, Brine CJ, Castle JE, Zikakis JP (1981) Chitin: New facets of research. Science 212:749–753

Berger LR, Reynolds DM (1958) The chitinase system of a strain of *Streptomyces griseus*. Biochim Biophys Acta 29:522–534

Bassler BL, Yu C, Lee YC, Roseman S (1991) Chitin utilization by marine bacteria. Degradation and catabolism of chitin oligosaccharides by *Vibrio furnissii*. J Biol Chem 266:24276–24286

Berg T, Schild S, Reidl J (2007) Regulation of the chitobiose-phosphotransferase system in *Vibrio cholerae*. Arch Microbiol 187:433–439

Bhattacharya D, Nagpure A, Gupta RK (2007) Bacterial chitinases: properties and potential. Crit Rev Biotechnol 27:21–28

Boraston AB (2005) The interaction of carbohydrate-binding modules with insoluble non-crystalline cellulose is enthalpically driven. Biochem J 385:479–484

Brameld KA, Goddard WA (1998) Substrate distortion to a boat conformation at subsite -1 is critical in the mechanism of family 18 chitinases. J Am Chem Soc 120:3571–3580

Clarke ND (1994) A proposed mechanism for the self-splicing of proteins. Proc Natl Acad Sci USA 91:11084–11088

Colson S, van Wezel GP, Craig M, Noens EE, Nothaft H, Mommaas AM, Titgemeyer F, Joris B, Rigali S (2008) The chitobiose-binding protein, DasA, acts as a link between chitin utilization and morphogenesis in *Streptomyces coelicolor*. Microbiology 154:373–382

Fuchs RL, McPherson SA, Drahos DJ (1986) Cloning of a *Serratia marcescens* gene encoding chitinase. Appl Environ Microbiol 51:504–509

Fukamizo T (2000) Chitinolytic enzymes: catalysis, substrate binding, and their application. Curr Protein Pept Sci 1:105–124

Georgelis N, Yennawar NH, Cosgrove DJ (2012) Structural basis for entropy-driven cellulose binding by a type-A cellulose-binding module (CBM) and bacterial expansin. Proc Natl Acad Sci USA 109:14830–14835

Ghinet MG, Roy S, Poulin-Laprade D, Lacombe-Harvey MÈ, Morosoli R, Brzezinski R (2010) Chitosanase from *Streptomyces coelicolor* A3(2): biochemical properties and role in protection against antibacterial effect of chitosan. Biochem Cell Biol 88:907–916

Gooday GW (1990a) The ecology of chitin degradation. Adv Mibrob Ecol 11:387–430

Gooday GW (1990b) Physiology of microbial degradation of chitin and chitosan. Biodegradation 1:177–190

Grover A (2012) Plant chitinases: genetic diversity and physiological roles. Crit Rev Plant Sci 31:57–73

Hall TM, Porter JA, Young KE, Koonin EV, Beachy PA, Leahy DJ (1997) Crystal structure of a Hedgehog autoprocessing domain: homology between Hedgehog and self-splicing proteins. Cell 91:85–97

Hamre AG, Lorentzen SB, Väljamäe P, Sørlie M (2014) Enzyme processivity changes with the extent of recalcitrant polysaccharide degradation. FEBS Lett 588:4620–4624

Hashimoto M, Ikegami T, Seino S, Ohuchi N, Fukada H, Sugiyama J, Shirakawa M, Watanabe T (2000) Expression and characterization of the chitin-binding domain of chitinase A1 from *Bacillus circulans* WL-12. J Bacteriol 182:3045–3054

Hiramatsu S, Fujie M, Usami S, Sakai K, Yamada T (2000) Two catalytic domains of Chlorella virus CVK2 chitinase. J Biosci Bioeng 89:252–257

Hirono I, Yamashita M, Aoki T (1998) Note: Molecular cloning of chitinasegenes from *Vibrio anguillarum* and *V. parahaemolyticus*. J Appl Microbiol 84:1175–1178

Horn SJ, Sørbotten A, Synstad B, Sikorski P, Sørlie M, Vårum KM, Eijsink VG (2006) Endo/exo mechanism and processivity of family 18 chitinases produced by *Serratia marcescens*. FEBS J 273:491–503

Howard MB, Ekborg NA, Taylor LE 2nd, Weiner RM, Hutcheson SW (2004) Chitinase B of *Microbulbifer degradans* 2-40 contains two catalytic domains with different chitinolytic activities. J Bacteriol 186:1297–1303

Hult EL, Katouno F, Uchiyama T, Watanabe T, Sugiyama J (2005) Molecular directionality in crystalline beta-chitin: hydrolysis by chitinases A and B from *Serratia marcescens* 2170. Biochem J 388:851–856

Hunt DE, Gevers D, Vahora NM, Polz MF (2008) Conservation of the chitin utilization pathway in the Vibrionaceae. Appl Environ Microbiol 74:44–51

Ito T, Katayama T, Hattie M, Sakurama H, Wada J, Suzuki R, Ashida H, Wakagi T, Yamamoto K, Stubbs KA, Fushinobu S (2013) Crystal structures of a glycoside hydrolase family 20 lacto-*N*-biosidase from *Bifidobacterium bifidum*. J Biol Chem 288:11795–11806

Itoh T, Hibi T, Fujii Y, Sugimoto I, Fujiwara A, Suzuki F, Iwasaki Y, Kim JK, Taketo A, Kimoto H (2013) Cooperative degradation of chitin by extracellular and cell surface-expressed chitinases from *Paenibacillus* sp. strain FPU-7. Appl Environ Microbiol 79:7482–7490

Itoh T, Hibi T, Suzuki F, Sugimoto I, Fujiwara A, Inaka K, Tanaka H, Ohta K, Fujii Y, Taketo A, Kimoto H (2016) Crystal structure of chitinase ChiW from Paenibacillus sp. str. FPU-7 reveals a novel type of bacterial cell-surface-expressed multi-modular enzyme machinery. PLoS ONE 11:e0167310

Itoh T, Sugimoto I, Hibi T, Suzuki F, Matsuo K, Fujii Y, Taketo A, Kimoto H (2014) Overexpression, purification, and characterization of *Paenibacillus* cell surface-expressed chitinase ChiW with two catalytic domains. Biosci Biotechnol Biochem 78:624–634

Jee JG, Ikegami T, Hashimoto M, Kawabata T, Ikeguchi M, Watanabe T, Shirakawa M (2002) Solution structure of the fibronectin type III domain from *Bacillus circulans* WL-12 chitinase A1. J Biol Chem 277:1388–1397

Kawase T, Yokokawa S, Saito A, Fujii T, Nikaidou N, Miyashita K, Watanabe T (2006) Comparison of enzymatic and antifungal properties between family 18 and 19 chitinases from *S. coelicolor* A3(2). Biosci Biotechnol Biochem 70:988–998

Keyhani NO, Roseman S (1999) Physiological aspects of chitin catabolism in marine bacteria. Biochim Biophys Acta 1473:108–122

Kitaoku Y, Fukamizo T, Numata T, Ohnuma T (2017) Chitin oligosaccharide binding to the lysin motif of a novel type of chitinase from the multicellular green alga, *Volvox carteri*. Plant Mol Biol 93:97–108

Kusaoke H, Shinya S, Fukamizo T, Kimoto H (2017) Biochemical and biotechnological trends in chitin, chitosan, and related enzymes produced by *Paenibacillus* IK-5 Strain. Int J Biol Macromol 104:1633–1640

Lacombe-Harvey MÈ, Brzezinski R, Beaulieu C (2018) Chitinolytic functions in actinobacteria: ecology, enzymes, and evolution. Appl Microbiol Biotechnol

Li H, Greene LH (2010) Sequence and structural analysis of the chitinase insertion domain reveals two conserved motifs involved in chitin-binding. PLoS ONE 5:e8654

Li X, Roseman S (2004) The chitinolytic cascade in Vibrios is regulated by chitin oligosaccharides and a two-component chitin catabolic sensor/kinase. Proc Natl Acad Sci USA 101:627–631

Lombard V, Golaconda Ramulu H, Drula E, Coutinho PM, Henrissat B (2014) The Carbohydrate-active enzymes database (CAZy) in 2013. Nucleic Acids Res 42:D490–D495

Macdonald SS, Blaukopf M, Withers SG (2015) *N*-Acetylglucosaminidases from CAZy family GH3 are really glycoside phosphorylases, thereby explaining their use of histidine as an acid/base catalyst in place of glutamic acid. J Biol Chem 290:4887–4895

Madhuprakash J, Dalhus B, Rani TS, Podile AR, Eijsink VGH, Sørlie M (2018) Key residues affecting transglycosylation activity in family 18 chitinases: insights into donor and acceptor subsites. Biochemistry 57:4325–4337

Mallakuntla MK, Vaikuntapu PR, Bhuvanachandra B, Das SN, Podile AR (2017) Transglycosylation by a chitinase from *Enterobacter cloacae* subsp. cloacae generates longer chitin oligosaccharides. Sci Rep 7:5113

Meena S, Gothwal RK, Krishna Mohan M, Ghosh P (2014) Production and purification of a hyperthermostable chitinase from *Brevibacillus formosus* BISR-1 isolated from the Great Indian Desert soils. Extremophiles 18:451–462

Meibom KL, Li XB, Nielsen AT, Wu CY, Roseman S, Schoolnik GK (2004) The *Vibrio cholerae* chitin utilization program. Proc Natl Acad Sci U S A. 101:2524–2529

Meibom KL, Blokesch M, Dolganov NA, Wu CY, Schoolnik GK (2005) Chitin induces natural competence in *Vibrio cholerae*. Sci 310:1824–1827.

Merzendorfer H, Zimoch L (2003) Chitin metabolism in insects: structure, function and regulation of chitin synthases and chitinases. J Exp Biol 206:4393–4412

Miyashita K, Fujii T, Saito A (2000) Induction and repression of a *Streptomyces lividans* chitinase gene promoter in response to various carbon sources. Biosci Biotechnol Biochem 64:39–43

Monreal J, Reese ET (1969) The chitinase of *Serratia marcescens*. Can J Microbiol 15:689–696

Ohno T, Armand S, Hata T, Nikaidou N, Henrissat B, Mitsutomi M, Watanabe T (1996) A modular family 19 chitinase found in the prokaryotic organism *Streptomyces griseus* HUT 6037. J Bacteriol 178:5065–5070

Payne CM, Baban J, Horn SJ, Backe PH, Arvai AS, Dalhus B, Bjørås M, Eijsink VG, Sørlie M, Beckham GT, Vaaje-Kolstad G (2012) Hallmarks of processivity in glycoside hydrolases from crystallographic and computational studies of the *Serratia marcescens* chitinases. J Biol Chem 287:36322–36330

Perrakis A, Tews I, Dauter Z, Oppenheim AB, Chet I, Wilson KS, Vorgias CE (1994) Crystal structure of a bacterial chitinase at 2.3 A resolution. Structure 2:1169–1180

Rathore AS, Gupta RD (2015) Chitinases from bacteria to human: properties, applications, and future perspectives. Enzyme Res. 2015:791907

Rombouts FM, Phaff HJ (1976) Lysis of yeast cell walls. Lytic β-(1→6)-glucanase from B*acillus circulans* WL-12. Eur J Biochem 63:109–120

Saito A, Ebise H, Orihara Y, Murakami S, Sano Y, Kimura A, Sugiyama Y, Ando A, Fujii T, Miyashita K (2013) Enzymatic and genetic characterization of the DasD protein possessing *N*-acetyl-β-D-glucosaminidase activity in *Streptomyces coelicolor* A3(2). FEMS Microbiol Lett 340:33–40

Saito A, Fujii T, Yoneyama T, Redenbach M, Ohno T, Watanabe T, Miyashita K (1999) High-multiplicity of chitinase genes in *Streptomyces coelicolor* A3(2). Biosci Biotechnol Biochem 63:710–718

Saito A, Shinya T, Miyamoto K, Yokoyama T, Kaku H, Minami E, Shibuya N, Tsujibo H, Nagata Y, Ando A, Fujii T, Miyashita K (2007) The dasABC gene cluster, adjacent to dasR, encodes a novel ABC transporter for the uptake of N, N'-diacetylchitobiose in Streptomyces coelicolor A3(2). Appl Environ Microbiol 73:3000–3008

Schaefer J, Kramer KJ, Garbow JR, Jacob GS, Stejskal EO, Hopkins TL, Speirs RD (1987) Aromatic cross-links in insect cuticle: detection by solid-state 13C and 15 N NMR. Science 235:1200–1204

Schneewind O, Missiakas DM (2012) Protein secretion and surface display in Gram-positive bacteria. Philos Trans R Soc Lond B Biol Sci 367:1123–1139

Schrempf H (2001) Recognition and degradation of chitin by streptomycetes. Antonie Van Leeuwenhoek 79:285–289

Shimosaka M, Fukumori Y, Narita T, Zhang X, Kodaira R, Nogawa M, Okazaki M (2001) The bacterium *Burkholderia gladioli* strain CHB101 produces two different kinds of chitinases belonging to families 18 and 19 of the glycosyl hydrolases. J Biosci Bioeng 91:103–105

Shirota K, Sato T, Sekiguchi J, Miyauchi K, Mochizuki A, Matsumiya M (2008) Purification and characterization of chitinase isozymes from a red algae, *Chondrus verrucosus*. Biosci Biotechnol Biochem 72:3091–3099

Sikorski P, Sørbotten A, Horn SJ, Eijsink VG, Vårum KM (2006) *Serratia marcescens* chitinases with tunnel-shaped substrate-binding grooves show endo activity and different degrees of processivity during enzymatic hydrolysis of chitosan. Biochemistry 45:9566–9574

Suzuki K, Sugawara N, Suzuki M, Uchiyama T, Katouno F, Nikaidou N, Watanabe T (2002) Chitinases A, B, and C1 of *Serratia marcescens* 2170 produced by recombinant *Escherichia coli*: enzymatic properties and synergism on chitin degradation. Biosci Biotechnol Biochem 66:1075–1083

Suzuki K, Taiyoji M, Sugawara N, Nikaidou N, Henrissat B, Watanabe T (1999) The third chitinase gene (chiC) of *Serratia marcescens* 2170 and the relationship of its product to other bacterial chitinases. Biochem J 343:587–596

Świątek MA, Tenconi E, Rigali S, van Wezel GP (2012a) Functional analysis of the *N*-acetylglucosamine metabolic genes of *Streptomyces coelicolor* and role in control of development and antibiotic production. J Bacteriol 194:1136–1144

Świątek MA, Urem M, Tenconi E, Rigali S, van Wezel GP (2012b) Engineering of *N*-acetylglucosamine metabolism for improved antibiotic production in *Streptomyces coelicolor* A3(2) and an unsuspected role of NagA in glucosamine metabolism. Bioengineered 3:280–285

Tanaka T, Fukui T, Imanaka T (2001) Different cleavage specificities of the dual catalytic domains in chitinase from the hyperthermophilic archaeon *Thermococcus kodakaraensis* KOD1. J Biol Chem 276:35629–35635

Tharanathan RN, Kittur FS (2003) Chitin-the undisputed biomolecule of great potential. Crit Rev Food Sci Nutr 43:61–87

Tews I, Terwisscha van Scheltinga AC, Perrakis A, Wilson KS, Dijkstra BW (1997) Substrate-assisted catalysis unifies two families of chitinolytic enzymes. J Am Chem Soc 119:7954–7959

Uchiyama T, Kaneko R, Yamaguchi J, Inoue A, Yanagida T, Nikaidou N, Regue M, Watanabe T (2003) Uptake of N, N'-diacetylchitobiose [(GlcNAc)2] via the phosphotransferase system is essential for chitinase production by Serratia marcescens 2170. J Bacteriol 185:1776–1782

Uchiyama T, Katouno F, Nikaidou N, Nonaka T, Sugiyama J, Watanabe T (2001) Roles of the exposed aromatic residues in crystalline chitin hydrolysis by chitinase A from *Serratia marcescens* 2170. J Biol Chem 276:41343–41349

Umemoto N, Ohnuma T, Osawa T, Numata T, Fukamizo T (2015) Modulation of the transglycosylation activity of plant family GH18 chitinase by removing or introducing a tryptophan side chain. FEBS Lett 589:2327–2333

Uni F, Lee S, Yatsunami R, Fukui T, Nakamura S (2012) Mutational analysis of a CBM family 5 chitin binding domain of an alkaline chitinase from *Bacillus* sp J813. Biosci Biotechnol Biochem 76:530–535

Vaaje-Kolstad G, Horn SJ, Sørlie M, Eijsink VG (2013) The chitinolytic machinery of *Serratia marcescens*-a model system for enzymatic degradation of recalcitrant polysaccharides. FEBS J 280:3028–3049

van Aalten DM, Komander D, Synstad B, Gåseidnes S, Peter MG, Eijsink VG (2001) Structural insights into the catalytic mechanism of a family 18 exo-chitinase. Proc Natl Acad Sci USA 98:8979–8984

Watanabe T, Ariga Y, Sato U, Toratani T, Hashimoto M, Nikaidou N, Kezuka Y, Nonaka T, Sugiyama J (2003) Aromatic residues within the substrate-binding cleft of *Bacillus circulans* chitinase A1 are essential for hydrolysis of crystalline chitin. Biochem J 376:237–244

Watanabe T, Ishibashi A, Ariga Y, Hashimoto M, Nikaidou N, Sugiyama J, Matsumoto T, Nonaka T (2001) Trp122 and Trp134 on the surface of the catalytic domain are essential for crystalline chitin hydrolysis by *Bacillus circulans* chitinase A1. FEBS Lett 494:74–78

Watanabe T, Ito Y, Yamada T, Hashimoto M, Sekine S, Tanaka H (1994) The roles of the C-terminal domain and type III domains of chitinase A1 from *Bacillus circulans* WL-12 in chitin degradation. J Bacteriol 176:4465–4472

Watanabe T, Oyanagi W, Suzuki K, Tanaka H (1990) Chitinase system of *Bacillus circulans* WL-12 and importance of chitinase A1 in chitin degradation. J Bacteriol 172:4017–4022

Chapter 8
Chitin Synthesis and Degradation in Fungi: Biology and Enzymes

Jinkui Yang and Ke-Qin Zhang

Abstract Chitin is one of the most important carbohydrates of the fungal cell wall, and is synthesized by chitin synthases. Chitin can be degraded by chitinases, which are important virulence factors in pathogenic fungi. Knowledge about the biosynthesis and degradation of chitin, and the enzymes responsible, has accumulated in recent years. In this review, we analyze the amino acid sequences of chitin synthases from several typical fungi. These enzymes can be divided into seven groups. While the different chitin synthases from a single fungus share a low degree of similarity, the same type of chitin synthase from different fungi shows high similarity. The number of chitinase genes in fungi display wide variation, from a single gene in *Schizosaccharomyces pombe*, to 36 genes in *Trichoderma virens*. Chitinases from different fungi can be divided into four groups. The functions of chitin synthases and chitinases in several typical fungi are summarized, and the crystal structures of chitinases and chitinase modification are also discussed.

Keywords Chitin · Chitin synthase · Chitinase · Crystal structure · Modification

8.1 Introduction

Chitin is a linear polymer of β-(1,4)-linked N-acetylglucosamine (GlcNAc) synthesized by the chitin synthase/regulator system. Chitin can then be enzymatically deacetylated to chitosan by chitin deacetylases. The fungal cell wall is a complex organelle that is a composite of glucan and chitin fibers held together by proteins and mannan (Adams 2004). The content and localization of chitin vary among fungi. The cell wall is composed of two layers; the internal one forms a scaffold and is composed of chitin and β-1,3-glucan, while the external one is usually formed

J. Yang · K.-Q. Zhang (✉)
State Key Laboratory for Conservation and Utilization of Bio-Resources in Yunnan, and Key Laboratory for Microbial Resources of the Ministry of Education, Yunnan University, 650091 Kunming, Yunnan, China
e-mail: kqzhang1@ynu.edu.cn

Q. Yang and T. Fukamizo (eds.), *Targeting Chitin-containing Organisms*, Advances in Experimental Medicine and Biology 1142,
https://doi.org/10.1007/978-981-13-7318-3_8

from other polysaccharides and glycoproteins. Glycoproteins often possess a remnant of the glycosylphosphatidylinositol anchor to attach via a β-1,6-glucan to β-1,3-glucan (Latgé 2007; Muszewska et al. 2017). The primary role of chitin appears to maintain structural integrity. However, other roles have been hypothesized, including epithelial adhesion, the linkage between the cell wall and capsule, and antifungal resistance (Goldman and Vicencio 2012). The cell wall is the outermost layer of fungi and is exposed to the surrounding environment, so is essential to maintain cellular structure, to protect the cells from environmental stresses (Latgé 2007), and for survival in the normal hypotonic conditions found in nature and in artificial media (Bulawa 1993).

8.2 Chitin Synthesis and Chitin Synthases

Chitin is the product of chitin synthases (Chs), which use the nucleotide UDP-GlcNAc as a sugar donor, and require a divalent metal ion for activity (Bulawa 1993). Chs have been studied in detail in yeasts and filamentous fungi such as *Aspergillus fumigatus*, *Botrytis cinerea*, *Candida albicans*, *Magnaporthe oryzae*, *Neurospora crassa,* and *Saccharomyces cerevisiae*. Filamentous fungi contain more Chs than yeasts (Table 8.1); for example, three and four Chs were found in *S. cerevisiae* and *C. albicans*, respectively, while *A. fumigatus*, *A. nidulans,* and *B. cinerea* have eight Chs, and the model fungus *N. crassa* and pathogenic fungi *Beauveria bassiana*, *M. oryzae*, and *T. harzianum* contain seven.

The Chs from *B. cinerea* and other filamentous fungi, such as *A. fumigatus*, *A. nidulans*, *M. oryzae,* and *N. crassa*, can be divided into seven groups (ChsI, ChsII,

Table 8.1 Members of the seven classes of chitin synthases in different fungi[a]

Fungus	I	II	III	IV	V	VI	VII	Total
S. cerevisiae	Chs1	Chs2		Chs3				3
C. albicans	Chs2, Chs8	Chs1		Chs3				4
A. fumigatus	ChsA	ChsB	ChsC, ChsG	ChsF	ChsE	ChsH (XP_755676)	ChsD	8
A. nidulans	ChsC	ChsA	ChsB, ChsF	ChsD	CsmA	CsmB	ChsG	8
B. bassiana	Chs3	Chs2	Chs1	Chs4	Chs5	Chs6	ChsD	7
B. cinerea	ChsI	ChsII	ChsIIIa, ChsIIIb	ChsIV	ChsV	ChsVI	ChsVII	8
M. oryzae	Chs3	Chs2	Chs1	Chs4	Chs5	Chs6	ChsD	7
N. crassa	Chs3	Chs2	Chs1	Chs4	ChsC	Chs6	ChsD	7
T. harzianum	Chs3	Chs2	Chs1	Chs4	ChsC	Chs6	ChsD	7

[a]Standard genetic nomenclature for *S. cerevisiae* and *C. albicans* has been used to designate CHS proteins from all fungi. The enzymes have been assigned to classes based on the classification proposed by Roncero (2002)

ChsIII, ChsIV, ChsV, ChsVI, and ChsVII), and some fungi contain two members of ChsIII, for example *A. fumigatus*, *A. nidulans* and *B. cinerea*. Moreover, the conserved QRRRW signature motif, which locates in the characteristic catalytic domain pfam03142 (Chitin_synth_2), is found in all fungal Chs (Fajardo-Somera et al. 2015). Interestingly, the Chs from the same fungus share a low degree of similarity, for example, ChsA from *A. fumigatus* shares 40.4, 37.4, 8.8, 6.5, 8.1, 37.9, and 9.4% similarity with ChsB, ChsC, ChsD, ChsF, ChsE, ChsG, and ChsVII (ChsH), respectively. Similarly, ChsI from *B. cinerea* shares 43.5, 38.4, 8.6, 9.0, 8.7, and 7.1% similarity with ChsII, ChsIII, ChsIV, ChsV, ChsVI, and ChsVII. However, the same type of Chs from different filamentous fungi share a high degree of similarity, for example, ChsI from *B. cinerea* shares 71.0, 67.7, 65.6, 68.3, and 66.6% similarities with the homologous Chs from *A. fumigatus*, *T. harzianum*, *B. bassiana*, *N. crassa,* and *M. oryzae*, respectively. Similarly, ChsVII from *B. cinerea* shares 57.8–71.6% similarity with orthologs from other fungi.

A phylogenetic tree of Chs from different fungi was constructed according to their amino acid sequences, and these enzymes clustered into three clades (A, B, and C) (Fig. 8.1). Clade A can be divided into three subclades (A-I, A-II, and A-III). A-I contains ChsI from *B. cinerea*; ChsA from *A. fumigatus*; Chs3 from *B. bassiana, M. oryzae,* and *N. crassa*; Chs1 from *S. cerevisiae*; and Chs2 and Chs8 from *C. albicans*. Subclade A-II contains ChsII from *B. cinerea*, ChsB from *A. fumigatus*, Chs2 from *S. cerevisiae,* and Chs1 from *C. albicans*, as well as Chs2 from other fungi. Interestingly, *B. cinerea* and *A. fumigatus* contain two members of ChsIII, i.e., ChsIIIa and ChsIIIb (*B. cinerea*), and ChsC and ChsG (*A. fumigatus*); they are categorized into subclade A-III. Chs3 from *B. bassiana* and *N. crassa*, as well as Chs1 from *M. oryzae*, are also clustered in this branch. Clade B can also be separated into three subclades (B-I, B-II, and B-III). B-I contains ChsIV from *B. cinerea*, ChsF from *A. fumigatus*, Chs4 from other fungi, and Chs3 from *S. cerevisiae* and *C. albicans*. ChsE from *A. fumigatus* and ChsV from *B. cinerea* and other fungi are categorized in subclade B-II. XP_755676 (ChsH) from *A. fumigatus*, and ChsVI from *B. cinerea* and other fungi, are in subclade B-III. Clade C consists of ChsVII from *A. fumigatus*, and ChsD from *B. cinerea* and other fungi. This phylogenetic analysis was consistent with previous reports (Roncero 2002; Ruiz-Herrera et al. 2002; Choquer et al. 2007).

The multiplicity of Chs enzymes suggests that they may have redundant roles in cell wall synthesis, and the roles of Chs have been studied in filamentous fungi and yeasts such as *S. cerevisiae* and *C. albicans*. In *S. cerevisiae*, ScChs1, a class I enzyme, replenishes chitin in the birth/bud scar after cytokinesis and hence has a repair function. The class II enzyme ScChs2 is involved in the synthesis of the primary septum, and the class IV enzyme ScChs3 is responsible both for the formation of the chitin ring when the bud emerges, and for the chitin in the lateral cell wall (Roncero 2002). Meanwhile, eight Chs genes—*ChsC, ChsA, ChsB, ChsF, ChsD, ChsG, CsmA,* and *CsmB* (Table 8.1)—were identified in the fungus *A. nidulans*, and encode class I, II, IIIa, IIIb, IV, V, VI, and VII enzymes, respectively (Horiuchi 2009). While AnChsB appears to have crucial roles in hyphal tip growth,

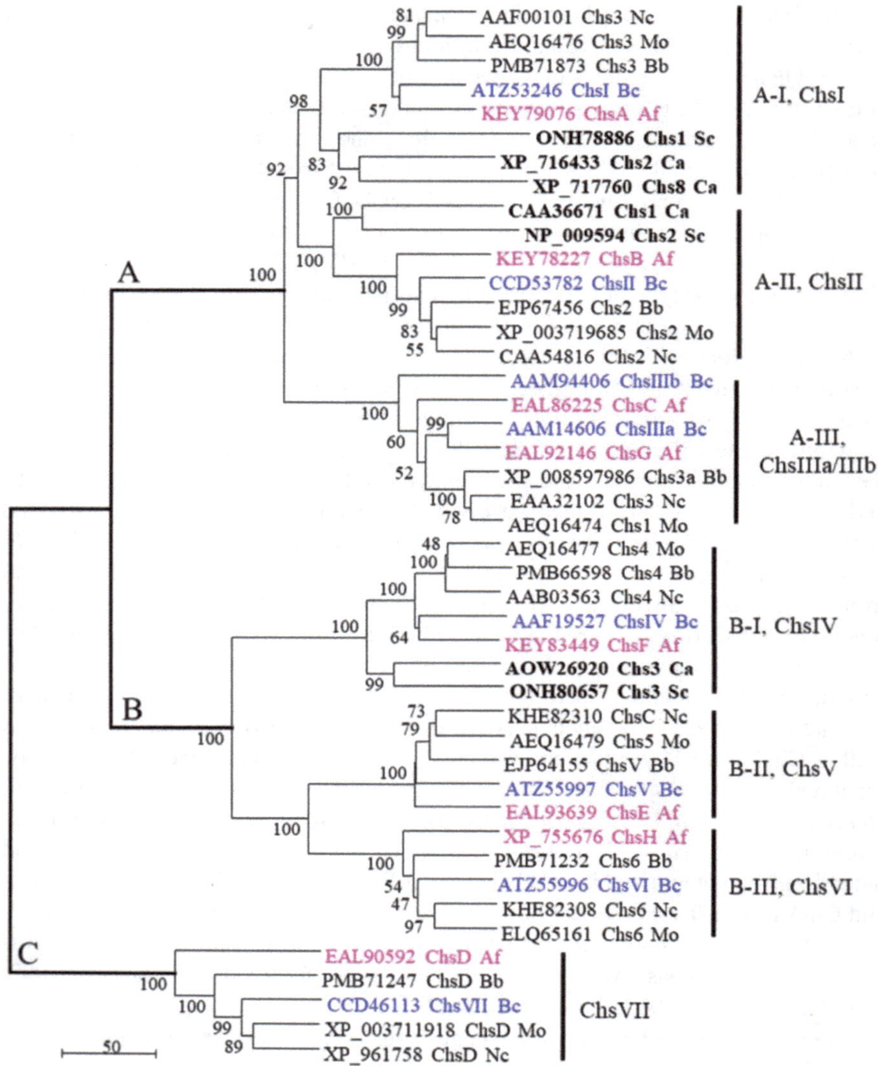

Fig. 8.1 The phylogenetic tree was constructed based on the amino acid sequence of Chs from different fungi. The GenBank numbers of these homologous Chs were given in front of the fungal name. Af, *A. fumigatus*; Bb, *B. bassiana*; Bc, *B. cinerea*; Ca, *C. albicans*; Mo, *M. oryzae*; Nc, *N. crassa*; Sc, *S. cerevisiae*. Numbers around nodes indicated the bootstrap value. The bar marker showed the genetic distance

AnChsA and AnChsC have overlapping functions in septum formation and condition (Motoyama et al. 1997). AnCsmA and AnCsmB perform compensatory roles in hyphal growth and could be involved in regulating septal pore formation (Yamada et al. 2005). Recently, the function of seven Chs in the growth and development of *N. crassa* was investigated. Analyses of deletion mutants suggested that Chs-6 has a role primarily in hyphal extension and ascospore formation, Chs-5 in aerial hyphae, conidia and ascospore formation, Chs-3 in perithecia development, and Chs-7 in all of the aforementioned (Fajardo-Somera et al. 2015). Moreover, seven *Chs* genes have been identified in plant pathogenic fungus *M. oryzae*. Three of them (*Chs1*, *Chs6*, and *Chs7*) were found to be important for plant infection; the Δ*Chs1* and Δ*Chs7* mutants were significantly reduced in virulence, and the Δ*Chs6* mutant was nonpathogenic. *Chs1* plays a specific role in conidiogenesis, and most Δ*Chs1* conidia had no septum or spore tip mucilage. The Δ*Chs6* mutant was reduced in hyphal growth and conidiation, and it failed to penetrate and grow invasively in plant cells. The Δ*Chs7* mutant was defective in appressorium penetration and invasive growth (Kong et al. 2012).

In summary, individual *Chs* genes play diverse roles in hyphal growth, conidiogenesis, appressorium development, and pathogenesis in yeasts and filamentous fungi (Horiuchi 2009; Kong et al. 2012; Lenardon et al. 2010).

8.3 Chitin Degradation and Chitinases

Chitinases are enzymes belonging to glycoside hydrolase family 18 (GH18) and 19 (GH19) and are responsible for the hydrolysis of beta-1,4-linkages in chitin (Cantarel et al. 2009). GH18 and GH19 contain chitinases, which display differences in amino acid sequences, domains, and 3D protein structures (Henrissat 1991). The GH18 family is widely distributed, and found in viruses, bacteria, plants, fungi, and animals (Li 2006), while chitinases from the GH19 family are found mainly in plants, but have also been described in bacteria, viruses, and nematodes (Geng et al. 2002; Honda et al. 2008).

The number of chitinase genes shows wide variation in fungal genomes, from a single gene in *Schizosaccharomyces pombe* up to 36 genes in *T. virens* (Gruber and Seidl-Seiboth 2012). For example, eighteen chitinase sequences were found in the *T. reesei* genome, and the chitinase domain composition was used to classify them into subgroups A, B, and C (Seidl et al. 2005). Meanwhile, 12 and 19 chitinases were found in the model fungi *A. nidulans* and *N. crassa*, respectively, and 14 chitinase sequences were retrieved from the plant-pathogenic fungus *M. oryzae* (Gruber and Seidl-Seiboth 2012). Recently, 24 genes belonging to GH18 were identified in the entomopathogenic fungus *Metarhizium anisopliae*; these putative chitinases were classified based on domain organization and phylogenetic analysis into the previously described A, B, and C chitinase subgroups, and into a new subgroup D (Junges et al. 2014). Moreover, 12 GH18 genes were identified in the genome of the mycoparasitic fungus *Clonostachys rosea*, and phylogenetic analysis revealed that *C. rosea* contains eight genes in subgroup A, two genes in subgroup B, and two genes in subgroup C (Tzelepis et al. 2015).

The amino acid sequences of chitinases from four representative fungi including *A. nidulans*, *M. oryzae*, *N. crassa,* and *T. reesei* were retrieved from their genomes;

the orthologs of chitinases 18-15 and 18-18 of *T. reesei* were also downloaded for the fungus *T. harzianum*. A neighbor-joining phylogenetic tree was constructed using MEGA 7.0 software. These chitinases were classified into four groups (Fig. 8.2), which was slightly different from the analysis in *T. reesei*. Group A contains 25 chitinases, and can be divided into three subfamilies. Subfamily A-I contains three chitinases from *T. reesei* including 18-5, 18-6, and 18-7; A-II also contains two chitinases from *T. reesei* (18-2 and 18-3), and three chitinases from *T. reesei* are gathered in subfamily A-III. Chitinases from the other three fungi were also distributed in these three subfamilies. Group B consists of 16 members and can also be divided into three subfamilies. B-I contains four chitinases, they are specific for *M. oryzae* and *N. crassa*, and B-II contains two chitinases from *T. reesei* (18-12 and 18-17), as well as one each from *A. nidulans*, *M. oryzae*, and *N. crassa*. Subfamily B-III contains four chitinases from *T. reesei*, including 18-13, 18-14, 18-16, and 18-18. Group C contains 18 chitinases and can be divided into two subfamilies. C-I includes chitinases from *A. nidulans*, *M. oryzae*, and *N. crassa*, while four chitinases from *T. reesei* (18-1, 18-8, 18-9, and 18-10) are in subfamily C-II. In addition, chitinase 18-15 from *T. reesei* and its homologue from *T. harzianum* formed a single group, named group D. Our analysis is consistent with previous reports, although chitinase 18-18 from *T. reesei* was clustered in group A by Seidl et al. (2005), and 18-15 was not included in any of the trees in previous analyses. In fact, chitinase 18-15 from *T. reesei* shares a high degree of similarity to orthologues from different *Trichoderma* spp. (Seidl et al. 2005) and entomopathogenic fungi such as *Metarhizium anisopliae*, *M. robertsii*, *M. acridum*, and *B. bassiana* (Junges et al. 2014).

The biological functions of chitinases are very diverse and include roles in yeast and filament morphogenesis, autolysis, acquisition of chitin for nutritional purposes, and mycoparasitism. In *S. cerevisiae*, the chitinase ScCts1p is essential for chitin degradation during cell division, and deletion of the *Cts1* gene leads to a defect in cell separation and the formation of multicellular aggregates (Kuranda and Robbins 1975). Another chitinase, ScCts2p, is involved in ascus formation (Giaever et al. 2002). Filamentous fungi contain expanded chitinase gene families, so it is difficult to reveal the functions of individual chitinases in these fungi because of the need to investigate multiple deletion mutants to observe phenotypic alterations. The functions of chitinases in filamentous fungi have, however, been investigated by gene expression, disruption, and transcriptional analysis. For example, an endochitinase Bbchit1 was purified from the fungus *B. bassiana* and the gene *Bbchit1* was overexpressed in this fungus. Overproduction of Bbchit1 enhanced the virulence of *B. bassiana* toward aphids, as indicated by significantly lower 50% lethal concentrations (LC_{50}) and 50% lethal times of the transformants compared to the wild-type (WT) strain (Fang et al. 2005). Moreover, a chitinase gene *Lpchi1* (an orthologue of *T. reesei* Chi18-5) was isolated from the nematophagous fungus *Lecanicillium psalliotae* and expressed in *Pichia pastoris* GS115; the recombinant chitinase could degrade chitinous components of eggs of the root knot nematode *Meloidogyne incognita* and then significantly influence the development of the eggs (Gan et al. 2007a, b). Recently, the function of 10 genes encoding chitinases in

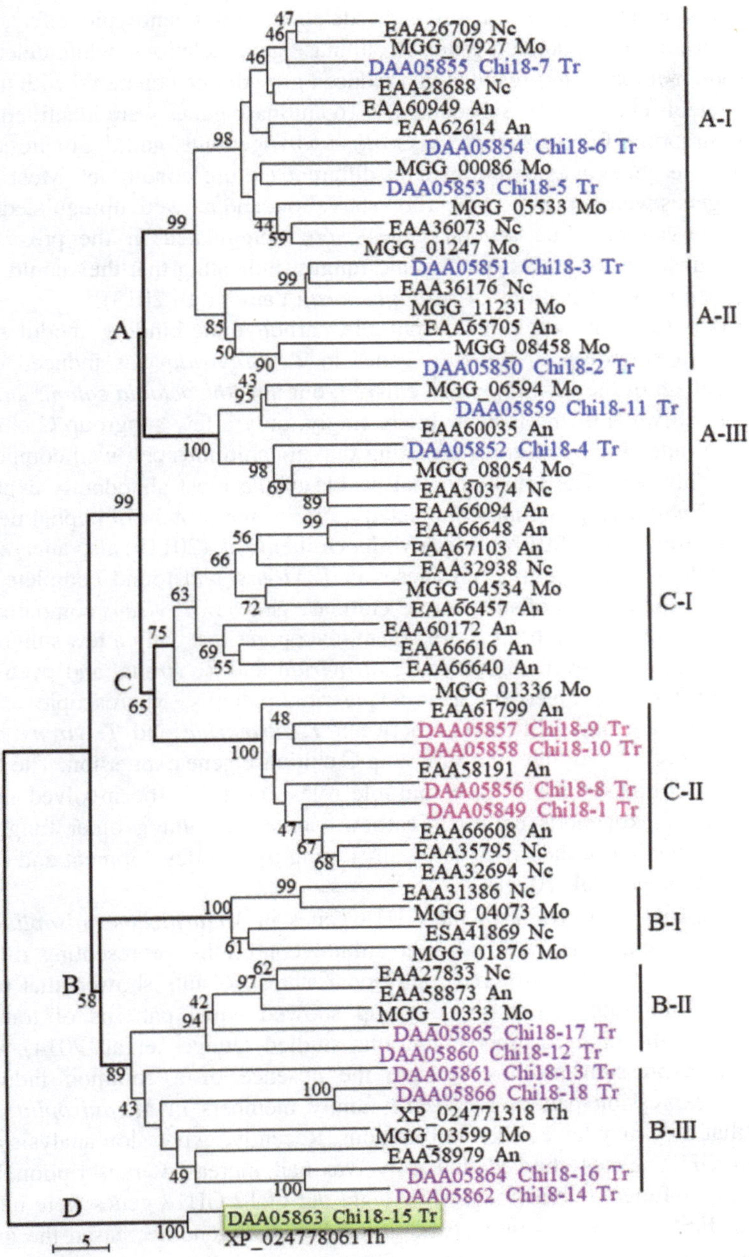

Fig. 8.2 The phylogenetic tree was constructed based on the amino acid sequence of chitinases from four representative fungi including *A. nidulans* (An), *M. oryzae* (Mo), *N. crassa* (Nc), and *T. reesei* (Tr), as well the orthologs of *T. reesei* chitinases 18-15 and 18-18 form the fungus *T. harzianum* (Th). The GenBank numbers of these homologous chitinases were given in front of the fungal name. Numbers around nodes indicated the bootstrap value. The bar marker showed the genetic distance

N. crassa was investigated using gene deletion. No phenotypic effects were detected for any of the studied group A chitinase gene deletions, while deletion of the B group member *chit-1* resulted in a reduced growth rate compared with the WT strain (Tzelepis et al. 2012). Subsequently, 16 chitinase genes were identified in the nematode-trapping fungus *Arthrobotrys oligospora* genome, and the expression of nine chitinase genes was determined in different culture conditions. Most of the chitinase genes were repressed by carbon starvation, and all were upregulated under nitrogen starvation, while several genes were upregulated in the presence of chitinous substrates or a plant-pathogenic fungus, indicating that they could play a role in biocontrol applications of *A. oligospora* (Yang et al. 2013).

Subgroup C chitinases contain multiple carbohydrate-binding modules. The expression of subgroup C chitinase genes in *T. atroviride* was induced during mycoparasitism of the fungal prey *B. cinerea*, but not *Rhizoctonia solani*, and only by fungal cell walls of the former. Interestingly, only a few subgroup C chitinase genes were inducible by chitin, suggesting that nonchitinous cell wall components can act as inducers. The transcriptional profile of the most abundantly expressed subgroup C chitinase gene, *tac6*, indicated a role of the protein in hyphal network formation (Gruber et al. 2011a). Meanwhile, Gruber et al. (2011b) also analyzed the gene regulation of subgroup C chitinases in *T. virens*, and found completely different expression profiles of subgroup C chitinase genes in *T. virens* compared with *T. atroviride*, although both fungi are potent mycoparasites. Only a few subgroup C chitinase orthologues were found in *T. atroviride* and *T. virens*, and even those showed substantially divergent gene expression patterns. Microscopic analysis revealed morphogenetic differences between *T. atroviride* and *T. virens*, which could be connected to differential subgroup C chitinase gene expression. The fungal subgroup C chitinases could have multiple roles and might be involved in both degradation of exogenous chitinous carbon sources, including other fungal cell walls, and recycling of their own cell walls during hyphal development and colony formation (Gruber et al. 2011b).

The transcriptional profile of the GH18 genes in *Metarhizium anisopliae* was determined by real-time PCR in eight culture conditions, representing different stages of development or different nutritional states. Results showed that not all members of the same chitinase subgroup showed equal patterns of transcript expression in the eight distinct conditions studied (Junges et al. 2014). These differential expression profiles indicate the absence of a common induction/repression expression pattern for GH18 family members in *M. anisopliae*, suggesting that they may have different functions. Recently, expression analysis of the *C. rosea* GH18 genes showed that only two had increased transcription levels during fungal–fungal interactions, while eight out of 12 GH18 genes were induced by chitin. Furthermore, deletion of the group C *chiC2* gene decreased the growth inhibitory activity of *C. rosea* culture filtrates toward *B. cinerea* and *R. solani*, although biocontrol of *B. cinerea* by *C. rosea* was not affected. In addition, the conidiation rate was significantly lower in the Δ*chiC2* strains compared to the WT strain (Tzelepis et al. 2015).

In summary, the functions of chitinases have been investigated in various fungi. These enzymes play diverse, but important, roles in cell wall remodeling during different stages of the fungal life cycle, influencing cell wall integrity, cell separation, mating, and stress resistance. Furthermore, chitinases might also be involved in defense against competing fungi and thus could contribute to overall fitness in complex environments such as soil or plant microbiomes (Langner and Göhre 2016).

8.4 Structure of Chitinases and Their Complexes with Inhibitors

Chitinases from pathogenic fungi have been shown to be virulence factors, and can play important roles in the infection of hosts. Structural studies of chitinases and chitinase–inhibitor complexes can provide crucial information on the modes of substrate binding, the specificity of chitinase inhibitors, and the mechanism of the hydrolysis reaction, and are also valuable for the development of chemotherapeutics with fungicidal, insecticidal, or anti-inflammatory potential.

The fungus *Coccidioides immitis* is the causative agent of coccidioidomycosis, one of the most widespread endemic diseases in the USA. The primary *Coccidioides* antigen was identified as a chitinase, CiX1 (Johnson and Pappagianis 1992). The X-ray crystal structure of chitinase CiX1 from *C. immitis* has been solved to 2.2 Å resolution (PDB entry 1D2 K). Like other members of GH18, this 427-residue protein has an eight-stranded β/α-barrel structure. Although lacking an N-terminal chitin anchoring domain, the enzyme closely resembles the chitinase from the proteobacterium *Serratia marcescens*. Among the conserved features are three *cis* peptide bonds, all involving conserved active site residues. The active site is formed from conserved residues, including tryptophans 47, 131, 315, 378, tyrosines 239 and 293, and arginines 52 and 295 (*C. immitis* CiX1 residue numbering). Glu171 is the catalytic acid in the hydrolytic mechanism (Hollis et al. 2000).

Aspergillus fumigatus produces a wide range of chitinolytic enzymes, and chitinase ChiB1 (AfChiB1) is the major example and inducible (Escott et al. 1998). The structures of AfChiB1 and AfChiB1–argifin/argadin complexes were characterized (argifin and argadin are cyclic pentapeptides that inhibit GH18 chitinases). AfChiB1 shares 66% identity to CiX1 from *C. immitis*, and their molecular structures are similar. Thus, AfChiB1 (PDB entry 1W9P) also has a $(\beta\alpha)_8$ fold, although it lacks helix α1, in common with other chitinases from GH18. Asp175 and Glu177 at the end of β4 form part of the GH18 chitinase DXE motif, with Glu177 as the catalytic acid (Rao et al. 2005). Residues affected by argifin binding are mostly conserved in the GH18 chitinases AfChiB1 and CiX1. The guanyl-urea moiety stacks on a conserved tryptophan (Trp384 in AfChiB1) and interacts with the glutamate and aspartate of the DXE motif (Glu177 and Asp175 in AfChiB1). On the opposite side of the active site, the guanyl-urea moiety also interacts with a conserved tyrosine (Tyr245 in AfChiB1) (Rao et al. 2005). Similarly, argadin binds the two chitinases (AfChiB1 and CiX1) in a similar orientation and position. The cyclized aspartic β-semialdehyde stacks with the same conserved tryptophan (Trp384 in AfChiB1) as the argifin guanyl-urea moiety. The argadin histidine side chain penetrates the active site to establish contacts with the DXE motif. On the opposite side of the active site, a conserved cluster of residues (Tyr245, Asp246, and Arg301 in AfChiB1) establishes three hydrogen bonds observed in chitinase–argadin complexes. Interestingly, two water molecules are observed in the complexes with chitinase and argadin at equivalent positions, mediating hydrogen bonds between argadin and the protein backbone (AfChiB1 Trp137, and Asp246) (Rao et al. 2005).

Clonostachys rosea is a mycoparasitic fungus, active against several plant-pathogenic fungi, such as *B. cinerea*. Meanwhile, *C. rosea* is reported as a potential agent for the biocontrol of nematodes. A chitinase (CrChi1) was isolated from *C. rosea* that could degrade the eggs of the root-knot nematode *Meloidogyne incognita* (Gan et al. 2007b). Subsequently, the structures of CrChi1 (PDB entry 3G6L) and a CrChi1–caffeine complex (PDB entry 3G6 M) were resolved. Like other GH18 chitinases, CrChi1 has a DXE motif at the end of strand β4, with Glu174 as the catalytic residue in the middle of the open end of the $(\beta\alpha)_8$ barrel. Two caffeine molecules were shown to bind to CrChi1 in subsites −1 and +1 in the substrate-binding domain. Chitinase CrChi1 showed sequence similarities ranging from 24.2 to 54.8% to five other structurally characterized chitinases from different organisms including bacteria, fungi, and human. The amino acid residues around the substrate-binding site and the catalytic center were highly conserved, the core parts of all these structures had similar folds, and their main differences were in the N- and C-terminal domains (Yang et al. 2010).

From the above analysis, it is clear that the structures of chitinases from different organisms are conserved, especially in the amino acid residues corresponding to the substrate-binding domain and the catalytic domain. Therefore, they probably share a common catalytic mechanism. Moreover, comparison of chitinase–inhibitor (such as argifin/argadin) complexes shows that although there is high sequence conservation of the residues in the active site, and the majority of protein–inhibitor contacts are made through these residues, subtle changes in residues near the active site can have significant effects. It should be possible to exploit these subtle differences to design inhibitor derivatives that specifically inhibit a particular chitinase, and develop interesting targets for inhibition with selective compounds from fungicidal, insecticidal, and chemotherapeutic perspectives.

8.5 Chitinase Modification

Chitinases are considered key hydrolytic enzymes in pathogenic fungi, and they play an important role in biological control. Several researchers have attempted to improve the virulence of fungi by structural modification of chitinases, such as fusing one or more chitin-binding domains (ChBDs) and constructing fusion proteins with protease and chitinase activities. ChBDs are found mainly at the N-termini of plant chitinases, but in bacterial and fungal chitinases they can be located either at the C- or N-terminal end (Junges et al. 2014). Only a few fungal chitinases have been shown to contain a ChBD. Previous studies have shown that ChBDs exhibit remarkably high specificity for chitin, and the binding activity is reversible. The ChBD is a tunnel-like structure, which facilitates chitinase binding, thus allowing the efficient degradation of chitin (Limón et al. 2001; Kowsari et al. 2013). The investigation of chitinase modification has focused on the mycoparasitic pathogens *Trichoderma* spp. and the insect pathogen *B. bassiana*.

T. harzianum is one of the most potent biocontrol agents against a wide range of economically important aerial and soilborne plant pathogens. Chitinases are key hydrolytic enzymes in the lysis of cell walls of fungi, and thus they play an important role in biological control. Among *Trichoderma* chitinases, Chit42 is essential for biocontrol activities against phytopathogenic fungi, and Chit42 does not contain a ChBD. In 2001, hybrid chitinases with stronger chitin-binding capacity were constructed by fusing to Chit42 a ChBD from *Nicotiana tabacum* ChiA chitinase and the cellulose-binding domain from cellobiohydrolase II of *T. reesei*. The chimeric chitinases had similar activities toward the soluble substrate, but higher hydrolytic activity than the native chitinase toward high molecular mass insoluble substrates such as ground chitin or chitin-rich fungal cell walls (Limón et al. 2001). Subsequently, a chimeric chitinase with improved enzyme activity was produced in *T. harzianum* by fusing a ChBD from *T. atroviride* chitinase 18-10 to Chit42. The improved chitinase displayed a 1.7-fold higher specific activity toward the insoluble chitin than Chit42. Moreover, Chit42-ChBD transformants showed higher antifungal activity toward seven phytopathogenic fungal species (Kowsari et al. 2013).

Entomopathogenic fungi, such as *B. bassiana*, infect a host insect by penetrating the insect cuticle and can produce extracellular proteases and chitinases that degrade these proteinaceous and chitinous components, allowing hyphal penetration through the cuticle and access to the insect hemolymph. Two chitinase genes, *Bbchit1* and *Bbchit2*, have been cloned from *B. bassiana*, neither of which has a ChBD (Fang et al. 2005). In 2007, several *B. bassiana* hybrid chitinases were produced where the chitinase Bbchit1 was fused to ChBDs derived from plant, bacterial, or insect sources. A hybrid chitinase containing a ChBD from the silkworm *Bombyx mori* chitinase fused to Bbchit1 showed the greatest ability among the hybrids to bind to chitin. This hybrid chitinase gene (encoding Bbchit1–BmChBD) was then placed under the control of a fungal constitutive promoter and transformed into *B. bassiana*. Insect bioassays showed a 23% of reduction in time to death in the transformant compared to the WT strain (Fan et al. 2007). Subsequently, a fusion gene of *Bbchit1* linked to *CDEP1* (a cuticle degrading protease gene) was overexpressed in *B. bassiana*, and transformants secreting the fusion protein (CDEP1–Bbchit1) penetrated the cuticle significantly faster than the WT or transformants overexpressing either *Bbchit1* or *CDEP1* alone. Moreover, expression of CDEP1–Bbchit1 resulted in a 60.5% reduction in LC_{50} against *Myzus persicae*, more than twice the reduction obtained by overexpression of Bbchit1 (Fang et al. 2009).

In summary, engineered chitinases show increased ability in chitin hydrolysis compared with native chitinases. Furthermore, when the engineered chitinase is overexpressed in fungi, such as *T. harzianum* and *B. bassiana*, the genetically modified fungi may show an increase in virulence compared with strains with constitutively expressed native chitinase genes. This approach provides a method for improving the virulence of pathogenic fungi and developing high-efficiency agents to control plant diseases.

8.6 Conclusions and Perspectives

Chitin is a structurally important component of the fungal cell wall. The synthesis of chitin is mediated by chitin synthase, an integral membrane enzyme that catalyzes the transfer of GlcNAc from UDP-GlcNAc to a growing chitin chain. Chitinases can hydrolyze the β-(1-4) linkages in polymers of chitin; these enzymes are found in a wide variety of organisms, including viruses, bacteria, plants, and animals. In recent years, many fungal genomes were sequenced, and many chitin synthases and chitinases were identified. Filamentous fungi contain more genes encoding chitin synthase than yeasts, and these chitin synthases may be divided into seven groups. Similarly, more chitinase genes have been identified in filamentous fungi than yeasts, and these enzymes can be categorized into four groups. The specific roles of the chitin synthases and chitinases in several fungi have been studied using genetic, biochemical and transcriptional analyses. Chitin synthases play diverse roles in hyphal growth, conidiogenesis, appressorium development, and pathogenesis in yeasts and filamentous fungi (e.g., Horiuchi 2009; Lenardon et al. 2010; Kong et al. 2012), and chitinases are known to be involved in cell separation, hyphal growth and branching, sexual development, spore germination, and autolysis (e.g., Giaever et al. 2002; Fang et al. 2005; Gruber et al. 2011a, b; Tzelepis et al. 2012, 2015; Yang et al. 2013).

At present, the functions of chitin synthases and chitinases have been revealed in only a few fungi, such as *A. nidulans*, *B. cinerea*, *M. oryzae*, *S. cerevisiae,* and *T. reesei*. Where there are multiple chitin synthases and chitinases in a fungus, especially in filamentous fungi, it is difficult to illuminate the function of individual chitin synthases or chitinases in fungal growth and development; multiple deletion mutants needed to be constructed to observe phenotypic alterations, and it is difficult to construct such mutants because of the limited selectable markers available in fungi. However, the development of CRISPR technology might greatly facilitate research into these enzymes. Moreover, omics technologies, such as transcriptomics and proteomics, have also been applied to research on the chitin synthases and chitinases in fungi, and provide a good technical platform to reveal their functions (Junges et al. 2014).

Chitinases are important virulence factors in pathogenic fungi. Studies on the structures of chitinases and chitinase–inhibitor complexes provide crucial information on the reaction mechanism of the enzymes hydrolyzing chitin, which will provide a good basis for optimization and modification of chitinase structure, and improvement in chitin hydrolysis, thus enhancing the virulence of pathogenic fungi and developing high-efficiency agents to control plant diseases and other harmful fungi.

Acknowledgements The research described here is jointly supported by the NSFC-Yunnan Joint Fund (U1402265), the National Natural Science Foundation of China (approved nos. 31272093 and 31360019).

References

Adams DJ (2004) Fungal cell wall chitinases and glucanases. Microbiology 150:2029–2035

Bulawa CE (1993) Genetics and molecular biology of chitin synthesis in fungi. Annu Rev Microbiol 47:505–534

Cantarel BL, Coutinho PM, Rancurel C, Bernard T, Lombard V, Henrissat B (2009) The Carbohydrate-Active EnZymes database (CAZy): an expert resource for Glycogenomics. Nucleic Acids Res 37:D233–D238

Choquer M, Becker HF, Vidal-Cros A (2007) Identification of two group A chitinase genes in *Botrytis cinerea* which are differentially induced by exogenous chitin. Mycol Res 111:615–625

Escott GM, Hearn VM, Adams DJ (1998) Inducible chitinolytic system of *Aspergillus fumigatus*. Microbiology 144:1575–1581

Fajardo-Somera RA, Jöhnk B, Bayram Ö, Valerius O, Braus GH, Riquelme M (2015) Dissecting the function of the different chitin synthases in vegetative growth and sexual development in *Neurospora crassa*. Fungal Genet Biol 75:30–45

Fan Y, Fang W, Guo S, Pei X, Zhang Y, Xiao Y et al (2007) Increased insect virulence in *Beauveria bassiana* strains overexpressing an engineered chitinase. Appl Environ Microbiol 73:295–302

Fang W, Leng B, Xiao Y, Jin K, Ma J, Fan Y et al (2005) Cloning of *Beauveria bassiana* chitinase gene *Bbchit1* and its application to improve fungal strain virulence. Appl Environ Microbiol 71:363–370

Fang W, Feng J, Fan Y, Zhang Y, Bidochka MJ, Leger RJ et al (2009) Expressing a fusion protein with protease and chitinase activities increases the virulence of the insect pathogen *Beauveria bassiana*. J Invertebr Pathol 102:155–159

Gan Z, Yang J, Tao N, Liang L, Mi Q, Li J et al (2007a) Cloning of the gene *Lecanicillium psalliotae* chitinase Lpchi1 and identification of its potential role in the biocontrol of root-knot nematode *Meloidogyne incognita*. Appl Microbiol Biotechnol 76:1309–1317

Gan Z, Yang J, Tao N, Yu Z, Zhang KQ (2007b) Cloning and expression analysis of a chitinase gene Crchi1 from the mycoparasitic fungus *Clonostachys rosea* (syn. *Gliocladium roseum*). J Microbiol 45:422–430

Geng J, Plenefisch J, Komuniecki PR, Komuniecki R (2002) Secretion of a novel developmentally regulated chitinase (family 19 glycosyl hydrolase) into the perivitelline fluid of the parasitic nematode, *Ascaris suum*. Mol Biochem Parasitol 124:11–21

Giaever G, Chu AM, Ni L, Connelly C, Riles L, Veronneau S et al (2002) Functional profiling of the *Saccharomyces cerevisiae* genome. Nature 418:387–391

Goldman DL, Vicencio AG (2012) The chitin connection. MBio 3:pii:e00056–12

Gruber S, Kubicek CP, Seidl-Seiboth V (2011a) Differential regulation of orthologous chitinase genes in mycoparasitic *Trichoderma* species. Appl Environ Microbiol 77:7217–7226

Gruber S, Vaaje-Kolstad G, Matarese F, López-Mondéjar R, Kubicek CP, Seidl-Seiboth V (2011b) Analysis of subgroup C of fungal chitinases containing chitin-binding and LysM modules in the mycoparasite *Trichoderma atroviride*. Glycobiology 21:122–133

Gruber S, Seidl-Seiboth V (2012) Self versus non-self: fungal cell wall degradation in *Trichoderma*. Microbiology 158:26–34

Henrissat B (1991) A classification of glycosyl hydrolases based on amino acid sequence similarities. Biochem J 280:309–316

Hollis T, Monzingo AF, Bortone K, Ernst S, Cox R, Robertus JD (2000) The X-ray structure of a chitinase from the pathogenic fungus *Coccidioides immitis*. Protein Sci 9:544–551

Honda Y, Taniguchi H, Kitaoka M (2008) A reducing-end-acting chitinase from Vibrio proteolyticus belonging to glycoside hydrolase family 19. Appl Microbiol Biotechnol 78:627–634

Horiuchi H (2009) Functional diversity of chitin synthases of *Aspergillus nidulans* in hyphal growth, conidiophore development and septum formation. Med Mycol 47:S47–S52

Johnson SM, Pappagianis D (1992) The coccidioidal complement fixation and immunodiffusion-complement fixation antigen is a chitinase. Infect Immun 60:2588–2592

Junges A, Boldo JT, Souza BK, Guedes RL, Sbaraini N, Kmetzsch L et al (2014) Genomic analyses and transcriptional profiles of the glycoside hydrolase family 18 genes of the entomopathogenic fungus *Metarhizium anisopliae*. PLoS ONE 9:e107864

Kong LA, Yang J, Li GT, Qi LL, Zhang YJ, Wang CF et al (2012) Different chitin synthase genes are required for various developmental and plant infection processes in the rice blast fungus *Magnaporthe oryzae*. PLoS Pathog 8:e1002526

Kowsari M, Motallebi M, Zamani M (2013) Protein engineering of chit42 towards improvement of chitinase and antifungal activities. Curr Microbiol 68:495–502

Kuranda MJ, Robbins PW. Chitinase is required for cell separation during growth of *Saccharomyces cerevisiae*. J Biol Chem 266:19758–19767

Langner T, Göhre V (2016) Fungal chitinases: function, regulation, and potential roles in plant/pathogen interactions. Curr Genet 62:243–254

Latgé JP (2007) The cell wall: a carbohydrate armour for the fungal cell. Mol Microbiol 66:279–290

Lenardon MD, Munro CA, Gow NA (2010) Chitin synthesis and fungal pathogenesis. Curr Opin Microbiol 13:416–423

Li DC (2006) Review of fungal chitinases. Mycopathologia 161:345–360

Limón MC, Margolles-Clark E, Benítez T, Penttilä M (2001) Addition of substrate-binding domains increases substrate-binding capacity and specific activity of a chitinase from *Trichoderma harzianum*. FEMS Microbiol Lett 198:57–63

Motoyama T, Fujiwara M, Kojima N, Horiuchi H, Ohta A, Takagi M (1997) The *Aspergillus nidulans* genes *chsA* and *chsD* encode chitin synthases which have redundant functions in conidia formation. Mol Gen Genet 253:520–528

Muszewska A, Pilsyk S, Perlińska-Lenart U, Kruszewska JS (2017) Diversity of cell wall related proteins in human pathogenic fungi. J Fungi 4:pii:E6

Rao FV, Houston DR, Boot RG, Aerts JM, Hodkinson M, Adams DJ et al (2005) Specificity and affinity of natural product cyclopentapeptide inhibitors against *A. fumigatus*, human, and bacterial chitinases. Chem Biol 12:65–76

Roncero C (2002) The genetic complexity of chitin synthesis in fungi. Curr Genet 41:367–378

Ruiz-Herrera J, González-Prieto JM, Ruiz-Medrano R (2002) Evolution and phylogenetic relationships of chitin synthases from yeasts and fungi. FEMS Yeast Res 1:247–256

Seidl V, Huemer B, Seiboth B, Kubicek CP (2005) A complete survey of *Trichoderma* chitinases reveals three distinct subgroups of family 18 chitinases. FEBS J 272:5923–5939

Tzelepis GD, Melin P, Jensen DF, Stenlid J, Karlsson M (2012) Functional analysis of glycoside hydrolase family 18 and 20 genes in *Neurospora crassa*. Fungal Genet Biol 49:717–730

Tzelepis G, Dubey M, Jensen DF, Karlsson M (2015) Identifying glycoside hydrolase family 18 genes in the mycoparasitic fungal species *Clonostachys rosea*. Microbiology 161:1407–1419

Yamada E, Ichinomiya M, Ohta A, Horiuchi H (2005) The class V chitin synthase gene *csmA* is crucial for the growth of the *chsA/chsC* double mutant in *Aspergillus nidulans*. Biosci Biotechnol Biochem 69:87–97

Yang JK, Gan Z, Lou Z, Tao N, Mi Q, Liang L et al (2010) Crystal structure and mutagenesis analysis of chitinase CrChi1 from the nematophagous fungus *Clonostachys rosea* in complex with the inhibitor caffeine. Microbiology 156:3566–3574

Yang JK, Yu Y, Li J, Zhu W, Geng Z, Jiang D et al (2013) Characterization and functional analyses of the chitinase-encoding genes in the nematode-trapping fungus *Arthrobotrys oligospora*. Arch Microbiol 195:453–462

Chapter 9
Chitin in Arthropods: Biosynthesis, Modification, and Metabolism

Xiaojian Liu, Jianzhen Zhang and Kun Yan Zhu

Abstract Chitin is a structural constituent of extracellular matrices including the cuticle of the exoskeleton and the peritrophic matrix (PM) of the midgut in arthropods. Chitin chains are synthesized through multiple biochemical reactions, organized in several hierarchical levels and associated with various proteins that give their unique physicochemical characteristics of the cuticle and PM. Because, arthropod growth and morphogenesis are dependent on the capability of remodeling chitin-containing structures, chitin biosynthesis and degradation are highly regulated, allowing ecdysis and regeneration of the cuticle and PM. Over the past 20 years, much progress has been made in understanding the physiological functions of chitinous matrices. In this chapter, we mainly discussed the biochemical processes of chitin biosynthesis, modification and degradation, and various enzymes involved in these processes. We also discussed cuticular proteins and PM proteins, which largely determine the physicochemical properties of the cuticle and PM. Although rapid advances in genomics, proteomics, RNA interference, and other technologies have considerably facilitated our research in chitin biosynthesis, modification, and metabolism in recent years, many aspects of these processes are still partially understood. Further research is needed in understanding how the structural organization of chitin synthase in plasma membrane accommodate chitin biosynthesis, transport of chitin chain across the plasma membrane, and release of the chitin chain from the enzyme. Other research is also needed in elucidating the roles of chitin deacetylases in chitin organization and the mechanism controlling the formation of different types of chitin in arthropods.

Keywords Chitin biosynthesis · Chitin deacetylation · Chitin degradation · Cuticle · Peritrophic matrix

---

X. Liu · J. Zhang (✉)
Research Institute of Applied Biology, Shanxi University, Taiyuan, Shanxi 030006, China
e-mail: zjz@sxu.edu.cn

K. Y. Zhu (✉)
Department of Entomology, Kansas State University, 123 Waters Hall, Manhattan, KS 66506, USA
e-mail: kzhu@ksu.edu

Q. Yang and T. Fukamizo (eds.), *Targeting Chitin-containing Organisms*, Advances in Experimental Medicine and Biology 1142,
https://doi.org/10.1007/978-981-13-7318-3_9

9.1 Introduction

Arthropods are invertebrate animals (lacking a backbone) that belong to the phylum Arthropoda. These animals are characterized by their exoskeleton (external skeleton), segmented body, and paired jointed appendages. To date, over a million arthropod species have been described, accounting for more than 80% of all described living animal species. Based on the fossil record, Arthropoda aroused about 550–600 million years ago (Brusca 2000), and is considered one of the first animal groups to take the step from water to land and switch from aquatic to terrestrial life. Arthropoda is classified into five subphyla including Trilobitomorpha, Chelicerata, Crustacea, Hexapoda, and Myriapoda. Recent molecular and genetic data show that Hexapoda, comprising Insecta and three other noninsect hexapod classes (Protura, Diplura, and Collembola), are monophyletic, but Crustacea are not. The monophyletic Hexapoda and paraphyletic Crustacea form a single superclade called Pancrustacea (Regier et al. 2010).

Arthropods have remarkable abilities of rapidly adapting to and colonizing in various environments, which makes them one of the most successful and diverse animal groups. Many species of crustaceans including crabs, lobsters, shrimps, and crayfish are an important food for human consumption. Many insect species are important pollinators for crops and fruits. It is estimated that more than 10% of the value from the agricultural food production in the world attributes to the arthropod pollination (Gallai et al. 2009). On the other hand, many arthropod species (e.g., insects and mites) are devastating pests of agricultural crops and forests. It is estimated that approximately 18–26% of annual crop losses in the world is caused by the arthropod pests (Culliney 2014). Furthermore, many arthropod species (e.g., ticks, mosquitoes) are major groups of pathogen vectors for human and animal diseases. For example, the malaria mosquito (*Anopheles gambiae*) causes 445,000 human deaths in 2016 alone (World Health Organization 2017). The success of arthropods, whether as beneficial organisms or harmful pests, can be attributed to their unique characteristics, which have been acquired through several hundred million years of evolution. One such feature is the presence of the exoskeleton, the common feature of all arthropods, which provides extensive protection from the environment.

9.1.1 Structure, Formation, and Functions of Exoskeleton

The body wall of arthropods is known as integument, which consists of an outer noncellular structure (cuticle), a middle cellular layer (epidermis), and an inner unicellular layer (basement membrane) (Fig. 9.1). The nonliving cuticle is a continuous extracellular structure that is formed by the epidermis. The exoskeleton comprised of cuticle layers is a rigid body covering for supporting and protecting the soft tissues of arthropods. The histology of the arthropod cuticle has been

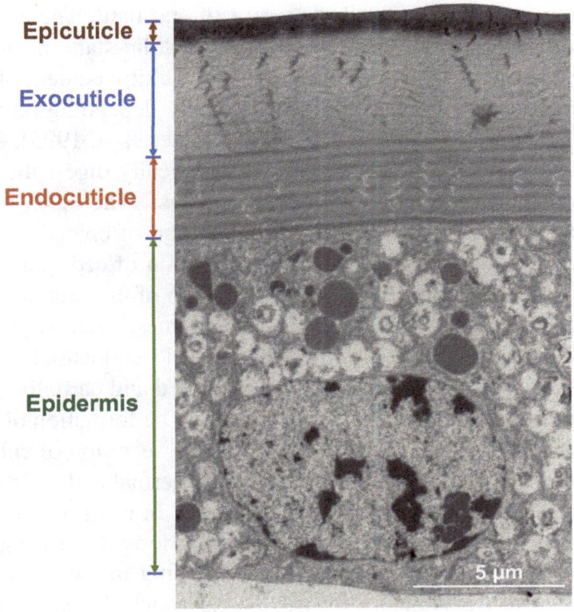

Fig. 9.1 Structure of the integument observed under transmission electron microscope (TEM) in the fifth-instar nymph of *Locusta migratoria* (basement membrane not shown)

extensively studied in the past two centuries. The cuticle is composed of two main layers: the outer chitin-free epicuticle and the inner chitinous procuticle (Fig. 9.1). The epicuticle is made up of two sublayers, an outer epicuticle (also called the envelope or cuticulin layer) and an inner layer (called epicuticle) (Locke 2001). The procuticle is subdivided into an upper exocuticle and a lower endocuticle (Neville 1975). In some insects, there is a layer of mesocuticle located between the endocuticle and the exocuticle (Andersen 1979; Noble-Nesbitt 1991; Barbakadze et al. 2006).

The epidermis is the outer cell layer of arthropods and is mostly single layered. The density and shape of epidermal cells may change substantially during the arthropod growth and development. The epidermal cells secrete cuticle components and enzymes involved in cuticle modification and digestion at the time of molting. At the beginning of the molting cycle, the smooth surface of the epidermal cell transforms into microvilli at the apices of which epicenters of cuticulin are secreted. The new cuticulin seals the epidermis and protects it from the molting fluid. The cuticulin layer becomes the outer epicuticle and an inner epicuticle is secreted as the cuticle is progressively elaborated. Oenocytes, named because of their pale amber color, is found in many anatomical locations, including the epidermis. They are often large and polyploid cells with an extensive tubular endoplasmic reticulum. The function of oenocytes is still not completely understood, but several studies indicate that they are responsible for lipid processing and deposition on the surface of the exoskeleton (Martins and Ramalho-Ortiago 2012; Makki et al. 2014; Yu et al. 2016).

The cuticle layers of arthropods vary substantially in the composition, mechanical properties, and function. The exocuticle is the most highly sclerotized structure and generally is very stiff and hard. The endocuticle is composed of soft flexible protein and chitin layers. Solid-state nuclear magnetic resonance and gravimetric analyses revealed that the chitin content of the exuvial dry mass may be as high as 40% but vary considerably depending on the insect species and cuticle type even in a single organism (Kramer et al. 1995). To accommodate the increase in body size, arthropods must periodically digest the old cuticle and produce new cuticle in the molting (ecdysis) process. The onset of molting is characterized by apolysis, which involves the separation of epidermal cells from the old cuticle by secretion of molting fluid and formation of ecdysial zoom (Locke and Huie 1979). The separation results in the formation of the exuvial space where new procuticle is laid down. Meanwhile, various enzymes, such as proteases and chitinases, which digest the main constituents of the old endocuticle, are secreted into the molting fluid. All the old endocuticle is digested and partially resorbed, which allows for the recycling of old cuticle components. The formation of new epicuticle starts after the ecdysial space opens as a result of the secretion of cuticle proteins and chitin fibers through the apical membranes of epidermal cells. Finally, the arthropods shed their exuvia and their new cuticle expands. In most insects, extensive deposition of new cuticle continues in the intermolt periods. Following the cuticle expansion, sclerotization of the cuticle occurs, resulting in the characteristic rigid exoskeleton.

The exoskeleton formed by the cuticle layers is a multifunctional structure of arthropods (Vincent and Wegst 2004). As the exoskeleton is an outer covering of the arthropod body, it helps stabilize the body shape, serves as muscle attachment sites, and allows for the locomotion and flight. It protects arthropods from various harms, including mechanical damage, radiation, desiccation, and invasion of pathogenic microorganisms. Different body colors of arthropods, some of which are contributed by the pigmentation of the cuticle, may function as camouflage or warning signals. The exoskeleton may also function as a sensory interface with the environment, which is involved in communications, either by providing pheromones or producing signaling body structures and colors.

9.1.2 Structure, Formation, and Functions of Peritrophic Matrix

In many arthropods, such as insects, the alimentary canal is divided into three main regions with specialized functions: the foregut, midgut, and hindgut, which are all formed by a single layer of epithelia. The epithelia of the foregut and hindgut are ectodermal in origin, whereas the midgut epithelium is endodermal. Most of the food digestion occurs in the midgut where the cells of the midgut are actively involved in the production and secretion of digestive enzymes. More than 250 years ago, Lyonet (1762) found a membrane surrounding the food bolus in a monograph on the anatomy of the goat moth caterpillar (*Cossus cossus*). Balbiani (1890) then named this anatomical structure as a peritrophic membrane due to its appearance like a membranous sack surrounding the gut content in the midgut. Peters (1992) proposed to substitute the term "membrane" with "envelope", taking into account that "membrane" is reserved in biology for lipid bilayers. This term was successively replaced by the peritrophic matrix (PM), which emphasizes that the PM is an apical extracellular matrix with surprisingly complex characteristics (Ramos et al. 1994). This similar structure has been found in many arthropods including species of the subphyla Chelicerata, Crustacea, Myriapoda, and in many insect orders (Peters 1992).

The PM is an invertebrate-unique, semi-permeable structure that lines the midgut lumen of most insects, except for hemipterans, thysanopterans, and adult lepidopterans, which feed only plant sap, nectar, or blood (Silva and Terra 1995), and the coleopteran families Carabidae and Dytiscidae, which exhibit extraintestinal digestion. In addition, PM appears to be absent in strepsipterans, raphidiopterans, and megalopterans (Peters 1992). However, some studies on the absence of a PM in particular insect species need to be viewed with caution. In the yellow fever mosquito (*Aedes aegypti*), for instance, PM is detectable only after female mosquitoes ingest a blood meal (Richards and Richards 1977). In other cases, PM is partly solubilized during fixation and can be easily detected only by dissection. The PM of the midgut in invertebrates is considered to be functionally similar to the mucous secretions of the digestive tract in vertebrates.

Although there is a great diversity of PM structures in different insect species, PMs have been commonly categorized into two types (Types I and II) based on the mode of the PM formation. Type I PM is secreted by the entire epithelium and formed simply by delamination from the surface of the midgut epithelium, whereas Type II PM is produced by a small number of specialized cells of the cardia, a valve-like organ at the junction between the foregut and the midgut. Type I PM is widespread in insects and particularly prevalent in lepidopteran larvae in which forms a "felt-like" material with a thickness of 0.5–1.0 mm. In contrast, Type II PM is more organized and contains one to three laminated layers which are found in primitive orders (e.g., Dermaptera and Isoptera) (Shao et al. 2001; Kato et al. 2006). Both types of PM are mainly composed of chitin microfibers embedded in the matrix of proteins, glycoproteins, and proteoglycans (Terra 2001). It is assumed that the chitin content in the PM generally accounts for 3–13% (w/w) depending on the respective species, whereas the proteins account for 20–55% (w/w) of the total PM mass (De Mets and Jeuniaux 1962; Ono and Kato 1968; Zimmermann et al. 1975; Becker 1980). Ultrastructural observations of PM suggest that chitin appears to form a flexible framework onto which the proteins are assembled to form a matrix structure (Wang and Granados 2000).

The PM is a physical barrier protecting the midgut epithelium from abrasive materials ingested by an arthropod, including abrasive food particles, pathogens, and certain toxins. It is also a biochemical barrier, sequestering and in some cases, inactivating ingested toxins. The PM separates the midgut cavity into two components including the endoperitrophic space (i.e., the gut lumen including the food bolus) and the ectoperitrophic space (i.e., the space between the PM and epithelium), which helps the midgut increase the efficiency in nutrient acquisition and reuse the hydrolytic enzymes (Bolognesi et al. 2001).

9.2 Biology of Chitin in Arthropods

9.2.1 Chemical Compositions and Basic Structure of Chitin

Chitin is composed of long, unbranched polysaccharide chains (polymers) of β-1,4-linked *N*-acetylglucosamines (GlcNAc), and is the second most abundant biological polymer after cellulose found in nature (Merzendorfer 2006). Chitin occurs in a multitude of organisms from fungi to mollusks and nematodes, but is certainly most prominent in the cuticle of the exoskeleton and the PM of the midgut in arthropods. In arthropods, chitin is arranged in unique ways and provides additional hydrogen-bonding opportunities to interact with various proteins and possibly tanning agents during the process of cuticle deposition (Fig. 9.2). In particular, the presence of the amino groups in chitin polymer is highly advantageous for cross-linking with other functional groups for conducting modification reactions (Zhu et al. 2016).

In arthropods, about 20 chitin polymers form a crystalline microfibril (also referred to as a rod or crystallite) of approximately 3 nm in diameter and 300 nm in length by hydrogen bonds between the amine and carbonyl groups of

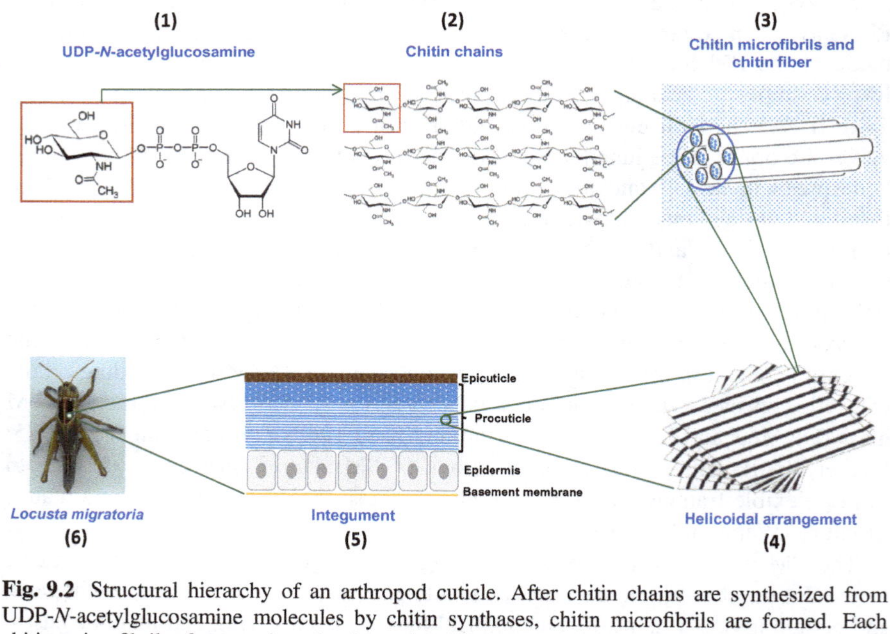

Fig. 9.2 Structural hierarchy of an arthropod cuticle. After chitin chains are synthesized from UDP-*N*-acetylglucosamine molecules by chitin synthases, chitin microfibrils are formed. Each chitin microfibril of approximately 3 nm in diameter and 300 nm in length is formed by approximately 20 chitin chains orientated in an antiparallel fashion in α-chitin and wrapped with proteins (1–3). Many chitin–protein microfibrils then form a larger chitin–protein fiber of about 60 nm in diameter (3). The bundles of chitin–protein fibers arrange parallel to each other to form a chitin–protein lamina. Many laminae are finally stacked and arranged in a helicoidal fashion (4). Chitin occurs as chitin–protein matrices in arthropod procuticle (5, 6)

N-acetylglucosamine of the adjacent chitin polymers (Fig. 9.2). Chitin microfibrils play a key role in the physical properties of the material such as its elasticity and chemical properties such as solubility. The X-ray diffraction analysis suggests that chitin is a polymorphic substance that occurs in three different forms known as α-, β-, and γ-chitin (Zhu et al. 2016). They differ mainly in the degree of hydration, the orientation of the sugar chains, and the number of chitin chains per unit cell (Rudall and Kenchington 1973; Kramer and Koga 1986).

In α-chitin, adjacent chitin polymers are arranged in an antiparallel fashion with each other, which makes it a highly ordered crystalline structure. This form is stabilized by a high number of hydrogen bonds of four hydroxyl groups and two amide groups in the repeating unit. The α-chitin is also the most abundant form found in the cell walls of fungi and the exoskeletons of arthropods. Studies on the crystalline α-chitin matrix of the lobster (*Homarus americanus*) show that everywhere in the carapace, the texture is optimized in such a way that the same crystallographic axis of the chitin matrix is parallel to the normal to the local tangent plane of the carapace. Notable differences in the texture are observed between hard mineralized parts and soft membranous parts (Raabe et al. 2005a, b, c, 2006). In crustaceans, it is heavily mineralized with calcium or magnesium salts, further increasing its strength (Bentov et al. 2016).

In β-chitin, the chains are arranged parallel to each other, whereas in γ-chitin, the chains are grouped in sets of three strands where two parallel chains are in one direction and the third one in the opposite direction (Ruddal 1963). Essentially, the arrangement of γ-chitin is a mixture of both α- and β-chitins. In contrast to α-chitin, both β- and γ-chitins are less tightly packed and can be swollen in water and dissolved in formic acid (Merzendorfer 2006). Because of its open structure, β-chitin is more facile to help forming hydrogels (Ianiro et al. 2014). Both β- and γ-chitins are, therefore, highly flexible and hydrated (Peters 1992). The β-chitin is found in squid pens, spines of diatoms, tubes of giant tubeworms, and possibly in the PM lining the midgut epithelial cells of insects, whereas the γ-chitin is found in the cocoons of some beetles. The precise structure of γ-chitin is not resolved to the same extent as the other two forms of chitin.

9.2.2 Higher Order Structure of Chitin in Cuticle

The architecture of chitin microfibrils was first described in the cuticle of crustaceans by Bouligand (1965). The structure was subsequently confirmed in insects (Neville and Luke 1969). A striking characteristic of chitin in arthropods is its well-defined hierarchy of the structural organization to form the exoskeleton and the PM (Fig. 9.2). As previously mentioned, each chitin microfibril is composed of about 20 single chitin chains wrapped with chitin-binding proteins (CBPs) (Fabritius et al. 2009). The average size of each chitin microfibril is about 3 nm in diameter and 300 nm in length. Many chitin microfibrils then form a larger chitin–protein fiber of about 60 nm in diameter. The bundles of chitin fibers finally arrange parallel to each other to form a horizontal sheet (or lamina). The laminae are stacked and arranged in parallel to the apical surface of the underlying epidermal cells.

The orientation is often different in successive levels throughout the thickness of the cuticle. New laminae are added continuously to the growing procuticle from the assembly zone between the epidermis and the cuticle during intermolt periods. These laminae can be assembled on top of one another in two ways. In the first pattern, each layer of fibers rotates anticlockwise through a constant angle in successive levels. This arrangement termed Bouligand structure is helicoidal and results in a series of thin lamellae in which the chitin fibrils assume a parabolic shape corresponding to each 180° stack (Bouligand 1972). This is called lamellate cuticle, which has a characteristic appearance in electron microscopy. In some cases, the intervening layers of the helicoidal cuticle are very thin so that the orientation appears to change suddenly from one layer to the next. This is called a pseudo-orthogonal or plywood-like arrangement. This is similar to a "cross-ply" laminate used in constructions. In addition, the vertical stacks of laminae are further stabilized by pore canals that traverse the procuticle in a cork-screw-like arrangement. The pore canal fibers made of chitin can be seen in the middle of these canals (Fabritius et al. 2009).

9.2.3 Higher Order Structure of Chitin in Peritrophic Matrix

In contrast to the cuticle, chitin in the PM exhibits great structural diversity, which likely reflects evolutionary adaptations to different food sources and physiological and/or immunological challenges. The chitin microfibrils in the PM are more hydrated, flexible, and the thickness and number of layers vary widely among different species (Hegedus et al. 2009). They do not form laminae as in the cuticle, but form loosely associated grid-like structures (Harper and Hopkins 1997; Harper et al. 1998). Chitin chains assemble to a condensed microfibril of 2–6 nm in diameter and up to 500 nm or even longer in length (Lehane 1997). About 20–400 chitin microfibrils are organized into a microfibrillar bundle with a diameter of about 20 nm (Kramer et al. 1985; Peters 1992). Microvilli are approximately 145 nm wide at their apex, and the interstitial spaces of the PM lattice are about 125 nm. This suggests that about 150 repeating units, or 300 GlcNAc residues, run between the nodes of the lattice.

According to the arrangement of microfibrillar textures, which can be visualized by ultrastructural microscopy, Peters (1992) proposed three structural types: orthogonal, hexagonal, and random felt-like arrangements. The PM consists of an organized lattice of chitin fibrils held together by chitin-binding proteins. Additional proteinaceous material is added to the forming meshwork, which becomes thicker and has a reduced pore size. The chitin fibers of the PM are thought to confer tensile strength to the PM (Peters 1992). According to the mode of delamination, PMs were categorized into type I and type II (Wigglesworth 1930). The PMs may be formed by either the entire midgut (Type I) or the cardia at the junction between the foregut and the midgut (Type II). Interestingly, there is no correlation between the PM types and the ultrastructural arrangement of the microfibers. It should be noted that there is a lack of structural studies on the type of chitin in the PM of insects.

9.3 Chitin Biosynthesis

9.3.1 Biochemical Processes and Important Enzymes in Chitin Biosynthesis

Although chitin biosynthesis is essential for arthropod growth and development, knowledge of this process is still fragmentary in insects and other arthropods (Cohen 2010). Candy and Kilby (1962) are the first to propose a chitin biosynthetic pathway for insects. The process starts with trehalose which is a sugar consisting of two molecules of glucose and ends with UDP-*N*-acetylglucosamine (UDP-GlcNAc), which is an essential substrate to form chitin polymers. The formation of an activated substrate UDP-GlcNAc requires several enzymes including trehalase, hexokinase, glucose-6-phosphate isomerase, glutamine-fructose-6-phosphate aminotransferase, glucosamine-6-phosphate *N*-acetyltransferase, phosphoglucosamine mutase, and UDP-*N*-acetylglucosamine pyrophosphorylase (UAP) (Cohen 1987; Merzendorfer and Zimoch 2003; Cohen 2010; Muthukrishnan et al. 2012). The pathway from UDP-GlcNAc to chitin was finally established in the southern armyworm (*Spodoptera eridania*) by using cell-free extracts (Jaworski et al. 1963). Many subsequent studies conducted in various insect species support this general pathway (Table 9.1).

9.3.1.1 Trehalase

Trehalose (α-D-glucopyranosyl-α-D-glucopyranoside) is a major hemolymph sugar and an indispensable substrate for energy production and chitin biosynthesis in insects. Trehalose in insects is synthesized by trehalose-6-phosphate synthase (TPS) and trehalose-6-phosphatase (TPP) (Becker et al. 1996; Chen et al. 2002). The reaction from trehalose to glucose is catalyzed by trehalase, the first enzyme in the chitin biosynthetic pathway in the organisms producing chitin (Becker et al. 1996; Thompson 2002). To date, trehalases have been purified from various insect species. They are divided into the soluble (Tre-1) and the membrane-bound (Tre-2) trehalases. The first insect trehalase gene encoding a soluble trehalase was cloned in the yellow mealworm (*Tenebrio molitor*) (Takiguchi et al. 1992), whereas a gene

Table 9.1 Representative studies on the enzymes and their coding genes in the reactions of insect chitin biosynthetic pathway

Reaction	Enzyme	Genes	References
Trehalose → β-D-glucose	Trehalase (TRE)	Two genes in most insects; three–four genes in *Nilaparvata lugens, Locusta migratoria, Harmonia axyridis,* and *Tribolium castaneum*	Takiguchi et al. (1992), Mitsumasu et al. (2005), Liu et al. (2016), Tang et al. (2016), Zhao et al. (2016), Shi et al. (2016b) and Tang et al. (2016)
β-D-glucose → Glucose-6-phosphate	Hexokinase	Few genes in insects	Muthukrishnan et al. (2012)
Glucose-6-phosphate → Fractose-6-phosphate	Glucose-6-phosphate isomerase	One gene in most insects	Yang et al. (2015)
Fractose-6-phosphate → Glucosamine-6-phosphate	Glutamine: fructose-6-phosphate aminotransferase (GFAT)	One–two genes in insects	Kato et al. (2006) and Huang et al. (2007)
Glucosamine-6-phosphate → *N*-acetylglucosamine-6-phosphate	Glucosamine-6-phosphate *N*-acetyltransferase	Presumably one gene in most insects	Kato et al. (2005)
N-acetylglucosamine-6-phosphate → *N*-acetylglucosamine-1-phosphate	Phosphoacetylglucosamine mutase	One gene in most insects	Palaka et al. (2017)
N-acetylglucosamine-1-phosphate →UDP-*N*-acetylglucosamine	UDP-*N*-acetylglucosamine pyrophosphorylase (UAP)	One gene in most insects; two genes in some locusts and beetles	Tonning et al. (2006), Arakane et al. (2011), Liu et al. (2013), Shi et al. (2016a)
UDP-*N*-acetylglucosamine→ chitin polymer	Chitin synthase (CHS)	Two genes in most insects; one gene in species lacking the peritrophic matrix	Tellam et al. (2000), Merzendorfer (2006), Wang et al. (2012), Bansal et al. (2012), Mansur et al. (2014), Tetreau et al. (2015a, b), Qu and Yang (2011, 2012) and Shirk et al. (2015)

encoding membrane-bound trehalase was sequenced in the domesticated silkmoth (*Bombyx mori*) (Mitsumasu et al. 2005).

In the beet armyworm (*Spodoptera exigua*), *SeTre-1* is expressed highly in the integument and Malpighian tubules, whereas *SeTre-2* is expressed in tracheae and fat bodies, which suggests that these two genes have different roles in chitin synthesis in the integument and the midgut. RNA interference (RNAi) experiments showed that knockdown of *SeTre-1* and *SeTre-2* reduced the chitin content of the integument and the midgut, respectively (Chen et al. 2010). However, bioinformatics analyses of fully sequenced insect genomes and transcriptomes revealed the presence of more than one soluble trehalase gene in all insect species examined to date (Liu et al. 2016; Shi et al. 2016b; Tang et al. 2016; Zhao et al. 2016). In the brown planthopper (*Nilaparvata lugens*), silencing each of the three genes encoding soluble trehalases by RNAi can affect chitin biosynthesis and degradation, resulting in molting deformities (Zhao et al. 2016). In the migratory locust (*Locusta migratoria*), there are two soluble trehalases (LmTreS1 and LmTreS2), one membrane-bound (LmTreM) and one membrane-bound-like trehalases (LmTreM-like). However, nymphs can successfully molt to adults after RNAi by injecting them with each of four double-stranded RNAs (dsRNAs) targeting these four trehalase genes (Liu et al. 2016). It appears that the roles of different trehalase genes in chitin biosynthesis may vary in different insect species.

9.3.1.2 Glutamine-Fructose-6-Phosphate Aminotransferase

Glutamine-fructose-6-phosphate aminotransferase (GFAT) catalyzes the formation of glucosamine 6-phosphate and is the first and rate-limiting enzyme of the hexosamine biosynthetic pathway. Silencing each of the two GFAT genes (*GFAT-1* and *GFAT-2*) by RNAi in the yellow fever mosquito (*Aedes aegypti*) suggested that *GFAT-1* may have a regulatory role in chitin biosynthesis in the midgut and this was reversed by the administration of glucosamine into the hemocoel (Kato et al. 2006). In the blood-sucking cattle tick (*Haemaphysalis longicornis*), silencing the expression of *GFAT* gene also resulted in less blood feeding, lower body weight gain and ultimately in death (Huang et al. 2007). GFAT is also sensitive to feedback inhibition by UDP-GlcNAc, indicating that the importance of this enzyme in regulating the flow of carbon into this pathway. However, the functional significance for the presence of two *GFAT* genes has not been explored so far.

9.3.1.3 UDP-*N*-Acetylglucosamine Pyrophosphorylase

UDP-GlcNAc is the activated form of GlcNAc needed for chitin biosynthesis and its formation is catalyzed by UDP-GlcNAc pyrophosphorylase (UAP) in the cell cytoplasm via the following reversible reaction: UTP + GlcNAc-1-P ↔ UDP-GlcNAc + ppi (Peneff et al. 2001). UAP is also important for glycosylation of proteins, sphingolipids and secondary metabolites with *N*-acetylglucosamine (GlcNAc) or GPI anchors which bridge proteins to the cell membrane, or for conjugation of 7-β-hydroxylated bile acids (Marschall et al. 1992; Eisenhaber et al. 2003). The cDNAs of *UAP* genes have been sequenced from several insect species. The importance of UAP for insect development has also been shown by studies in the fruit fly (*Drosophila melanogaster*). Mutants of *DmUAP* (also called *mummy* in this species) show various defects in trachea morphogenesis, cuticle formation, fasciculation of the central nervous system, dorsal closure, and eye development (Araujo et al. 2005; Schimmelpfeng et al. 2006).

Most insect species are known to possess only one *UAP* gene, whereas several species, including *L. migratoria*, the red flour beetle (*Tribolium castaneum*) and the Colorado potato beetle (*Leptinotarsa decemlineata*), are known to have two different *UAP* genes (*UAP1* and *UAP2*) (Arakane et al. 2011; Liu et al. 2013; Shi et al. 2016a). RNAi experiments showed that different *UAP* genes had different functions. In *T. castaneum*, both *TcUAP* genes are critical for insect survival, but only *TcUAP1* is required for chitin biosynthesis of the cuticle and the PM. RNAi of *TcUAP2* also led to a high mortality of the insect, but it may be due to defective glycosylation of proteins or secondary metabolites, whose functions are critical for insect survival (Arakane et al. 2011). Similarly, knockdown of *LdUAP1* by RNAi resulted in reduced chitin contents and impaired molting in *L. decemlineata*, whereas knockdown of *LdUAP2* affected the PM integrity and blocked the development and fat body depletion of the insect (Shi et al. 2016a). In contrast, *LmUAP1* appears to be responsible for chitin biosynthesis of both the integument and midgut in *L. migratoria*, whereas *LmUAP2* is not essential for the insect development at least in nymphal stage (Liu et al. 2013). Thus, the role of the *UAP2* gene varies in different insect species. Since most insect species have a single *UAP* gene, the presence of two genes in some insects may suggest that the two *UAP* paralogs have derived from a relatively recent gene duplication event. Indeed, phylogenetic analysis of all known UAP proteins indicates that the two UAPs in *L. migratoria* or *T. castaneum* are first tightly clustered, and then clustered with the UAPs from other insect species. The UAPs of mammals appear to be closely related to those of insects, whereas the UAPs from yeast and nematodes are much less related to those of mammals and insects.

9.3.1.4 Chitin Synthase

The last step in the chitin biosynthetic pathway is catalyzed by chitin synthase (CHS), which catalyzes the transfer of sugar moieties from activated sugar donors to specific acceptors in all chitin-containing organisms. CHSs belong to hexosyltransferases in the large family (Family 2) of glycosyltransferases (UDP-GlcNAc: chitin 4-β-*N*-acetylglucosaminyltransferase, EC 2.4.1.16), which also includes the closely related cellulose synthases (Coutinho et al. 2003). CHSs have been extensively studied in fungi (Valdivieso et al. 1999). Fungal CHSs are encoded by a large family of genes and as many as eight different *CHS* genes have been identified in a single fungal species (Munro and Gow 2001). Various fungal CHSs have been found to have different roles, including sporulation and cell division, and their expression varies throughout different developmental stages (Specht et al. 1996; Valdivieso et al. 1999; Munro and Gow 2001).

In contrast to fungi, insects appear to have fewer *CHS* genes. The first cDNA sequence encoding a CHS was reported in the Australian sheep blowfly (*Lucilia cuprina*) (Tellam et al. 2000). To date, two *CHS* genes (*CHS1* and *CHS2*, also known as *CHS-A* and *CHS-B*, respectively) have been identified in most insect species based on their differences in the cDNA sequences and tissue-specific expressions (Fig. 9.3). However, only one *CHS* gene has been found in several hemipteran insects, such as *N. lugens*, the soybean aphid (*Aphis glycines*), the pea aphid (*Acyrthosiphon pisum*), and the kissing bug (*Rhodnius prolixus*) as these insects do not have a PM (Bansal et al. 2012; Wang et al. 2012; Mansur et al. 2014). Nevertheless, an analogous structure to the PM, which is named perimicrovillar membrane (PMM), has been described in *R. prolixus* (Alvarenga et al. 2016). The PMM also contains chitin as shown by using different techniques. It is clear that *CHS1* is exclusively expressed in the epidermis underlying the cuticular exoskeleton and related ectodermal cells such as tracheal cells, whereas *CHS2* gene

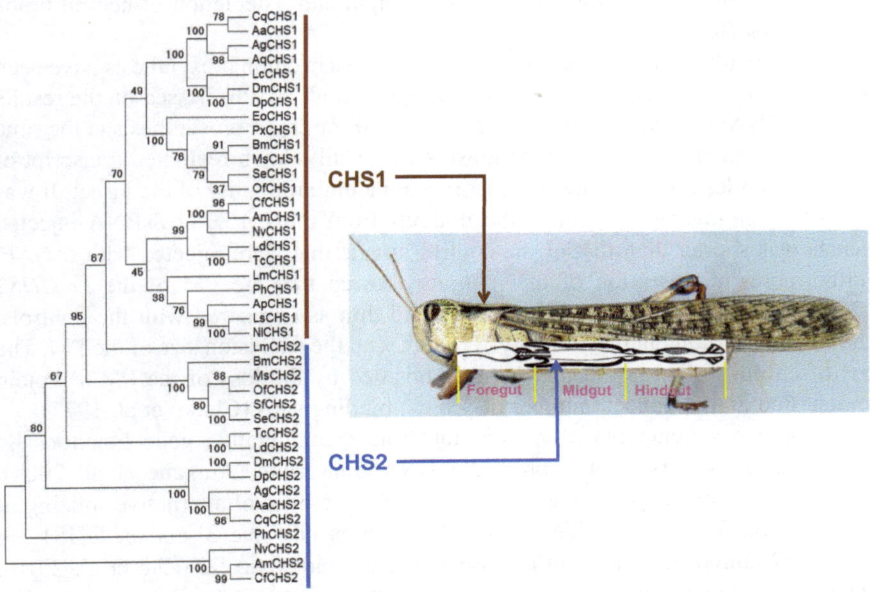

Fig. 9.3 A dendrogram showing the phylogenetic relationships of insect chitin synthases (CHSs). The tree was generated by MEGA 7 based on full-length amino acid sequences of chitin synthases from 24 different insect species, including *Acyrthosiphon pisum* (Ap), *Aedes aegypti* (Aa), *Anopheles gambiae* (Ag), *Anopheles quadrimaculatus* (Aq), *Apis mellifera* (Am), *Bombyx mori* (Bm), *Choristoneura fumiferana* (Cf), *Culex quinquefasciatus* (Cq), *Drosophila melanogaster* (Dm), *Drosophila pseudoobscura* (Dp), *Ectropis oblique* (Eo), *Laodelphax striatellus* (Ls), *Leptinotarsa decemlineata* (Ld), *Locusta migratoria* (Lm), *Lucilia cuprina* (Lc), *Manduca sexta* (Ms), *Nasonia vitripennis* (Nv), *Nilaparvata lugens* (Nl), *Ostrinia furnacalis* (Of), *Pediculus humanus corporis* (Ph), *Plutella xylostella* (Px), *Spodoptera exigua* (Se), *Spodoptera frugiperda* (Sf), and *Tribolium castaneum* (Tc)

is specifically expressed in gut epithelial cells that produce PM-associated chitin in insects (Merzendorfer and Zimoch 2003; Arakane et al. 2004, 2005; Hogenkamp et al. 2005; Zimoch et al. 2005; Ashfaq et al. 2007).

In *D. melanogaster*, embryos with the homozygous mutations of *CHS1* (also known as *kkv*) have a blimp phenotype and are unable to break open the egg shell (Ostrowski et al. 2002; Moussian et al. 2005). Much of the recent research on insect CHSs has been focused on their molecular properties and functional analyses. For example, RNAi experiments in many insect species have confirmed that *CHS1* is responsible for chitin synthesis in the cuticle and cuticular lining of the foregut, hindgut, and trachea, whereas *CHS2* is dedicated to chitin synthesis in the PM (Fig. 9.3). In *T. castaneum*, RNAi by using *TcCHS1*-specific dsRNA can disrupt all three types of molting events (larva–larva, larva–pupa, and pupa–adult) due to the reduced cuticular chitin synthesis. As expected, silencing of *TcCHS2* led to a complete absence of chitin in the midgut and also to the disruption of the PM function (Arakane et al. 2005, 2008; Agrawal et al. 2014; Kelkenberg et al. 2015). Because of the loss of the PM's capability to partition digestive enzymes and substrates, the larvae developed nutritional deficiencies resulting in a starvation phenotype with significantly reduced body weight and a depletion of neutral lipids in fat bodies (Kelkenberg et al. 2015).

Similar results of different roles of two *CHS* genes in chitin biosynthesis have been obtained in *L. migratoria* (Zhang et al. 2010; Liu et al. 2012b). Based on the results from the RNAi experiments, *LmCHS1* is required for chitin biosynthesis at the time just before and after molting. In contrast, significantly down-regulated transcript of *LmCHS2* can lead to a cessation of feeding and a high mortality of the insect. It was found that the average length of the midguts from the *LmCHS2* dsRNA-injected locusts was shorter than that of the control insects that were injected with ds*GFP*. Furthermore, microsection of the midguts showed that the PM of the *LmCHS2* dsRNA-injected nymphs was amorphous and thin as compared with the controls. Thus, the chitin content may directly correlate with the exclusion size of the PM. The loss of chitin fibrils may further be accompanied by the loss of the PM's protein content due to the reduced number of protein binding sites (Clarke et al. 1977).

Alternative splicing plays an important role in modulating gene functions by expanding the diversity of expressed mRNA transcripts (Arakane et al. 2004). Indeed, one of the unique features of *CHS1* is the presence of alternative splicing of its transcript. The polypeptides encoded by exons near the 3'-end of CHS1 are exactly 59 amino acids long in all insect species studied so far (Zhu et al. 2016). The amino acid sequences of the same species have significant sequence similarity/identity between the two alternative splicing variants. However, phylogenetic analysis of this region alone from multiple insect species of different orders indicates that two splicing variants of *CHS1* from same species are closely related, whereas CHS1 (CHS1A and CHS1B) and CHS2 from different species are clearly grouped into two different clusters (Arakane et al. 2004; Ashfaq et al. 2007; Zhang et al. 2010; Wang et al. 2012; Yang et al. 2013). Studies on *T. castaneum*, the tobacco hornworm (*Manduca sexta*), and the oriental fruit fly (*Bactrocera dorsalis*) have shown that the two alternatively spliced transcripts change during their developmental stages with *CHS1B* being prominent in the pupal stage (Arakane et al. 2004; Hogenkamp et al. 2005; Zimoch et al. 2005; Zhang et al. 2010; Yang et al. 2013). In general, *CHS1A* is expressed predominately in the integument, whereas *CHS1B* transcript seems to be enriched in the tracheae at least in some insect species such as *M. sexta*, *B. dorsalis,* and *L. migratoria.*

The specific functions of the two alternatively spliced transcripts of the *CHS1* gene have been investigated using RNAi. In several insects, silencing either of the two transcripts can result in molting failure and death (Arakane et al. 2005; Zhang et al. 2010; Wang et al. 2012; Yang et al. 2013). However, their phenotypes and timing of developmental arrest are not the same, suggesting that the two alternatively spliced transcripts may have distinctly different roles. In lepidopterans, such as *B. mori*, and *M. sexta*, and the Asian corn borer (*Ostrinia furnacalis*), four alternatively spliced transcripts of CHS1 have been identified; one near the 5′- end and one close to the 3′-end (Qu and Yang 2011, 2012; Tetreau et al. 2015a, b). In addition, an alternatively spliced transcript of *CHS2* gene has also been identified in the corn earworm (*Helicoverpa zea*) (Shirk et al. 2015). One of the two transcripts is predicted to result in a truncated protein, thus lacking any enzymatic activity. The functional significance of this observation and whether similar splicing of *CHS2* transcripts occurs in other insect species remain to be explored.

9.3.2 The Catalytic Mechanism of Insect Chitin Synthase

The CHS catalyzes the transfer of sugar moieties from activated sugar donors to specific acceptors, thereby forming a glycosidic bond. The domain structure and membrane topology of CHS show that CHS is a membrane-integral enzyme consisting of 8–9 transmembrane helices (TMHs) in the N-terminal domain (domain A), a second TMH in the catalytic domain (domain B), and two TMHs in the C-terminal domain (domain C). There are several respective consensus sequences as the CHS signature motifs including EDR and Q(Q/R)XRW. Some of the conserved residues in these motifs have been implicated to be essential for the catalytic mechanism as they may be involved in the substrate protonation (Zhu et al. 2002; Moussian et al. 2005; Merzendorfer 2006).

To date, there is no crystallographic analysis of the three-dimensional structure of CHS from any arthropod species. However, structures of a bacterial cellulose synthase (NodC) from *Sinorhizobium meliloti* have been analyzed in detail (Dorfmueller et al. 2014). NodC is a CHS-like enzyme, whose catalytic mechanism appears to be consistent with the role of a conserved aspartate acting as a catalytic base in the nucleophilic attack of the 4-OH group of the acceptor sugar. During the elongation, the sugar at +1 position rotates only once during the addition of two sugars. Sequence similarities among the catalytic domains of GTF2 glycosyltransferases strongly suggest similar tertiary structures and thus similar catalytic mechanisms. Because chitin fibrils are deposited on extracellular surfaces, the nascent chitin chains are translocated across the plasma membrane. The notable structural difference is the presence of a long tunnel at the active site of both cellulose synthase and insect CHS but a closed pocket in NodC. Such structural differences may suggest that both cellulose synthase and insect CHS can accommodate several sugars of the elongating chains of cellulose and chitin, respectively, whereas NodC can accommodate only a chain with 5 sugars.

Although chitin is a linear chain of β-1,4-linked GlcNAc residues, every single sugar is rotated by 180° with respect to its neighboring sugars. In order to facilitate simultaneous polymerization, the inverse orientation of the sugar moieties may require two active sites in close proximity. In principle, the formation of the active site(s) for chitin synthesis could depend on the oligomerization of different CHS molecules. In *M. sexta*, a trimeric CHS complex was detected by partial purification of MsCHS-2 from the larval midgut (Maue et al. 2009). Because of the oligomeric state, enzyme activity might be lost during the purification, which makes it impossible for measuring its activity.

9.3.3 Regulation of Chitin Biosynthesis

9.3.3.1 Ecdysone Involved in Regulation of Insect Chitin Biosynthesis

Molting and metamorphosis in insects are regulated by 20-hydroxyecdysone (20E) and modulated by juvenile hormone (JH) (Riddiford et al. 2001). As well demonstrated in holometabolous insects, especially in *D. melanogaster*, molting is initiated by 20E and its receptor complex, ecdysone receptor (EcR) and ultraspiracle (USP) (King-Jones and Thummel 2005). The ligand–receptor complex 20E-EcR/USP directly activates a small set of early-response genes such as *E74*, *E75*, and *Broad* (Yao et al. 1993). These very early expressed transcription factors regulate a much larger set of secondary response genes such as *βFTZ-F1*, *DHR3*, *DHR39*, and *E78B*, which then activates the late genes that play direct roles in several specific processes of molting and metamorphosis (Thummel 2002). These processes include chitin synthesis and degradation, and cuticle melanization and sclerotization. Current research shows that the process of chitin synthesis and degradation behaves as an ecdysone-induced response. These genes include *CHS1* and its paralog, *CHS2*, the soluble trehalase (*TRE-1*), glucose-6-P isomerase (*G6PI*), and *UAP* (Gagou et al. 2002; Tonning et al. 2006; Yao et al. 2010; Ampasala et al. 2011; Qu and Yang 2012; Yang et al. 2013).

In *Drosophila*, the expression of the *mmy* gene encoding UAP could be prematurely up-regulated by 20E in the epidermis, salivary gland, and proventriculus (Araújo et al. 2005; Schimmelpfeng et al. 2006; Tonning et al. 2006). In agreement with this result, *LmUAP1* in chitin biosynthetic pathway is a 20E late-response gene in *L. migratoria*. Specifically, injection of 20E can induce the expression of *LmUAP1* but suppression of *LmEcRcom* expression by RNAi can result in down-regulation of *LmUAP1* (Liu et al. 2018). Furthermore, the 20E treatment and dsRNA injection to target *EcR* gene showed increased and decreased expressions of two *L. migratoria* chitinase 5 (*CHT5*) genes, *LmCHT5-1* and *LmCHT5-2*, respectively. These results suggest that both of these duplicated *LmCHT5* genes are responsive to 20E (Li et al. 2015). In *S. exigua*, 20E can up-regulate several genes including *SeTre-1*, *SeG6PI*, *SeUAP*, *SeCHSA*, and *SeCHSB* in chitin biosynthetic pathway; thus, *EcR* plays a key role in the regulation of chitin biosynthesis via inducing their expression (Yao et al. 2010). Up-regulation of the *CHS* gene expression by 20E has also been observed in other insect species, including *D. melanogaster* (Gagou et al. 2002), *M. sexta* (Zimoch et al. 2005), and *O. furnacalis* (Qu and Yang 2012). The consensus sequences of the ecdysone-inducible Broad Complex (*BR-C*) and *E74A* early genes were predicted within the promoter region of *DmCHSB* and *OfCHSB* (Gagou et al. 2002).

9.3.3.2 miRNAs Involved in Regulation of Insect Chitin Biosynthesis

MicroRNAs (miRNAs) are small noncoding regulatory RNAs of approximately 21–24 nucleotides in length. The miRNAs have been shown to contribute to the posttranscriptional regulation of gene expression in both plants and animals (Carrington and Ambros 2003; Smibert and Lai 2008), and hundreds of miRNAs have been identified in various organisms (Wienholds and Plasterk 2005; Belles et al. 2012). The miRNAs play an essential role in regulating the development, cell differentiation, apoptosis, and other critical biological events (Bartel 2009). In *D. melanogaster*, Let-7 complex (miR-100, miR-125, and Let-7) and miR-34 play key roles in the regulation of metamorphosis (Sempere et al. 2003). Research has also show the roles of miRNAs in chitin biosynthesis. In *B. mori*, for example, a genome-wide analysis revealed trehalase as a target of miR-8 (Yu et al. 2008). In *N. lugens*, the conserved miRNAs, miR-8-5p, and miR-2a-3p, negatively regulate the membrane-bound trehalase (*TRE-2*) and phosphoacetylglucosamine mutase (PAGM) genes. Moreover, the levels of miR-8-5p and miR-2a-3p are directly regulated by *BR-C*, an early-response gene, in the 20E signaling pathway (Chen et al. 2013). In *L. migratoria*, injections of miR-71 and miR-263 agomirs can suppress the expression of *LmCHS1* and *Lm CHT10*, which consequently alters the chitin production of new and old cuticles and results in a molting-defective phenotype (Yang et al. 2016).

9.4 Structural Organization and Assembly of Chitin

9.4.1 Structural Role of Chitin/Deacetylated Chitin in Assembly of Insect Cuticle

In arthropods, chitin polymers are often modified by the deacetylation of a certain proportion of *N*-acetyl-β-D-glucosamine units to yield β-1,4-linked D-glucosamine units (i.e., deacetylated units) (Zhu et al. 2016). This process alters the physicochemical properties of arthropod cuticle but the degree of deacetylation varies a great deal among different arthropod species and different body parts of the same species. Naturally occurring chitin exhibits variable degrees of deacetylation ranges from 5 to 25%. Usually, the modified chitin containing more than 80% of deacetylated D-glucosamine residues is called chitosan. To date, no enzyme is known to synthesize chitosan directly from precursors; thus, chitosan can only arise by chemical or enzymatic deacetylation of preformed chitin.

Although the role of chitin deacetylation is still not well understood, it is conceivable that partial deacetylation may render the chitin matrix more resistant to hydrolysis by endochitinases, provide amino groups for cross-linking to proteins, and generate selective binding sites for cuticle or PM proteins. In arthropods, chitin deacetylation is achieved by chitin deacetylases (CDAs; EC 3.5.1.41). Indeed, interfering with the deacetylation by RNAi of *CDA* genes can result in abnormal tracheal tubes, cuticle and joint defects, molting failure, and mortality (Yu et al. 2016). Interestingly, chitosan also has antibacterial properties that might also provide some protection against the invasion of bacteria.

9.4.2 Chitin Deacetylases and Possible Roles in Cuticle Organization

The CDAs are metalloproteins belonging to a family of carbohydrate esterases that catalyze the *N*-deacetylation of different carbohydrate substrates including chitin, xylan, and peptidoglycan (Tsigos et al. 2000). CDAs are found in all organisms producing chitin, which include fungi, nematodes, and arthropods. They are also found in bacteria and plants, where they may have digestive or defense functions. The first insect cDNA encoding a CDA-like protein was characterized from a midgut cDNA expression library of the cabbage looper (*Trichoplusia ni*) (Guo et al. 2005). Since then, studies have shown that CDAs belong to a gene family ranging from 4 to 9 members in insects (Campbell et al. 2008; Dixit et al. 2008; Xi et al. 2014; Tetreau et al. 2015b).

Recent phylogenetic analysis of CDA sequences has shown that insect CDAs can be classified into five groups based on sequence similarity and domain diversity (Yu et al. 2016) (Fig. 9.4). Except for Group V, all the CDAs contain a single chitin-binding peritrophin-A domain (perA domain, also called chitin-binding domain 2 (CBD2) at the *N*-terminal region and a polysaccharide deacetylase-like catalytic domain (CE4) at the C-terminal region. Group I is composed of CDA1 and CDA2 and Group II is composed of CDA3. In addition to a CBD2 and an EC4, these CDAs have a low-density lipoprotein receptor class A domain (LDLa), but the identities of the amino acid sequences are only about 38% between Groups I and II. Groups III and IV include CDA4 and CDA5, respectively. These CDAs lack the LDLa domain but retain the CBD2 and CE4 domains. They differ in the length of the linker sequence between the domains. Group V includes CDA6, CDA7, CDA8, and

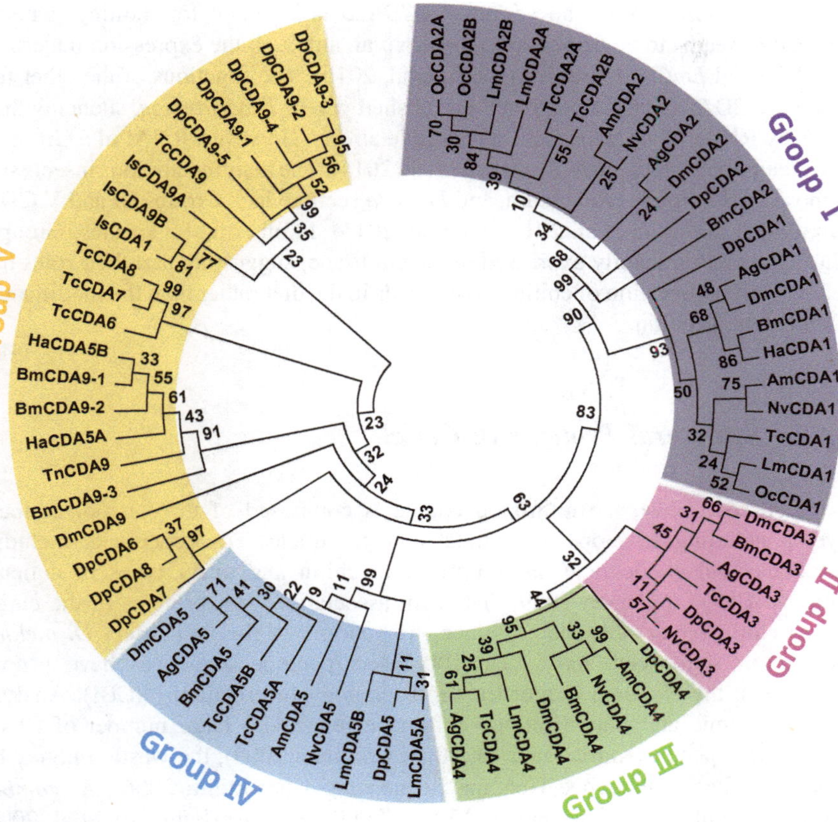

Fig. 9.4 Phylogenetic analysis of chitin deacetylases (CDAs) from 14 different arthropod species, including *Anopheles gambiae* (Ag), *Apis mellifera* (Am), *Bombyx mori* (Bm), *Choristoneura fumiferana* (Cf), *Daphnia pulex* (Dp), *Drosophila melanogaster* (Dm), *Helicoverpa armigera* (Ha), *Ixodes scapularis* (Is), *Locusta migratoria* (Lm), *Mamestra configurata* (Mc), *Nasonia vitripennis* (Nv), *Oxya chinensis* (Oc), *Tribolium castaneum* (Tc), and *Trichoplusia ni* (Tn). Different colors indicate different groups of insect CDAs

CDA9, which only possess a CE4 domain. The variety of CDAs further increases by the alternative splicing of *CDA* genes. For example, two *CDA2* alternative splicing variants (*CDA2a* and *CDA2b*) have been found in several insect species including *T. castaneum* (Arakane et al. 2009) and *L. migratoria* (Yu et al. 2016).

CDAs play important roles in insect growth and development. In *D. melanogaster*, the CDA Serpentine (Serp, CDA1) and Vermiform (Verm, CDA2) mutant embryos display elongated and tortuous tracheal tubes (Luschnig et al. 2006; Wang et al. 2006). In *T. castaneum*, multiple variants of different CDAs have been extensively studied. *TcCDA1* and *TcCDA2* are essential for the insect molting and survival (Arakane et al. 2009). *TcCDA2a* is needed for the establishment of the soft femoral-tibial joint cuticle, whereas *TcCDA2b* is involved in the formation of the hard elytra. In *L. migratoria*, the variant *LmCDA2a* is essential for molting, whereas *LmCDA2b* seems to be dispensable for survival, although the expression patterns of *LmCD2a* and *LmCD2b* are similar (Yu et al. 2016). The functions of the other four groups of CDAs have not been well established yet. In *T. castaneum*, silencing these genes by RNAi failed to produce any visible effects. However, RNAi of a Group IV CDA leads to mortality in *N. lugens* (Xi et al. 2014). The hemimetabolous insects such as those of Hemiptera, Anapleura, and *Locusta* seem to lack Groups III and V CDAs (Dixit et al. 2008; Xi et al. 2014; Yu et al. 2016). In other insect species, Group V CDAs are predominantly expressed in the gut tissue, suggesting that their roles may be involved in digestion of chitinous materials in the diet rather than the modification of endogenous chitin.

9.4.3 Structural Proteins in Cuticle

As discussed above, the arthropod cuticle is composed of three major cuticular layers, including envelope, epicuticle, and procuticle. The procuticle, including exocuticle and endocuticle, is comprised of chitin and many types of cuticular proteins (CPs) (Andersen 1979). The firm association with proteins in the cuticle makes chitin remarkably stable. Since the genome of the first insect *D. melanogaster* was sequenced, more than 100 insect genome sequences have become available at the National Center for Biotechnology Information (NCBI). Analyses of the genomic and transcriptomic data have revealed a large number of CPs in *Drosophila* species (Karouzou et al. 2007; Cornman 2009), the western honey bee (*Apis mellifera*) (Honeybee Genome Sequencing Consortium 2006), *A. gambiae* (Cornman et al. 2008; Cornman and Willis 2009), *B. mori* (Futahashi et al. 2008), and *L. migratoria* (Zhao et al. 2017). For instance, over 200 putative CP genes have been identified in *A. gambiae* (Cornman et al. 2008; Cornman and Willis 2009) and in *B. mori* (Futahashi et al. 2008), and more than 700 CPs are currently available at the cuticleDB website (http://bioinformatics2.biol.uoa.gr/cuticleDB/index.jsp).

The CPs are classified by using distinctive names derived from their specific sequence motifs and amino acid compositions (Table 9.2). Most CPs belong to the CPR family, which contains the Rebers and Riddiford Consensus that were first identified by Rebers and Riddiford (1988). The consensus sequences contain chitin-binding domain (CBD) with the amino acid residues responsible for the binding (Rebers and Willis 2001a, b; Togawa et al. 2004). The original Rebers and Riddiford motif was G-(x8)-G-x(6)-Y-x(2)-A-x-E-x-G-F-x(7)-P-x-P (R&R Consensus), where x represents any amino acid residue and the values in the parentheses indicate the number of amino acid residues (Rebers and Riddiford 1988). The CPR protein family can be further divided into three types, RR-1, RR-2, and RR-3. The CPRs with the RR-1 type domain is generally found in the soft cuticle, whereas RR-2 in hard cuticle (Andersen 1998; Willis 2010). However, a precise definition of the RR-3 type has not yet been established (Andersen 2000; Willis 2010). In *T. castaneum*, RNAi of *TcCPR18* or *TcCPR27*, both belonging to

Table 9.2 Insect cuticular proteins and their conserved structural features

Protein family	Conserved structural features	References
CPR	All with the R&R Consensus: G-x(8)-G-x(6)-Y-x(2)-A-x-E-x-G-F-x(7)-P-x-P: RR1: RR-1 motif, mainly in soft (flexible) cuticles; RR2: RR-2 motif, mainly in hard (rigid) cuticles; RR3: less well defined with no discriminating features	Rebers and Riddiford (1988), Andersen (1998, 2000), Andersen et al. (1997), Togawa et al. (2004, 2007), Willis (2010), Ioannidou et al. (2014), Zhao et al. (2017), Zhou et al. (2019) and Pan et al. (2018)
CPAP	CPAP1: With 1 CBD2 domain ($CX_{11-12}CX_5CX_{9-14}CX_{12-16}CX_{6-8}C$) CPAP3: With 3 CBD2 domains ($CX_{11-12}CX_5CX_{9-14}CX_{12-16}CX_{6-8}C$)	Jasrapuria et al. (2010, 2012), Tetreau et al. (2015b), Zhao et al. (2017), Zhou et al. (2019) and Pan et al. (2018)
Tweedle	Tweedle motif: Block I: $K_{XX}Y/F$; Block II: $K_{X4-5}FIKAP$; Block III: $KT_{XX}YVL$; Block IV: KPEVY/ HF_XKY	Guan et al. (2006), Cornman and Willis (2009), Cornman (2009), Willis (2010) and Song et al. (2016)
CPF/ CPFL	CPF: With 42–44 conserved amino acids residues: A-(LIV)-x-(SA)-(QS)-x-(SQ)-x-(IV)-(LV)-R-S-x-G-(NG)-x(3)-V-S-x-Y-(ST)-K-(TA)-(VI)-D-(TS)-(PA)-(YF)-S-SV-x-K-x-D-x-R-(VI)-(TS)-N-x-GA; CPFL: Carboxyl terminal similar to the CPFs, but lack the 42–44 conserved amino acids residues	Andersen et al. (1997), Togawa et al. (2007), Cornman (2009), Futahashi et al. (2008), Zhao et al. (2017), Zhou et al. (2019) and Pan et al. (2018)
CPLC	CPLCA, CPLCG, CPLCW, and CPLCP: Low-complexity families, all with the CPLC motif but with their own distinct sequence features (with alanine, glycine, tryptophan, and proline, respectively)	Comman and Willis (2009) and Zhou et al. (2019)
CPCFC	C-x(5)-C motifs repeated two or three times	Jensen et al. (1997), Willis (2010), Willis et al. (2012) and Vannini et al. (2015)
Apidermin	With GC-rich regions and AAPA/V, only found in *Apis mellifera*	Kucharski et al. (2007)
CPG	With glycine-rich regions, including GG repeats (GXGX, GGXG, or GGGX), only found in *Bombyx mori*	Futahashi et al. (2008)
CPH	Hypothetical CPs, only found in *Bombyx mori*	Futahashi et al. (2008)
Others	New and unclassified members	He et al. (2007), Asano et al. (2013), Dong et al. (2016) and Lu et al. (2018)

the RR-2 type, can lead to disorganized chitin laminae and pore canals as well as aberrant hind wings although RNAi of *TcCPR18* shows less pronounced defects (Arakane et al. 2012).

Several CPs contain peritrophin-A type chitin-binding motifs, which are named CP analogous to peritrophins (CPAPs). The CPAPs have been grouped into two different families, CPAP1 and CPAP3, based on whether these proteins contain either one or three CBD2 domains (Jasrapuria et al. 2012). In *T. castaneum*, the CPAP family genes are all expressed only in epithelial cells (Jasrapuria et al. 2012). The genes encoding proteins of the CPAP3 family are the ortholog of the "gasp" or "obstructor" genes originally reported in *D. melanogaster* (Barry et al. 1999; Behr and Hoch 2005). Silencing of these genes by RNAi revealed that these proteins played key roles in maintaining the structural integrity of the cuticle in different parts of an insect (Jasrapuria et al. 2012; Petkau et al. 2012; Pesch et al. 2015).

There are many other CPs with different motifs, such as CPT (or TWDL), CPF, and CPF-like (CPFL) proteins, in arthropod cuticle. The CPT family is an unconventional CP family. It is sometimes called TWDL family after the name of a dominant mutation, *TweedleD* (*TwdlD*), which alters the body shape in *D. melanogaster* (Guan et al. 2006). The Tweedle motif shows a preponderance of β-pleated and aromatic residues (tyrosine and phenylalanine) on one face within a sheet, which provides an optimal location for interaction with chitin (Iconomidou et al. 2005). The name of the CPF family was derived from a 51-residue conserved region identified first in *Tenebrio* and *Locusta* (Andersen et al. 1997). However, some CPF proteins with only a 42- to 44-residue conserved region have also been found in several insect species, and these proteins are also called CPF (Togawa et al. 2007). Nevertheless, two CPF recombinant proteins of *A. gambiae*, CPF1 and CPF3, which contain a 44-residue conserved motif at the C-terminus, are not able to bind chitin in vitro (Togawa et al. 2007). On the other hand, CPFL proteins have carboxyl terminal similar to that of the CPF proteins but lack the conserved 44 amino acid residues. These proteins are named CPG proteins due to their high content of glycine (Zhong et al. 2006; Futahashi et al. 2008). To facilitate the analysis of putative CPs, Ioannidou et al. (2014) developed an online tool, CutProtFam-Pred (http://aias.biol.uoa.gr/CutProtFam-Pred/home.php). This tool allows for relatively accurate identification and classification of the CPs based on sequence alone. To date, however, the mechanism determining the specific binding of different types of CPs to chitin has not been elucidated.

9.4.4 Chitin-Binding Proteins in Peritrophic Matrix

As described above, the arthropod PM is composed of chitin and various proteins. Despite different modes by which Type I and Type II PMs are formed, their protein components are similar. Tellam et al. (1999) proposed four classes of the PM-associated proteins based on whether or not the proteins can be easily removed from the PM. Class I proteins are loosely associated proteins which can be removed with physiological buffers; Class II proteins can be removed from the PM with mild detergents, such as sodium dodecyl sulfate; Class III proteins are released only with strong denaturants, such as urea; and finally, Class IV proteins cannot be removed by any harsh means.

Much research has been directed toward the Class III proteins also known as the PM proteins (PMPs). The PMPs largely determine the overall structure and physicochemical properties of the PM. However, PMPs vary in the composition of the PM leading to different characteristics, which reflects the adaptation to physiological and immunological demands of the organisms. To date, PMPs have been identified in several PM-derived proteomes using mass spectrometry (Campbell et al. 2008; Hu et al. 2012; Toprak et al. 2016). In addition, numerous genes encoding PMPs have been identified in various insect species (Tetreau et al. 2015b). The PMPs can bind to chitin because they have one or more copies of CBD motifs, which are referred to as peritrophin-A, B, or C domains (Tellam et al. 1999). The peritrophin-A domain, which is the predominant CBD in arthropods, is alternatively called CBD2. It has been found in numerous PMPs (Barry et al. 1999; Behr and Hoch 2005; Jasrapuria et al. 2010).

Depending on the extent of glycosylation, the PMPs can be subdivided into two large groups, non-mucin-like proteins and mucin-like proteins. Non-mucin-like proteins may be *N*-glycosylated to some degree at the asparagine (Asn) residues. The peritrophins 44 and 48 are the first reported non-mucin-like proteins from *L. cuprina*. Each of them exhibits a signal peptide for secretion and expressed exclusively by cardia cells of the larvae (Elvin et al. 1996; Schorderet et al. 1998). These two proteins are referred to as glycoproteins because putative asparagine (N)-linked glycosylation sites are present, but the degree and type of glycosylation are markedly different from the mucin-like peritrophins discussed below. The functions of the genes encoding non-mucin-like proteins have been explored by RNAi. Silencing the expression of most non-mucin-like genes do not result in nutritional deficiencies, developmental retardation or mortality. However, parenteral RNAi for *TcPMP1-C* can significantly reduce the number of eggs laid in *T. castaneum* (Agrawal et al. 2014). This may be due to slightly impaired digestion not affecting the nutritional state of the mother but rather the development of eggs, which is more sensitive to minor changes in energy stores.

The term mucin was originally used to describe the glycosylated proteins in mucous secretions lining the mammalian digestive tract and classified into membrane-bound or secreted mucins. Similarly, insects also have secreted mucins (PM associated) and membrane-bound mucins (microvilli associated). Mucin-like proteins typically contain proline (Pro), serine (Ser) and threonine (Thr) enriched motifs (mucin domains) that are presumably glycosylated. More than 80% of the Ser, Thr, and Pro residues are O-glycosylated, generating large glycoproteins and proteoglycans. Wang and Granados (1997) are the first to identify a PMP with a largely attached glycan in *T. ni* and use the terminology of the invertebrate intestinal mucin (IIM). The IIM has properties of both peritrophins and mucins that attract large amounts of water molecules. The high degree of glycosylation separates IIMs from other glycoproteins with a significantly smaller number of sugar attachments.

These findings are consistent with the reports showing that the glycans contribute up to 50% of the total mass of IIMs (Wang and Granados 1997; Shi et al. 2004; Agrawal et al. 2014). Among the 11 PMPs in *T. castaneum*, only 2 glycosylated TcPMPs (TcPMP3 and TcPMP5- B) have marked Ser/Thr-rich linker domains with long repeats (Agrawal et al. 2014). RNAi of these two genes resulted in a significant increase in PM permeability for fluorescein isothiocyanate (FITC)–dextrans, suggesting that their sugar moieties largely determine PM permeability.

9.4.5 Chitin Assembly in Cuticle and Peritrophic Matrix

Chitin assembly is also a complex process in the formation of insect cuticle and very little is known on the process. The *Drosophila* mutants of *kkv1* gene encoding CHS1 exhibit the loss of cuticle integrity, leading to defects in the formation of epidermal and tracheal cuticles. Because mutant embryos stretch to several times the size of wild-type embryos, it is called as blimp-like phenotype. In the PM, chitin microfibrils frequently are organized into microfibrillar bundles with diameters of about 20 nm (Kramer et al. 1985; Peters and Latka 1986). Several studies have demonstrated that the PM formation and function can be disrupted by RNAi of *CHS2*. In *L. migratoria*, for example, injection of *LmCHS2* dsRNA in nymphs can result in amorphous and thin midgut as compared with the controls, which ultimately leads to a cessation of feeding and a high mortality (Liu et al. 2012). RNAi of *CHS2* in *T. castaneum* also led to pronounced losses of the PM and its function (Kelkenberg et al. 2015).

Furthermore, silencing the group I *TcCDA* can result in chitin reduction, loss of laminar organization and loss of integrity of all adult cuticles, indicating that deacetylation is important for cuticle organization (Arakane et al. 2009). Similar results have also been observed in *L. migratoria*. The transmission electron microscopic analysis showed that the chitin laminae disappeared and the cuticle became thick in the insects injected with ds*LmCDA2* (Yu et al. 2016). Blimp-like phenotypes have also been reported for other genes, such as knickkopf (*knk*) and retroactive (*rtv*), which are indirectly associated with chitin synthesis. In the *knk* and *rtv* mutants of *Drosophila*, luminal chitin can still be formed, but the formation of filamentous textures is disrupted (Moussian et al. 2006). Another chitin-binding protein, Obstructor-A (CPAP3-A) may also be involved in the organization of chitin in the assembly zone, even though the chitin staining appears normal in mutants lacking this protein (Petkau et al. 2012; Pesch et al. 2015).

9.5 Chitin Degradation and Recycling

9.5.1 Biochemical Processes of Chitin Degradation

During the molting process, old cuticle is partially degraded by chitin-degrading enzymes before insect ecdysis. Chitinases (CHTs, EC 3.2.1.14) and β-*N*-

acetylglucosaminidases (NAGs, EC 3.2.1.52) are two primary types of enzymes responsible for degrading chitin. All insect CHTs that have been studied so far are endo-splitting glycosidases, which degrade chitin chains to low-molecular-weight chitooligosaccharides of different chain lengths. The chitooligosaccharides are then cleaved from the nonreducing end into monomeric *N*-acetylglucosamine (GlcNAc) by NAGs. Currently, very little is known about the reactions that strip away the matrix-associated proteins to expose the chitin microfibrils to CHTs. In the gut and molting fluid, there is a large assortment of proteases, which are believed to assist in this process.

The presence of CHT and NAG together can result in a synergistic effect on chitin catabolism in the molting fluid of *M. sexta* (Fukamizo and Kramer 1985). The rate of hydrolysis is up to six times higher than the sum of the rates observed with either enzyme alone. Nonetheless, such an enhancement of the catalytic activity by the binary enzyme system can be significantly influenced by the concentration ratio of CHT to NAG. This implies that the regulation of both enzymes is under strict hormonal and/or developmental control. Indeed, the peak of 20E (the most active molting hormone) coincides precisely with the timing of the onset of apolysis. Injection of 20E into *B. mori*, *M. sexta*, and *L. migratoria* can increase the expression of *CHT* genes, suggesting that *CHT* genes are responsive to 20E (Kramer et al. 1993; Zheng et al. 2003; Li et al. 2015). In *L. migratoria*, the transcript levels of *LmCht5-1* and *LmCht5-2* increase within 2 h, reach the peak at 6 h, and maintain high at 12 h after the injection of 20E (Li et al. 2015). Furthermore, such an induction of *LmCht5-2* expression is unaffected by cycloheximide, a protein biosynthesis inhibitor, suggesting that the effect of 20E is directly at the gene transcriptional level.

9.5.2 Diversity of Chitin Degrading Enzymes in Arthropods

The CHTs belong to the family of glycoside hydrolase 18 (GH18), an ancient gene family widely found in archaea, prokaryotes, and eukaryotes. A large number of *CHT* and *CHT*-like genes were identified by a genome-wide search in *D. melanogaster* (Zhu et al. 2004). During the last several years, a rapid increase in the number of arthropod genome sequences has provided valuable resources for a bioinformatics-based investigation of *CHT* genes. The number of *CHT* genes in insect genomes ranges from a low of seven in aphids to as many as 24 in *T. castaneum* (Zhu et al. 2004, 2008a; Nakabachi et al. 2010; Zhang et al. 2011; Pan et al. 2012; Merzendorfer 2013; Tetreau et al. 2015a). These genes differ significantly in size, developmental and tissue expression patterns, and functions. Phylogenetic comparisons of their deduced proteins have led to a classification of eight groups (I–VIII) based on their catalytic domains in insects (Zhang et al. 2011). Recent studies, however, have revealed additional groups (IX, X, and h), which results in a total of 11 groups of CHT and CHT-like proteins in insects (Fig. 9.5). Among the 11 groups, each of eight groups (II, III, VI–X, and h) is represented by a

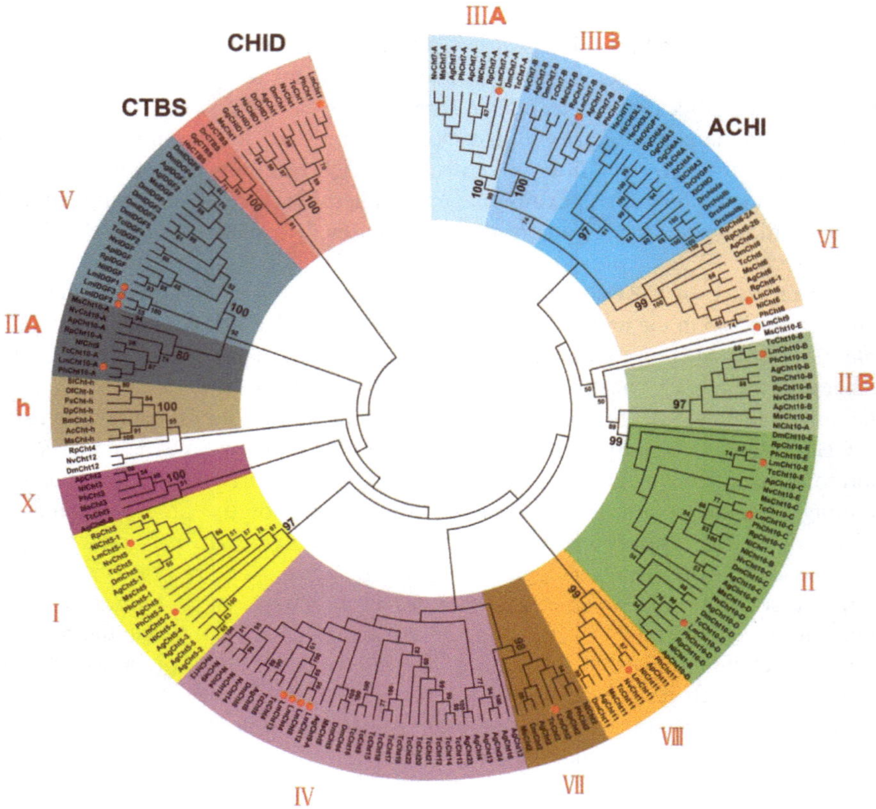

Fig. 9.5 Phylogenetic analysis of chitinases and chitinase-like proteins from 15 different animal species, including *Acyrthosiphon pisum* (Ap), *Ancylostoma ceylanicum* (Ac), *Anopheles gambiae* (Ag), *Danio rerio* (Dr), *Drosophila melanogaster* (Dm), *Gallus gallus* (Gg), *Homo sapiens* (Hs), *Locusta migratoria* (Lm), *Manduca sexta* (Ms), *Nasonia vitripennis* (Nv), *Nilaparvata lugens* (Nl), *Pediculus humanus corporis* (Ph), *Rhodnius prolixus* (Rp), *Tribolium castaneum* (Tc), and *Xenopus tropicalis* (Xt). Different colors indicate different groups of chitinases and chitinase-like proteins

single gene in each species, whereas each of the remaining three groups (I, IV, and V) has multiple CHT and CHT-like proteins. In Group I, there seems to be an expansion of *CHT5* genes (3, 4, or 5 in mosquito lineages and 2 in *L. migratoria*) (Zhang et al. 2011; Li et al. 2015). On the other hand, Group h CHTs appear to be Lepidoptera-specific (Tetreau et al. 2015a).

The domain organization and conserved motifs of insect CHTs and CHT-like proteins have been extensively reviewed (Arakane and Muthukrishnan 2010; Merzendorfer 2013). Domain analyses of the deduced proteins have shown that all the proteins encoded by these genes have a multiple-domain organization which includes 1–5 catalytic domains (GH18 domain, some of them may be catalytically inactive), 0–7 chitin-binding domains (also called CBD2), 0 or 1 leader signal peptide or transmembrane-spanning domain, and perA and serine/threonine-rich linker regions that are glycosylated. Members of most groups are thought to have chitinolytic activity. They contain the consensus sequence, DWEYP, which is considered an essential characteristic for a putative CHT protein. Group V CHT-like proteins lack this motif and; therefore, they do not have chitinolytic activity. The Group V proteins are named as imaginal disk growth factors (IDGFs) because they promote the growth of imaginal disk cells (Kawamura et al. 1999) or hemocyte agglutinating activity (Kanost et al. 1994).

Different *CHT* and *CHT*-like genes often show different developmental stage- and tissue-dependent expression patterns. In *A. gambiae*, for example, the *CHT* genes encoding proteins belonging to groups I, II, III, V, VI, VII, and VIII are expressed at almost all developmental stages from eggs through adult stages with the different expression levels, whereas the genes encoding the proteins belonging to Group IV display high complexity of expression patterns. The CHTs in groups I, II, and III are found in the molting fluid (Koga et al. 1992; Qu et al. 2014) but are expressed mainly in the integument. In *Tribolium*, it seems that all the Group IV genes are expressed in the larval gut tissues, but not in the carcass (whole body after gut and head are removed) (Arakane and Muthukrishnan 2010). In contrast, Group IV genes show diverse expression profiles in *A. gambiae*.

The distinctive functions of *CHT* genes belonging to different groups have been explored using RNAi approaches in several insect species (Zhu et al. 2008b; Zhang et al. 2012a, b; Li et al. 2015; Xi et al. 2015). In *T. castaneum*, *TcCHT5* (Group I), *TcCHT10* (Group II), *TcCHT7* (Group III), and *TcIDGF2* (Group V) are essential for insect molting and development (Zhu et al. 2008b). In *L. migratoria*, RNAi-mediated suppression of *LmCht5-1* transcript can cause severe molting defects and lethality from fifth-instar nymphs to adults, but such effects are not observed in similar RNAi studies on *LmCht5-2*, suggesting that *LmCht5-1* is essential for the development and survivorship of the locust (Li et al. 2015).

9.5.3 N-*Acetylglucosaminidases*

The enzymes primarily responsible for the production of GlcNAc monomers for chitin recycling are NAGs, which belong to CAZy glycoside hydrolase family 18 (GH18). Their genes or proteins have been identified or purified, respectively, from a variety of sources or tissues in insects, including molting fluid, hemolymph, integument, and gut (Koga et al. 1982; Nagamatsu et al. 1995; Reynolds and Samuels 1996; Zen et al. 1996; Filho et al. 2002; Leonard et al. 2006; Tomiya et al. 2006; Kokuho et al. 2007; Yang et al. 2008). To date, a total of four groups of NAGs have been identified by phylogenetic analysis, which include chitinolytic NAGs (Group I), chitinolytic NAGs (Group II), N-glycan processing NAGs (Group III), and hexosaminidases (Group IV) (Nagamatsu et al. 1995; Hogenkamp et al. 2008; Yang et al. 2008; Rong et al. 2013). Group I NAGs have been shown to possess catalytic activity in degradation of chitin oligosaccharides and are found in abundance in insects. The first crystal structure of NAG-1 (Hex-1) was determined from *O. furnacalis* (Liu et al. 2012). This enzyme exists as a homodimer and has a deeper and larger substrate-binding pocket capable of binding oligosaccharides than those of the human and bacterial enzymes.

The functions of *NAG1* gene have been explored in several insect species. In *T. castaneum*, *L. migratoria*, and *O. furnacalis*, down-regulation of *NAG1* transcript by RNAi can result in severely interrupted development (Hogenkamp et al. 2008; Liu et al. 2012; Rong et al. 2013). However, knockdown of *TcNAG2* transcript in *T. castaneum* only resulted in lower mortality, and the effects were less severe (Hogenkamp et al. 2008). RNAi of *OfHex2* in *O. furnacalis* caused nonlethal but severe abnormalities of the larval abdomen, and pupal and adult appendages (Liu et al. 2013). These effects appear to be consistent with their roles in the degradation of chitooligosaccharides. The third group of NAGs is called FDLs based on prior nomenclature of an orthologous gene from *D. melanogaster*, which has "fused lobes" phenotype (Leonard et al. 2006). Unlike other NAG genes, insect FDL genes can encode a broad spectrum of NAGs that function in glycan degradation. For example, *DmHEX3* may be involved in sperm–egg recognition and in fertilization (Cattaneo et al. 2006). The fourth group of NAGs is grouped along with hexosaminidases from mammalian and other sources. There is only limited evidence for their role in chitin metabolism, and therefore this group will not be discussed in this chapter.

9.5.4 Recycling of Chitin

A recycling pathway for GlcNAc has been characterized in *E. coli* (Park 2001). One of the components of this pathway is the enzyme *N*-acetylglucosamine kinase (EC 2.7.1.59), which catalyzes the transfer of a phosphate group from ATP to GlcNAc, producing GlcNAc-6-P. Murine and human *N*-acetylglucosamine kinase homologs have been identified (Hinderlich et al. 2000; Berger et al. 2002). However, it is unknown if a recycling mechanism exists for GlcNAc in insects. If such an enzyme exists in insects, it is conceivable that GlcNAc could reenter the chitin synthesis pathway as *N*-acetylglucosamine-6-phosphate (Zhu et al. 2016).

9.6 Concluding Remarks

Chitin biosynthesis, modification, assembly, degradation, and turnover are highly complex processes in arthropods. Although rapid advances in genomics, proteomics, RNAi, and other technologies have considerably facilitated the research in recent years, many of these processes are still partially understood. Muthukrishnan et al. (2018) recently identified many critical research questions in order to better understand these processes. Among them, the following questions appear to be most critical. How does the structural organization of CHS in plasma membrane accommodate chitin biosynthesis, transport of chitin chain across the plasma membrane, and release of the chitin chain from CHS? What role do chitin deacetylase(s) play in chitin organization within the cuticle? What is the mechanism controlling the formation of different types of chitin (i.e., α-, β-, and γ-chitin)? How do different CPs affect the chitin properties? Given the current rapid paces in research focusing on these processes, it is expected that rapid progress to address these questions will be made in the near future.

Acknowledgements We acknowledge that many relevant studies could not be cited due to space restrictions. We thank Dr. Qing Yang for her invitation to write this chapter. Relevant research conducted in the authors' laboratories was supported by the grants from the National Natural Science Foundation of China (Grant Nos. 31730074, 31672364) and the Kansas Agricultural Experiment Station, Manhattan, Kansas (KS 362, KS471). This manuscript has contribution no. 19-139-B from the Kansas Agricultural Experiment Station, Kansas State University, Manhattan, Kansas, USA.

References

Agrawal S, Kelkenberg M, Begum K, Steinfeld L, Williams CE, Kramer KJ et al (2014) Two essential peritrophic matrix proteins mediate matrix barrier functions in the insect midgut. Insect Biochem Mol Biol 49:24–34

Alvarenga ES, Mansur JF, Justi SA, Figueira-Mansur J, Dos Santos VM, Lopez SG et al (2016) Chitin is a component of the *Rhodnius prolixus* midgut. Insect Biochem Mol Biol 69:61–70

Ampasala DR, Zheng SC, Zhang DY, Ladd T, Doucet D, Krell PJ et al (2011) An epidermis-specific chitin synthase cDNA in *Choristoneura fumiferana*: cloning, characterization, developmental and hormonal-regulated expression. Arch Insect Biochem 76:83–96

Andersen SO (1979) Insect cuticle. Annu Rev Entomol 24:29–61

Andersen SO (1998) Amino acid sequence studies on endocuticular proteins from the desert locust, *Schistocerca gregaria*. Insect Biochem Mol Biol 28:421–434

Andersen SO (2000) Studies on proteins in post-ecdysial nymphal cuticle of locust, *Locusta migratoria*, and cockroach, *Blaberus craniifer*. Insect Biochem Mol Biol 30:569–577

Andersen SO, Rafn K, Roepstorff P (1997) Sequence studies of proteins from larval and pupal cuticle of the yellow mealworm, *Tenebrio moliter*. Insect Biochem Mol Biol 27:121–131

Arakane Y, Baguinon M, Jasrapuria S, Chaudhari S, Doyungan A, Kramer KJ et al (2011) Both UDP *N*-acetylglucosamine pyrophosphorylases of *Tribolium castaneum* are critical for molting, survival and fecundity. Insect Biochem Mol Biol 41:42–50

Arakane Y, Dixit R, Begum K, Park Y, Specht CA, Merzendorfer H et al (2009) Analysis of functions of the chitin deacetylase gene family in *Tribolium castaneum*. Insect Biochem Mol Biol 39:355–365

Arakane Y, Hogenkamp DG, Zhu YC, Kramer KJ, Specht CA, Beeman RW et al (2004) Characterization of two chitin synthase genes of the red fl our beetle, *Tribolium castaneum*, and alternate exon usage in one of the genes during development. Insect Biochem Mol Biol 34:291–304

Arakane Y, Lomakin J, Gehrke SH, Hiromasa Y, Tomich JM, Muthukrishnan S et al (2012) Formation of rigid, non-flight forewings (elytra) of a beetle requires two major cuticular proteins. PLoS Genet 8:e1002682

Arakane Y, Muthukrishnan S (2010) Insect chitinase and chitinase-like proteins. Cell Mol Life Sci 67:201–216

Arakane Y, Muthukrishnan S, Kramer KJ, Specht CA, Tomoyasu Y, Lorenzen MD et al (2005) The *Tribolium* chitin synthase genes *TcCHS1* and *TcCHS2* are specialized for synthesis of epidermal cuticle and midgut peritrophic matrix. Insect Mol Biol 14:453–463

Arakane Y, Specht CA, Kramer KJ, Muthukrishnan S, Beeman RW (2008) Chitin synthases are required for survival, fecundity and egg hatch in the red fl our beetle, *Tribolium castaneum*. Insect Biochem Mol Biol 38:959–962

Araújo SJ, Aslam H, Tear G, Casanova J (2005) mummy/cystic encodes an enzyme required for chitin and glycan synthesis, involved in trachea, embryonic cuticle and CNS development analysis of its role in *Drosophila* tracheal morphogenesis. Dev Biol 288:179–193

Asano T, Taoka M, Shinkawa T, Yamauchi Y, Isobe T, Sato D (2013) Identification of a cuticle protein with unique repeated motifs in the silkworm, *Bombyx mori*. Insect Biochem Mol Biol 43:344–351

Ashfaq M, Sonoda S, Tsumuki H (2007) Developmental and tissue-specifi c expression of CHS1 from *Plutella xylostella* and its response to chlorfluazuron. Pestic Biochem Physiol 89:20–30

Balbiani EG (1890) E'tudes anatomiques et histologiques sur le tube digestif des Crytops. Arch Zool Exp Gen 8:1–82

Bansal R, Mian MA, Mittapalli O, Michel AP (2012) Characterization of a chitin synthase encoding gene and effect of diflubenzuron in soybean aphid, *Aphis glycines*. Int J Biol Sci 8:1323–1334

Barbakadze N, Enders S, Gorb S, Arzt E (2006) Local mechanical properties of the head articulation cuticle in the beetle *Pachnoda marginata* (Coleoptera, Scarabaeidae). J Exp Biol 209:722–730

Barry MK, Triplett AA, Christensen AC (1999) A peritrophin-like protein expressed in the embryonic tracheae of *Drosophila melanogaster*. Insect Biochem Mol Biol 29:319–327

Bartel DP (2009) MicroRNAs: target recognition and regulatory functions. Cell 136:215–233

Becker A, Schloder P, Steele JE, Wegener G (1996) The regulation of trehalose metabolism in insects. Experientia 52:433–439

Becker B (1980) Effects of polyoxin D on in vitro synthesis of peritrophic membranes in *Calliphora erythrocephala*. Insect Biochem 10:101–106

Behr M, Hoch M (2005) Identification of the novel evolutionary conserved obstructor multigene family in invertebrates. FEBS Lett 579:6827–6833

Belles X, Cristino AS, Tanaka ED, Rubio M, Piulachs MD (2012) Insect MicroRNAs: from molecular mechanisms to biological roles. In: Gilbert LI (ed) Insect molecular biology and biochemistry. Academic, New York, pp 30–56

Bentov S, Aflalo ED, Tynyakov J, Glazer L, Sagi A (2016) Calcium phosphate mineralization is widely applied in crustacean mandibles. Sci Rep 6:22118

Berger M, Chen H, Reutter W, Hinderlich S (2002) Structure and function of *N*-acetylglucosamine kinase. Eur J Biochem 269:4212–4218

Bolognesi R, Ribeiro AF, Terra WR, Ferreira C (2001) The peritrophic membrane of *Spodoptera frugiperda*: secretion of peritrophins and role in immobilization and recycling digestive enzymes. Arch Insect Biochem Physiol 47:62–75

Bouligand Y (1965) On a twisted fibrillar arrangement common to several biologic structures. Comptes Rendus Hebdomadaires des Seances de l'Academie des Sciences, Serie D 261:4864–4867

Bouligand Y (1972) Twisted fibrous arrangements in biological-materials and cholesteric mesophases. Tissue Cell 4:189–217

Brameld KA, Shrader WD, Imperiali B, Gddard WA III (1998) Substrate assistance in the mechanism of family 18 chitinases: theoretical studies of potential intermediates and inhibitors. J Mol Biol 280:913–923

Brusca RC (2000) Unrevealing the history of arthropod diversification. Ann Mo Bot Gard 87:13–25

Campbell PM, Cao AT, Hines ER, East PD, Gordon KHJ (2008) Proteomic analysis of the peritrophic matrix from the gut of the caterpillar, *Helicoverpa armigera*. Insect Biochem Mol Biol 38:950–958

Candy DJ, Kilby BA (1962) Studies on chitin synthesis in the desert locust. J Exp Biol 39:129–140

Carrington JC, Ambros V (2003) Role of microRNAs in plant and animal development. Science 301:336–338

Cattaneo F, Pasini ME, Intra J, Matsumoto M, Briani F, Hoshi M et al (2006) Identification and expression analysis of *Drosophila melanogaster* genes encoding β-hexosaminidases of the sperm plasma membrane. Glycobiology 16:786–800

Chen J, Liang ZK, Liang YK, Pang R, Zhang WQ (2013) Conserved microRNAs miR-8-5p and miR-2a-3p modulate chitin biosynthesis in response to 20-hydroxyecdysone signaling in the brown planthopper, *Nilaparvata lugens*. Insect Biochem Mol Biol 43:839–848

Chen J, Tang B, Chen H, Yao Q, Huang X, Chen J et al (2010) Different functions of the insect soluble and membrane-bound trehalase genes in chitin biosynthesis revealed by RNA interference. PLoS ONE 5:e10133

Chen Q, Ma E, Behar KL, Xu T, Haddad GG (2002) Role of trehalose phosphate synthase in anoxia tolerance and development in *Drosophila melanogaster*. J Biol Chem 277:3274–3279

Clarke L, Temple GH, Vincent JF (1977) The effects of a chitin inhibitor-dimilin- on the production of peritrophic membrane in the locust, *Locusta migratoria*. J Insect Physi 23:241–246

Cohen E (1987) Chitin biochemistry: synthesis and inhibition. Annu Rev Entomol 32:71–93

Cohen E (2010) Chitin biochemistry: synthesis, hydrolysis and inhibition. Adv Insect Physiol 38:5–74

Cornman RS (2009) Molecular evolution of *Drosophila* cuticular protein genes. PLoS ONE 4: e8345

Cornman RS, Togawa T, Dunn WA, He N, Emmons AC, Willis JH (2008) Annotation and analysis of a large cuticular protein family with the R&R Consensus in *Anopheles gambiae*. BMC Genom 18:9–22

Cornman RS, Willis JH (2009) Annotation and analysis of low-complexity protein families of *Anopheles gambiae* that are associated with cuticle. Insect Mol Biol 18:607–622

Coutinho PM, Deleury E, Davies GJ, Henrissat B (2003) An evolving hierarchical family classification for glycosyltransferases. J Mol Bio 328:307–317

Culliney TW (2014) Chapter 8: Crop losses to arthropods. In: Pimentel D, Peshin R (eds) Integrated pest management: pesticide problems, vol. 3, Springer, pp 201–225

De Mets R, Jeuniaux C (1962) Composition of peritrophic membrane. Arch Int Physiol Biochim 70:93–96

Dixit R, Arakane Y, Specht CA, Richard C, Kramer KJ, Beeman RW et al (2008) Domain organization and phylogenetic analysis of proteins from the chitin deacetylase gene family of *Tribolium castaneum* and three other species of insects. Insect Biochem Mol Biol 38:440–451

Dong Z, Zhang W, Zhang Y, Zhang X, Zhao P, Xia Q (2016) Identification and characterization of novel chitin-binding proteins from the larval cuticle of silkworm, *Bombyx mori*. J Proteome Res 15:1435

Dorfmueller HC, Ferenbach AT, Borodkin VS, van Aalten DM (2014) A structural and biochemical model of processive chitin synthesis. J Biol Chem 289:23020–23028

Eisenhaber B, Maurer-Stroh S, Novatchkova M, Schneider G, Eisenhaber F (2003) Enzymes and auxiliary factors for GPI lipid anchor biosynthesis and posttranslational transfer to proteins. BioEssays 25:367–385

Elvin CM, Vuocolo T, Pearson RD, East IJ, Riding GA, Eisemann CH et al (1996) Characterization of a major peritrophic membrane protein, peritrophin-44, from the larvae of *Lucilia cuprina*: cDNA and deduced amino acid sequences. J Biol Chem 271:8925–8935

Fabritius HO, Sachs C, Triguero PR, Raabe D (2009) Influence of structural principles on the mechanics of a biological fiber-based composite material with hierarchical organization: the exoskeleton of the lobster *Homarus americanus*. Adv Mater Sci 21:391–400

Filho BP, Lemos FJ, Secundino NF, Pascoa V, Pereira ST, Pimenta PF (2002) Presence of chitinase and beta-*N*-acetylglucosaminidase in the *Aedes aegypti*: a chitinolytic system involving peritrophic matrix formation and degradation. Insect Biochem Mol Biol 32:1723–1729

Fukamizo T, Kramer KJ (1985) Mechanism of chitin oligosaccharide hydrolysis by the binary enzyme chitinase system in insect moulting fluid. Insect Biochem 15:1–7

Futahashi R, Okamoto S, Kawasaki H, Zhong YS, Iwanaga M, Mita K et al (2008) Genome-wide identification of cuticular protein genes in the silkworm, *Bombyx mori*. Insect Biochem Mol Biol 38:1138–1146

Gagou ME, Kapsetaki M, Turberg A, Kafetzopoulos D (2002) Stage-specific expression of the chitin synthase *DmeChSA* and *DmeChSB* genes during the onset of *Drosophila* metamorphosis. Insect Biochem Mol Biol 32:141–146

Gallai N, Salles J-M, Settele J, Vaissière BE (2009) Economic valuation of the vulnerability of world agriculture confronted with pollinator decline. Ecol Econ 68:810–821

Guan X, Middlebrooks BW, Alexander S, Wasserman SA (2006) Mutation of TweedleD, a member of an unconventional cuticle protein family, alters body shape in *Drosophila*. Proc Natl Acad Sci USA 103:16794–16799

Guo W, Li G, Pang Y, Wang P (2005) A novel chitin-binding protein identified from the peritrophic membrane of the cabbage looper, *Trichoplusia ni*. Insect Biochem Mol Biol 35:1224–1234

Harper MS, Hopkins TL (1997) Peritrophic membrane structure and secretion in European corn borer larvae (*Ostrinia nubilalis*). Tissue Cell 29:463–475

Harper MS, Hopkins TL, Czapla TH (1998) Effect of wheat germ agglutinin on formation and structure of the peritrophic membrane in European corn borer (*Ostrinia nubilalis*) larvae. Tissue Cell 30:166–176

He N, Botelho JM, Mcnall RJ, Belozerov V, Dunn WA, Mize T et al (2007) Proteomic analysis of cast cuticles from *Anopheles gambiae* by tandem mass spectrometry. Insect Biochem Mol Biol 37:135–146

Hegedus D, Erlandson M, Gillott C, Toprak U (2009) New insights into peritrophic matrix synthesis, architecture, and function. Annu Rev Entomol 54:285–302

Hinderlich S, Berger M, Schwarzkopf M, Effertz K, Reutter W (2000) Molecular cloning and characterization of murine and human *N*-acetylglucosamine kinase. Eur J Biochem 267:3301–3308

Hogenkamp DG, Arakane Y, Kramer KJ, Muthukrishnan S, Beeman RW (2008) Characterization and expression of the β-*N*-acetylhexosaminidase gene family of *Tribolium castaneum*. Insect Biochem Mol Biol 38:478–489

Hogenkamp DG, Arakane Y, Zimoch L, Merzendorfer H, Kramer KJ, Beeman RW et al (2005) Chitin synthase genes in *Manduca sexta*: characterization of a gut-specific transcript and differential tissue expression of alternately spliced mRNAs during development. Insect Biochem Mol Biol 35:529–540

Hu X, Chen L, Xiang X, Yang R, Yu S, Wu X (2012) Proteomic analysis of peritrophic membrane (PM) from the midgut of fifth-instar larvae, *Bombyx mori*. Mol Biol Rep 39:3427–3434

Huang X, Tsuji N, Miyoshi T, Motobu M, Islam MK, Alim MA et al (2007) Characterization of glutamine: fructose-6-phosphate aminotransferase from the ixodid tick, *Haemaphysalis longicornis*, and its critical role in host blood feeding. Int J Parasitol 37:383–392

Ianiro A, Giosia MD, Fermani S, Samor C, Barbalinardo M, Valle F et al (2014) Customizing properties of β-chitin in squid pen (gladius) by chemical treatments. Marine Drugs 12:5979–5992

Iconomidou VA, Willis JH, Hamodrakas SJ (2005) Unique features of the structural model of 'hard' cuticle proteins: implications for chitin-protein interactions and cross-linking in cuticle. Insect Biochem Mol Biol 35:553–560

Ioannidou ZS, Theodoropoulou MC, Papandreou NC, Willis JH, Hamodrakas SJ (2014) CutProtFam-Pred: detection and classification of putative structural cuticular proteins from sequence alone, based on profile hidden Markov models. Insect Biochem Mol Biol 52:51–59

Jasrapuria S, Arakane Y, Osman G, Kramer KJ, Beeman RW, Muthukrishnan S (2010) Genes encoding proteins with peritrophin A-type chitin-binding domains in *Tribolium castaneum* are grouped into three distinct families based on phylogeny, expression and function. Insect Biochem Mol Biol 40:214–227

Jasrapuria S, Specht CA, Kramer KJ, Beeman RW, Muthukrishnan S (2012) Gene families of cuticular proteins analogous to peritrophins (CPAPs) in *Tribolium castaneum* have diverse functions. PLoS ONE 7:e49844

Jaworski E, Wang L, Margo G (1963) Synthesis of chitin in cell-free extracts of *Prodenia eridania*. Nature 198:790

Jensen UG, Rothmann A, Skou L, Andersen SO, Roepstorff P, Højrup P (1997) Cuticular proteins from the giant cockroach, *Blaberus craniifer*. Insect Biochem Mol Biol 27:109–120

Kanost MR, Zepp MK, Ladendorff NE, Andersson LA (1994) Isolation and characterization of a hemocyte aggregation inhibitor from hemolymph of *Manduca sexta* larvae. Arch Insect Biochem Physiol 27:123–136

Karouzou MV, Spyropoulos Y, Iconomidou VA, Cornman RS, Hamodrakas SJ, Willis JH (2007) *Drosophila* cuticular proteins with the R&R Consensus: annotation and classification with a new tool for discriminating RR-1 and RR-2 sequences. Insect Biochem Mol Biol 37:754–760

Kato N, Mueller CR, Fuchs JF, Wessely V, Lan Q, Christensen BM (2006) Regulatory mechanisms of chitin biosynthesis and roles of chitin in peritrophic matrix formation in the midgut of adult *Aedes aegypti*. Insect Biochem Mol Biol 36:1–9

Kato N, Mueller CR, Wessely V, Lan Q, Christensen BM (2005) Mosquito glucosamine-6-phosphate N-acetyltransferase: cDNA, gene structure and enzyme kinetics. Insect Biochem Mol Biol 35:637–646

Kawamura K, Shibata T, Saget O, Peel D, Bryant PJ (1999) A new family of growth factors produced by the fat body and active on *Drosophila* imaginal disc cells. Development 126:211–219

Kelkenberg M, Odman-Naresh J, Muthukrishnan S, Merzendorfer H (2015) Chitin is a necessary component to maintain the barrier function of the peritrophic matrix in the insect midgut. Insect Biochem Mol Biol 56:21–28

King-Jones K, Thummel CS (2005) Nuclear receptors-a perspective from *Drosophila*. Nat Rev Genet 6:311–323

Koga D, Funakoshi T, Mizuki K, Ide A, Kramer KJ, Zen KC et al (1992) Immunoblot analysis of chitinolytic enzymes in integument and molting fluid of the silkworm, *Bombyx mori*, and the tobacco hornworm, *Manduca sexta*. Insect Biochem Mol Biol 22:305–311

Koga D, Mai MS, Dziadik-Turner C, Kramer KJ (1982) Kinetics and mechanism of exochitinase and β-*N*-acetylhexosaminidase from the tobacco hornworm, *Manduca sexta* L. (Lepidoptera: Sphingidae). Insect Biochem 12:493–499

Kokuho T, Yasukochi Y, Watanabe S, Inumuru S (2007) Molecular cloning and expression of two novel β-*N*-acetylglucosaminidases from silkworm *Bombyx mori*. Biosci Biotechnol Biochem 71:1626–1635

Kramer KJ, Corpuz L, Choi HK, Muthukrishnan S (1993) Sequence of a cDNA and expression of the gene encoding epidermal and gut chitinases of *Manduca sexta*. Insect Biochem Mol Biol 23:691–701

Kramer KJ, Dziadik-Turner C, Koga D (1985) Chitin metabolism in insects. In: Kerkut GA, Gilbert LI (eds) Comprehensive insect physiology, biochemistry, and pharmacology, vol 3. Pergamon Press, Oxford, pp 75–115

Kramer KJ, Hopkins TL, Schaefer J (1995) Applications of solids NMR to the analysis of insect sclerotized structures. Insect Biochem Mol Biol 25:1067–1080

Kramer KJ, Koga D (1986) Insect chitin: physical state, synthesis, degradation and metabolic regulation. Insect Biochem 16:851–877

Kucharski R, Maleszka J, Maleszka R (2007) Novel cuticular proteins revealed by the honey bee genome. Insect Biochem Mol Biol 37:128–134

Lehane MJ (1997) Peritrophic matrix structure and function. Annu Rev Entomol 42:525–550

Leonard R, Rendic D, Rabouille C, Wilson IB, Preat T, Altmann F (2006) The *Drosophila* fused lobes gene encodes an *N*-acetylglucosaminidase involved in *N*-glycan processing. J Biol Chem 281:4867–4875

Li D, Zhang J, Wang Y, Liu X, Ma E, Sun Y et al (2015) Two chitinase 5 genes from *Locusta migratoria*: molecular characteristics and functional differentiation. Insect Biochem Mol Biol 58:46–54

Liu T, Zhang HT, Liu FY, Wu QY, Shen X, Yang Q (2011) Structural determinants of an insect β-N-acetyl-D-hexosaminidase specialized as a chitinolytic enzyme. J Biol Chem 286:4049–4058

Liu XJ, Zhang HH, Li S, Zhu KY, Ma EB, Zhang JZ (2012) Characterization of a midgut-specific chitin synthase gene (*LmCHS2*) responsible for biosynthesis of chitin of peritrophic matrix in *Locusta migratoria*. Insect Biochem Mol Biol 42:902–910

Liu XJ, Li F, Li DQ, Ma EB, Zhang WQ, Zhu KY et al (2013) Molecular and functional analysis of UDP-*N*-acetylglucosamine pyrophosphorylases from the migratory locust, *Locusta migratoria*. PLoS One 8:e71970

Liu XJ, Sun YW, Cui M, Ma EB, Zhang JZ (2016) Molecular characteristics and functional analysis of trehalase genes in *Locusta migratoria*. Scientia Agricultura Sinica 49:4375–4386

Liu XJ, Sun YW, Li DQ, Li S, Ma EB, Zhang JZ (2018) Identification of *LmUAP1* as a 20-hydroxyecdysone response gene in the chitin biosynthesis pathway from the migratory locust, *Locusta migratoria*. Insect Sci 25:211–221

Locke M (2001) The Wigglesworth lecture: insects for studying fundamental problems in biology. J Insect Physiol 47:495–507

Locke M, Huie P (1979) Apolysis and the turnover of plasma membrane plaques during cuticle formation in an insect. Tissue Cell 11:277–291

Lu JB, Luo XM, Zhang XY, Pan PL, Zhang CX (2018) An ungrouped cuticular protein is essential for normal endocuticle formation in the brown planthopper. Insect Biochem Mol Biol 100:1–9

Luschnig S, Batz T, Armbruster K, Krasnow MA (2006) Serpentine and vermiform encode matrix proteins with chitin binding and deacetylation domains that limit tracheal tube length in *Drosophila*. Curr Biol 16:186–194

Lyonet P (1762) Trait'e Anatomique de la Chenille qui ronge le bois de Saule. Haye, Holland, La

Makki R, Cinnamon E, Gould AP (2014) The development and functions of oenocytes. Annu Rev Entomol 59:405–425

Mansur JF, Alvarenga ES, Figueira-Mansur J, Franco TA, Ramos IB, Masuda H et al (2014) Effects of chitin synthase double-stranded RNA on molting and oogenesis in the Chagas disease vector *Rhodnius prolixus*. Insect Biochem Mol Biol 51:110–121

Marschall HU, Matern H, Wietholtz H, Egestad B, Matern S, Sjövall J (1992) Bile acid *N*-acetylglucosaminidation *in vivo* and *in vitro* evidence for a selective conjugation reaction of 7 beta-hydroxylated bile acids in humans. J Clin Invest 89:1981–1987

Martins GF, Ramalho-Ortiago JM (2012) Oenocytes in insects. Inverteb Surviv J 9:139–152

Maue L, Meissner D, Merzendorfer H (2009) Purification of an active, oligomeric chitin synthase complex from the midgut of the tobacco hornworm. Insect Biochem Mol Biol 39:654–659

Merzendorfer H (2006) Insect chitin synthases: a review. J Comput Physiol B 176:1–15

Merzendorfer H (2013) Chitin synthesis inhibitors: old molecules and new developments. Insect Sci 20:121–138

Merzendorfer H, Zimoch L (2003) Chitin metabolism in insects: structure, function and regulation of chitin synthases and chitinases. J Exp Biol 206:4393–4412

Mitsumasu K, Azuma M, Niimi T, Yamashita O, Yaginuma T (2005) Membrane-penetrating trehalase from silkworm *Bombyx mori*: molecular cloning and localization in larval midgut. Insect Mol Biol 14:501–508

Moussian B, Schwarz H, Bartoszewski S, Nüsslein-Volhard C (2005) Involvement of chitin in exoskeleton morphogenesis in *Drosophila melanogaster*. J Morphol 264:117–130

Moussian B, Tång E, Tonning A, Helms S, Schwarz H, Nüsslein-Volhard C et al (2006) *Drosophila* Knickkopf and Retroactive are needed for epithelial tube growth and cuticle differentiation through their specific requirement for chitin filament organization. Development 133:163–171

Munro CA, Gow NA (2001) Chitin synthesis in human pathogenic fungi. Med Mycol 39(Suppl. 1):41–53

Muthukrishnan S, Arakane Y, Yang Q, Zhang C-X, Zhang J, Zhang W et al (2018) Future questions in insect chitin biology: a microreview. Arch Insect Biochem Physiol 98:e21454. https://doi.org/10.1002/arch.21454

Muthukrishnan S, Merzendorfer H, Arakane Y, Kramer KJ (2012) Chitin metabolism in insects. In: Biol Insect Biochem Mol (ed) Gilbert LI. Elsevier, San Diego, pp 193–235

Nagamatsu Y, Yanagisawa I, Kimoto M, Okamoto E, Koga D (1995) Purification of a chitooligosaccharidolytic β-*N*-acetylglucosaminidase from *Bombyx mori* larvae during metamorphosis and the nucleotide sequence of its cDNA. Biosci Biotechnol Biochem 59:219–225

Nakabachi A, Shigenobu S, Miyagishima S (2010) Chitinase-like proteins encoded in the genome of the pea aphid, *Acyrthosiphon pisum*. Insect Mol Biol 19:175–185

Neville AC (1975) Biology of the arthropod cuticle. Springer, Berlin

Neville AC, Luke BM (1969) A two-system model for chitin-protein complexes in insect cuticles. Tissue Cell 1:689–707

Noble-Nesbitt J (1991) Cuticular permeability and its control. In: Binnington K, Retnakaran A (eds) Physiology of the insect epidermis Melbourne. CSIRO, Australia, pp 240–251

Ono M, Kato S (1968) Amino acid composition of the peritrophic membrane in the silkworm, *Bombyx mori* L. Bull Sericult Exp Stn Jpn 23:1–8

Ostrowski S, Dierick HA, Bejsovec A (2002) Genetic control of cuticle formation during embryonic development of *Drosophila melanogaster*. Genetics 161:171–182

Palaka BK, Sapam TD, Ilavarasi AV, Chowdhury S, Sk Rajendiran, Khan MB et al (2017) Molecular cloning and characterization of phosphoacetylglucosamine mutase from *Bombyx mori*. J Entomol Zool Studies 5:1166–1178

Pan PL, Ye YX, Lou YH, Lu JB, Cheng C, Shen Y, Moussian B, Zhang CX (2018) A comprehensive omics analysis and functional survey of cuticular proteins in the brown planthopper. Proc Natl Acad Sci USA 115:5175–5180

Pan Y, Lu P, Wang Y, Yin L, Ma H, Ma G et al (2012) In silico identification of novel chitinase-like proteins in the silkworm, *Bombyx mori*, genome. J Insect Sci 12:150

Park JT (2001) Identification of a dedicated recycling pathway for an hydro-*N*-acetylmuramic acid and *N*-acetylglucosamine derived from *Escherichia coli* cell wall murein. J Bacteriol 183:3842–3847

Peneff C, Ferrari P, Charrier V, Taburet Y, Monnier C, Zamboni V et al (2001) Crystal structures of two human pyrophosphorylase isoforms in complexes with UDPGlc(Gal)NAc: role of the alternatively spliced insert in the enzyme oligomeric assembly and active site architecture. EMBO J 20:6191–6202

Pesch YY, Riedel D, Behr M (2015) Obstructor A organizes matrix assembly at the apical cell surface to promote enzymatic cuticle maturation in *Drosophila*. J Biol Chem 290:10071–10082

Peters W (1992) Peritrophic membranes, vol 30. Zoophysiology. Springer, Berlin

Peters W, Latka I (1986) Electron microscopic localization of chitin using colloidal gold labelled with wheat germ agglutinin. Histochemistry 84:155–160

Petkau G, Wingen C, Jussen LC, Radtke T, Behr M (2012) Obstructor-A is required for epithelial extracellular matrix dynamics, exoskeleton function, and tubulogenesis. J Biol Chem 287:21396–21405

Qu M, Ma L, Chen P, Yang Q (2014) Proteomic analysis of insect molting fluid with a focus on enzymes involved in chitin degradation. J Proteome Res 13:2931–2940

Qu M, Yang Q (2011) A novel alternative splicing site of class A chitin synthase from the insect *Ostrinia furnacalis* - Gene organization, expression pattern and physiological significance. Insect Biochem Mol Biol 41:923–931

Qu M, Yang Q (2012) Physiological significance of alternatively spliced exon combinations of the single-copy gene class A chitin synthase in the insect *Ostrinia furnacalis* (Lepidoptera). Insect Mol Biol 21:395–404

Raabe D, Al-Sawalmih A, Romano P, Sachs C, Brokmeier HG, Yi SB et al. (2005a) Structure and crystallographic texture of arthropod bio-composites. Icotom 14: Texture of Materials, Pts 1 and 2: 495–497 and 1665–1674

Raabe D, Romano P, Sachs C (2005b) The crustacean exoskeleton as an example of a structurally and mechanically graded biological nanocomposite material. Acta Mater 53:4281–4292

Raabe D, Romano P, Sachs C, Al-Sawalmih A, Brokmeier HG, Yi SB et al (2005c) Discovery of a honeycomb structure in the twisted plywood patterns of fibrous biological nanocomposite tissue. J Cryst Growth 283:1–7

Raabe D, Romano P, Sachs C, Fabritius H, Al-Sawalmih A, Yi S-B et al (2006) Microstructure and crystallographic texture of the chitin-protein network in the biological composite material of the exoskeleton of the lobster *Homarus americanus*. Mater Sci Eng, A 421:143–153

Ramos A, Mahowald A, Jacobs-Lorena M (1994) Peritrophic matrix of the black fly *Simulium vittatum*: formation, structure, and analysis of its protein components. J Exp Zool 268:269–281

Rebers JE, Riddiford LI (1988) Structure and expression of a *Manduca sexta* larval cuticle gene homologous to *Drosophila* cuticle genes. J Mol Biol 203:411–423

Rebers JE, Willis JH (2001a) A conserved domain in arthropod cuticular proteins binds chitin. Insect Biochem Mol Biol 31:1083–1094

Reger JF (1971) Fine structure of the surface coat of midgut epithelial cells in the homopteran *Phylloscis atra* (Fulgorid). J Submicrosc Cytol 3:353–358

Regier JC, Shultz JW, Zwick A, Hussey A, Ball B, Wetzer R et al (2010) Arthropod relationships revealed by phylogenomic analysis of nuclear protein-coding sequences. Nature 463:1079–1083

Rebers JE, Willis JH (2001b) A conserved domain in arthropod cuticular proteins binds chitin. Insect Biochem Mol Biol 31:1083–1094

Reynolds SE, Samuels RI (1996) Physiology and biochemistry of insect moulting fluid. Adv Insect Physiol 26:157–232

Richards AG, Richards PA (1977) The peritrophic membranes of insects. Annu Rev Entomol 22:219–240

Riddiford L, Cherbas P, Truman JW (2001) Ecdysone receptors and their biological actions. Vitam Horm 60:1–73

Rong S, Li D, Zhang X, Li S, Zhu KY et al (2013) β-*N*-acetylglucosaminidase gene is essential for larval-larval and larval-adult molting in *Locusta migratoria*. Insect Sci 20:109–119

Ruddal KM (1963) The chitin–protein complexes of insect cuticles. Adv Insect Physiol 1:257–313

Rudall KM, Kenchington W (1973) The chitin system. Biol Rev 48:597–636

Schimmelpfeng K, Strunk M, Stork T, Klambt C (2006) Mummy encodes an UDP-*N*-acetylglucosamine-dipohosphorylase and is required during *Drosophila* dorsal closure and nervous system development. Mech Dev 123:487–499

Schorderet S, Pearson RD, Vuocolo T, Eisemann C, Riding GA, Tellam RL (1998) cDNA and deduced amino acid sequences of a peritrophic membrane glycoprotein, 'peritrophin-48', from the larvae of *Lucilia cuprina*. Insect Biochem Mol Biol 28:99–111

Sempere LF, Sokol NS, Dubrovsky EB, Berger EM, Ambros V (2003) Temporal regulation of microRNA expression in *Drosophila melanogaster* mediated by hormonal signals and Broad-Complex gene activity. Dev Biol 259:9–18

Shao L, Devenport M, Jacobs-Lorena M (2001) The peritrophic matrix of hematophagous insects. Arch Insect Biochem Physiol 47:119–125

Shi X, Chamankhah M, Visal-Shah S, Hemmingsen SM, Erlandson M, Braun L et al (2004) Modeling the structure of the type I peritrophic matrix: characterization of a *Mamestra configurata* intestinal mucin and a novel peritrophin containing 19 chitin binding domains. Insect Biochem Mol Biol 34:1101–1115

Shi JF, Fu J, Mu LL, Guo WC, Li GQ (2016a) Two *Leptinotarsa* uridine diphosphate *N*-acetylglucosamine pyrophosphorylases are specialized for chitin synthesis in larval epidermal cuticle and midgut peritrophic matrix. Insect Biochem Mol Biol 68:1–12

Shi ZK, Liu X, Xu Q, Qin Z, Wang S, Zhang F et al (2016b) Two novel soluble trehalase genes cloned from *Harmonia axyridis* and regulation of the enzyme in a rapid changing temperature. Comput Biochem Physiol 198:10–18

Shirk PD, Perera OP, Shelby KS, Furlong RB, LoVullo ED, Popham HJR (2015) Unique synteny and alternate splicing of the chitin synthases in closely related heliothine moths. Gene 574:121–139

Silva CP, Terra WR (1995) An α-glucosidase from perimicrovillar membranes of *Dysdercus peruvianus* (Hemiptera: Pyrrhocoridae) midgut cells. Purification and properties. Insect Biochem Mol Biol 25:487–494

Smibert P, Lai EC (2008) Lessons from microRNA mutants in worms, flies and mice. Cell Cycle 7:2500–2508

Song TQ, Yang ML, Wang YL, Liu Q, Wang HM, Zhang J et al (2016) Cuticular protein LmTwdl1 is involved in molt development of the migratory locust. Insect Sci 23:520–530

Specht CA, Liu Y, Robbins PW, Bulawa CE, Iartchouk N, Winter KR et al (1996) The chsD and chsE genes of *Aspergillus nidulans* and their roles in chitin synthesis. Funfal Genet Bio 20:153–167

Takiguchi M, Niimi T, Su ZH, Yaginuma T (1992) Trehalase from male accessory gland of an insect, *Tenebrio molitor*, cDNA sequencing and developmental profile of the gene expression. Biochem J 288:19–22

Tang B, Wei P, Zhao L, Shi Z, Shen Q, Yang M et al (2016) Knockdown of five trehalase genes using RNA interference regulates the gene expression of the chitin biosynthesis pathway in *Tribolium castaneum*. BMC Biotechnol 16:67

Tellam RL, Vuocolo T, Johnson SE, Jarmey J, Pearson RD (2000) Insect chitin synthase cDNA sequence, gene organization and expression. Eur J Biochem 267:6025–6043

Tellam RL, Wijffels G, Willadsen P (1999) Peritrophic matrix proteins. Insect Biochem Mol Biol 29:87–101

Terra WR (2001) The origin and functions of the insect peritrophic membrane and peritrophic gel. Arch Insect Biochem Physiol 47:47–61

Tetreau G, Cao XL, Chen YR, Muthukrishnan S, Jiang H, Blissard GW et al (2015a) Overview of chitin metabolism enzymes in *Manduca sexta*: identification, domain organization, phylogenetic analysis and gene expression. Insect Biochem Mol Biol 62:114–126

Tetreau G, Dittmer NT, Cao X, Agrawal S, Chen YR, Muthukrishnan S et al (2015b) Analysis of chitin-1 binding proteins from *Manduca sexta* provides new insights into evolution of peritrophin A type chitin-binding domains in insects. Insect Biochem Mol Biol 62:27–41

Thompson SN (2002) Trehalose – the insect 'blood' sugar. Adv Insect Physiol 31:205–285

Thummel CS (2002) Ecdysone-regulated puff genes 2000. Insect Biochem Mol Biol 32:113–120

Togawa T, Dunn WA, Emmons AC, Willis JH (2007) CPF and CPFL, two related gene families encoding cuticular proteins of *Anopheles gambiae* and other insects. Insect Biochem Mol Biol 37:675–688

Togawa T, Natkato H, Izumi S (2004) Analysis of the chitin recognition mechanim of cuticle proteins from the soft cuticle of the silkworm, *Bombyx mori*. Insect Biochem Mol Biol 34:1059–1067

Tomiya N, Narang S, Park J, Abdul-Rahman B, Choi O, Singh S et al (2006) Purification, characterization, and cloning of a *Spodoptera frugiperda* Sf9 β-*N*-acetylhexosaminidase that hydrolyzes terminal *N*-acetylglucosamine on the *N*-glycan core. J Biol Chem 281:19545–19560

Tonning A, Helms S, Schwarz H, Uv AE, Moussian B (2006) Hormonal regulation of mummy is needed for apical extracellular matrix formation and epithelial morphogenesis in *Drosophila*. Development 133:331–341

Toprak U, Erlandson M, Baldwin D, Karcz S, Wan L, Coutu C et al (2016) Identification of the *Mamestra configurata* (Lepidoptera: Noctuidae) peritrophic matrix proteins and enzymes involved in peritrophic matrix chitin metabolism. Insect Sci 23:656–674

Tsigos I, Martinou A, Kafetzopoulos D, Bouriotis V (2000) Chitin deacetylases: new, versatile tools in biotechnology. Trends Biotechnol 18:305–312

Valdivieso MH, Duran A, Roncero C (1999) Chitin synthases in yeast and fungi. Experientia Supplementrum 87:55–69

Vannini L, Bowen JH, Reed TW, Willis JH (2015) The CPCFC cuticular protein family: Anatomical and cuticular locations in *Anopheles gambiae* and distribution throughout Pancrustacea. Insect Biochem Mol Biol 65:57–67

Vincent JF, Wegst UG (2004) Design and mechanical properties of insect cuticle. Arthropod Struct Dev 33:187–199

Wang P, Granados RR (1997) An intestinal mucin is the target substrate for a baculovirus enhancin. Proc Natl Acad Sci USA 94:6977–6982

Wang P, Granados RR (2000) Calcofluor disrupts the midgut defense system in insects. Insect Biochem Mol Biol 30:135–143

Wang S, Jayaram SA, Hemphala J, Senti KA, Tsarouhas V, Jin H et al (2006) Septatejunction-dependent luminal deposition of chitin deacetylases restricts tube elongation in the *Drosophila* trachea. Curr Biol 16:180–185

Wang Y, Fan HW, Huang HJ, Xue J, Wu WJ, Bao YY et al (2012) Chitin synthase 1 gene and its two alternative splicing variants from two sap-sucking insects, *Nilaparvata lugens* and *Laodelphax striatellus* (Hemiptera: Delphacidae). Insect Biochem Mol Biol 42:637–646

Wienholds E, Plasterk RH (2005) MicroRNA function in animal development. FEBS Lett 579:5911–5922

Wigglesworth VB (1930) The formation of the peritrophic membrane in insects, with special reference to the larvae of mosquitoes. Q J Microsc Sci 73:593–616

Willis JH (2010) Structural cuticular proteins from arthropods: annotation, nomenclature, and sequence characteristics in the genomics era. Insect Biochem Mol Biol 40:189–204

Willis JH, Papandreou NC, Iconomidou VA, Hamodrakas SJ (2012) Cuticular proteins. In: Li Gilbert (ed) Insect molecular biology and biochemistry. Academic, San Diego, pp 134–166

World Health Organization (2017) World malaria report 2017. Switzerland, World Health Organization, Geneva, p 160

Xi Y, Pan PL, Ye YX, Yu B, Zhang CX (2014) Chitin deacetylase family genes in the brown planthopper, *Nilaparvata lugens* (Hemiptera: Delphacidae). Insect Mol Biol 23:695–705

Xi Y, Pan PL, Ye YX, Yu B, Xu HJ, Zhang CX (2015) Chitinase-like gene family in the brown planthopper, *Nilaparvata lugens*. Insect Biochem Mol Biol 24:29–40

Yang ML, Wang YL, Jiang F, Song TQ, Wang HM, Liu Q et al (2016) miR-71 and miR-263 Jointly regulate target genes chitin synthase and chitinase to control locust molting. PLoS Genet 12:e1006257

Yang Q, Liu T, Liu FY, Qu MB, Qian XH (2008) A novel β-*N*-acetyl-D-hexosaminidase from the insect *Ostrinia furnacalis* (Guenée). FEBS J 275:5690–5702

Yang WJ, Wu YB, Chen L, Xu KK, Xie YF, Wang JJ (2015) Two chitin biosynthesis pathway genes in *Bactrocera dorsalis* (Diptera: Tephritidae): molecular characteristics, expression patterns, and roles in larval-pupal transition. J Econ Entomol 108:2433–2442

Yang WJ, Xu KK, Cong L, Wang JJ (2013) Identification, mRNA expression, and functional analysis of chitin synthase 1 gene and its two alternative splicing variants in oriental fruit fly, *Bactrocera dorsalis*. Int J Biol Sci 9:331–342

Yao Q, Zhang DW, Tang B, Chen J, Chen J, Lu L et al (2010) Identification of 20-hydroxyecdysone late-response genes in the chitin biosynthesis pathway. PLoS ONE 5:e14058

Yao TP, Forman BM, Jiang Z, Cherbas L, Chen JD, McKeown M et al (1993) Functional ecdysone receptor is the product of EcR and Ultraspiracle genes. Nature 366:476–479

Yu RR, Liu WM, Li DQ, Zhao XM, Ding GW, Zhang M et al (2016) Helicoidal organization of chitin in the cuticle of the migratory locust requires the function of the chitin deacetylase2 enzyme (LmCDA2). J Biol Chem 291:24352–24363

Yu X, Zhou Q, Li SC, Luo Q, Cai Y, Lin WC et al (2008) The silkworm (*Bombyx mori*) microRNAs and their expressions in multiple developmental stages. PLoS ONE 3:e2997

Zen KC, Choi HK, Nandigama K, Muthukrishnan S, Kramer KJ (1996) Cloning, expression and hormonal regulation of an insect β-*N*-acetylglucosaminidase gene. Insect Biochem Mol Biol 26:435–444

Zhang J, Liu X, Li D, Sun Y, Guo Y, Ma E et al (2010) Silencing of two alternative splicing derived mRNA variants of chitin synthase 1 gene by RNAi is lethal to the oriental migratory locust, *Locusta migratoria manilensis* (Meyen). Insect Biochem Mol Biol 40:824–833

Zhang J, Zhang X, Arakane Y, Muthukrishnan S, Kramer KJ, Ma E et al (2011) Comparative genomic analysis of chitinase and chitinase-like genes in the African malaria mosquito (*Anopheles gambiae*). PLoS ONE 6:e19899

Zhang D, Chen J, Yao Q, Pan Z, Chen J, Zhang W (2012a) Functional analysis of two chitinase genes during the pupation and eclosion stages of the beet armyworm *Spodoptera exigua* by RNA interference. Arch Insect Biochem Physiol 79:220–234

Zhang X, Zhang J, Park Y, Zhu KY (2012b) Identification and characterization of two chitin synthase genes in African malaria mosquito, *Anopheles gambiae*. Insect Biochem Mol Biol 42:674–682

Zhao LN, Yang MM, Shen QD, Shi ZK, Wang SG, Tang B (2016) Knockdown of three trehalases regulating trehalose and chitin metabolism in the rice brown planthopper *Nilaparvata lugens*. Sci Rep 6:27841

Zhao XM, Gou X, Qin ZY, Li DQ, Wang Y, Ma EB et al (2017) Identification and expression of cuticular protein genes based on *Locusta migratoria* transcriptome. Sci Rep 7:45462

Zheng YP, Retnakaran A, Krell PJ, Arif BM, Primavera M, Feng QL (2003) Temporal, spatial and induced expression of chitinase in the spruce budworm, *Choristoneura fumiferana*. J Insect Physiol 49:241–247

Zhong Y-S, Mita K, Shimada T, Kawasaki H (2006) Glycine-rich protein genes, which encode a major component of the cuticle protein genes in *Bombyx mori*. Insect Biochem Mol Biol 36:99–110

Zhou Y, Badgett MJ, Orlando R, Willis JH (2019) Proteomics reveals localization of cuticular proteins in *Anopheles gambiae*. Insect Biochem Mol Biol 104:91–105

Zhu KY, Merzendorfer H, Zhang W, Zhang J, Muthukrishnan S (2016) Biosynthesis, turnover, and functions of chitin in insects. Annu Rev Entomol 61:177–196

Zhu Q, Arakane Y, Banerjee D, Beeman RW, Kramer KJ, Muthukrishnan S (2008a) Domain organization and phylogenetic analysis of the chitinase-like family of proteins in three species of insects. Insect Biochem Mol Biol 38:452–466

Zhu Q, Arakane Y, Beeman RW, Kramer KJ, Muthukrishnan S (2008b) Functional specialization among insect chitinase family genes revealed by RNA interference. Proc Natl Acad Sci USA 105:6650–6655

Zhu Q, Deng Y, Vanka P, Brown SJ, Muthukrishnan S, Kramer KJ (2004) Computational identification of novel chitinase-like proteins in the *Drosophila melanogaster* genome. Bioinformatics 20:161–169

Zhu YC, Specht CA, Dittmer NT, Muthukrishnan S, Kanost MR, Kramer KJ (2002) Sequence of a cDNA and expression of the gene encoding a putative epidermal chitin synthase of *Manduca sexta*. Insect Biochem Mol Biol 32:1497–1506

Zimoch L, Hogenkamp DG, Kramer KJ, Muthukrishnan S, Merzendorfer H (2005) Regulation of chitin synthesis in the larval midgut of *Manduca sexta*. Insect Biochem Mol Biol 35:515–527

Zimmermann U, Mehlan D, Peters W (1975) Investigations on the transport function and structure of peritrophic membranes; V - Amino acid analysis and electron microscopic investigations of the peritophic membranes of the blowfly *Calliphora erythrocephala* Mg. Comp Biochem Physiol 51B:181–186

Chapter 10
Nematode Chitin and Application

Qi Chen and Deliang Peng

Abstract Plant-parasitic nematode infection is a global problem for agriculture and forestry. There is clearly a need for novel nematicides, because of the pitifully small repertoire of nematicides available and the effectiveness of losing or environmental prohibition of these nematicides. Chitin is the essential component of nematode eggshell and pharynx. The disturbance of chitin synthesis or hydrolysis led to nematode embryonic lethal, laying defective eggs or moulting failure. Thus, the key components in the chitin metabolic process are promising targets for anti-nematode agent's development. In this chapter, we focus on chitin and chitin synthase of nematodes, chitinases and their roles in nematode survival and application of chitin in nematode control.

Keywords Plant-parasitic nematode · Chitin · Chitinase · Chitin synthase

10.1 Introduction of Plant-Parasitic Nematodes

Nematodes are roundworms that constitute the phylum Nematoda. Only a small fraction of nematode genera contains plant-parasitic or animal-parasitic species, and the majority of nematodes are free-living (Siddique and Grundler 2018; van Megan et al. 2009). Plant-parasitic nematodes (PPNs) are devastating pathogens of plants that cause considerable yield losses of food and fibre crops, with an estimated value of $157 billion per year (Abad et al. 2008). Presently, more than 4100 PPN species have been described (Nicol et al. 2011). However, the full worldwide yield losses caused by PPNs are probably underestimated, particularly in developing countries, because growers and farmers are often unaware of the presence of phytoparasitic nematodes. Additionally, the symptoms such as reduced growth, stunting, choruses, mid-day wilting, leaf drop, small fruit, yellowing, curling and twisting of leaves and

---

Q. Chen · D. Peng (✉)
Institute of Plant Protection, Chinese Academy of Agricultural Sciences, Beijing 100193, China
e-mail: dlpeng@ippcaas.cn

Q. Yang and T. Fukamizo (eds.), *Targeting Chitin-containing Organisms*, Advances in Experimental Medicine and Biology 1142,
https://doi.org/10.1007/978-981-13-7318-3_10

stems, galls, stubby roots caused by PPNs are often non-specific, making it difficult to attribute crop losses to nematode damage (Siddique and Grundler 2018).

The first-stage and second-stage juveniles of PPNs (J1 and J2) occur in the egg. The infective second-stage juveniles hatched from eggs and moved to host plants attracted by gradients of chemical cues (Dusenbery 1997). After the invasion of the host plant roots, a permanent feeding site is established within the root around them. Nematodes feed, grow and three times moult to the adult stage. Adult males emerged from the root, while the females remain in the roots and laying eggs. The eggs are surrounded by an eggshell whose strength is provided by a chitinous layer (Bird and Self 1995).

Plant-parasitic nematodes possess a hollow spear called a stylet with which the nematode pierces cell wall to suck cytoplasmic contents from plant cell and release both proteinaceous (effectors) and non-proteinaceous molecules. The hollow stylet is connected to three enlarged, specialized oesophageal gland cells, which produce the effector molecules that are secreted in the host tissues to facilitate parasitism. Each of the three oesophageal glands consists of a single cell that contains an unusually long cytoplasmic extension ending in an ampulla. The effector proteins are synthesized in the gland cell and transported to the ampulla in membrane-bound granules. The ampulla, in turn, is connected to the lumen of the oesophagus by a valve. Some of the genes encoding oesophageal secretions are likely to have been acquired from prokaryotic microbes via horizontal gene transfer (van Megen et al. 2009; Siddique and Grundler 2018).

Plant-parasitic nematodes feed on a range of plant tissues including roots, flowers, stems and leaves, the majority of species feed on roots. Based on their feeding habits, PPNs can be broadly categorized as either endoparasitic or ectoparasitic nematodes. The most damaging to agriculture and most economically important secondary endoparasitic nematodes are root-knot nematodes (Meloidogyne spp.) and cyst nematodes (Heterodera spp. and Globodera spp.) (Siddique and Grundler 2018).

Once infective second-stage juveniles of root-knot and cyst nematode hatched, it migrates through the soil towards a host plant where it invades the plant root tissue near the tip and migrates towards the vascular cylinder, where root-knot nematodes (RKNs) induce the formation of 5–7 giant cells, cyst nematodes (CNs) induce the formation of a syncytium,. In the case of RKNs, proliferation of the tissue surrounding the nematode and the giant cells leads to the formation of a typical gall. The establishment of feeding sites (giant cells and syncytia) enables RKNs and CNs to take in large quantities of nutrients from the plant, facilitates nematode growth, and induces a pathologically disturbed allocation of photosynthetic products that reduce plant growth and yield (Siddique and Grundler 2018).

10.2 Nematode Cuticle

The nematode cuticle is a multilayered, proteinaceous structure secreted by the hypodermis, which has a variety of important functions in nematode biology. The protein of the cuticle is a type of collagen, the same class of protein as found in tendons of human beings. The cuticle maintains the body shape, provides a strong layer against which muscles can act during locomotion, and protects the nematode from the external environment. The cuticle is overlaid with a fuzzy coating material —the surface coat (SC). Substances on the surface of the cuticle are in direct contact with the outside world and therefore, have important roles in the interactions of various life stages with their environment, including the host and microorganisms. The eggs are generally the most resistant nematode life stage and many have a remarkable capacity for survival in stressful environments. The eggshell is, therefore, a most important protective barrier (Curtis et al. 2011).

10.3 Eggshell

Most nematode eggs are very similar to each other, morphologically and in size (average of 53–133 μm in length and 17–79 μm in width), irrespective of the size of the adult. They are ellipsoidal in shape with a transparent shell (Curtis et al. 2011). The eggshells of plant-parasitic nematodes typically consist of three layers, an outer uterine layer, a middle chitinous layer and an inner lipid layer (Bird and Bird 1991). In plant-parasitic nematodes, the composition of the uterine layer resembles the gelatinous matrix (GM) secreted from various organs (Curtis et al. 2011). The chitinous layer is often the thickest layer and provides structural rigidity to the eggshell. Protein is frequently present in this layer; it has been estimated that the eggshell of *Globodera rostochiensis* contains 59% protein and 9% chitin while that of *Meloidogyne javanica* contains 50% protein and 30% chitin (Clarke et al. 1967; Bird and Mcclure 1976). It is believed that the subdivision of the chitinous layers of cyst nematodes in distinctive outer and inner components reflects differences in their chemical composition. Chitin is synthesized by chitin synthase and the gene encoding this enzyme has been shown to be expressed in the laying adult females and fertilized eggs of various nematodes such as *Caenorhabditis elegans*, *Meloidogyne artiellia*, *Ascaris suum*, *Brugia malayi* and *Dirofilaria immitis* (Veronico et al. 2001; Harris and Fuhrman 2002; Curtis et al. 2011).

The inner lipid layer is responsible for the extreme impermeability of the nematode eggshell. In *Meloidogyne javanica*, it is formed in the middle region of the uterus, where proline-containing proteins are incorporated into both lipid and chitinous layers, to complete the synthesis of the eggshell. Nematode eggs are permeable to chemicals prior to the formation of the lipid layer, and also when this layer is broken down before hatching. Egg permeability changes are central to the hatching of cyst nematodes(Bird and Mcclure 1976; Perry and Clarke 1981, Curtis et al. 2011). This inference is corroborated by treatment with the biological nematicide DiTera® (fermentation product of killed *Myrothecium verrucaria*), which induces significant inhibition of hatching in *Globodera rostochiensis*, apparently inhibiting the eggshell permeability change by competitively blocking Ca^{2+}-binding sites on the eggshell (Twomey et al. 2000). DiTera® does not inhibit hatching in *Meloidogyne incognita* implying that it does not affect J2 directly during the hatching process or inhibit the action of enzymes. It has been suggested that the breakdown of the lipid layer is enzymatic, and in J2s of *Meloidogyne incognita* and *Meloidogyne javanica*, secretions have been shown to emanate from the amphids, secretory–excretory pore, and from around the mouth, while still inside the egg. An increase in size of the nucleolus of the dorsal pharyngeal glands was demonstrated in J2s of *Globodera rostochiensis* inside eggs that were stimulated with root diffusate, suggesting that the oesophageal glands were being activated. Leucine aminopeptidase activity was identified in the supernatant of eggs of *Heterodera glycines* (Curtis et al. 2011; Atkinson et al. 1987; Premachandran et al.1988). The eggs within *Heterodera glycines* cysts internally contaminated with fungus were observed that it had no inner lipid layer and believed that fungal lipases might have contributed to the disruption of both inner and outer lipid layers (Curtis et al. 2011).

Due to both the lipid layer and the juvenile epicuticle are derived directly from the secondary vitelline membrane of the embryo, both probably share the same protein moieties. Cross-reactivity of polyclonal and monoclonal antibodies produced against *Meloidogyne incognita* J2 within eggshells support this suggestion (Sharon et al. 2009; Curtis et al. 2011).

10.4 Chitin and Chitin Synthase

The presence of chitin has been documented in a variety of nematode species and tissues (Veronico et al. 2001). Chitin is a component of the eggshells of many nematodes, including the plant-parasitic nematodes *Meloidogyne javanica* (Bird and Mcclure 1976), *Meloidogyne artiellia* (Fanelli et al. 2005), *Meloidoderita kirjanovae* (Bird and Mcclure 1976), *Globodera rostochiensis* (Clarke et al. 1967), *Heterodera glycines* (Burgwyn et al. 2003), and *Tylenchulus semipenetrans* (Bird and Mcclure 1976), and has also been detected in the gelatinous matrix of RKN *Meloidogyne javanica* (Spiegel and Cohn 1985). Chitin is also present in the eggshells of many animal-parasitic nematodes (APNs) including *Ascaris suum* (Dubinský et al. 1986), *Brugia malayi* (Fuhrman et al. 1992), *Onchocerca volvulus* (Wu et al. 2001) and *Acanthocheilonema viteae* (Adam et al. 1996). In the model organism *Caenorhabditis elegans*, chitin is present in the eggshell. Synthesis of chitin in the eggshell of *Caenorhabditis elegans* is initiated at fertilization and chitin is deposited before the zygote exits the spermatheca. The synthesis of eggshell chitin prevents polyspermy and is essential for further embryonic development (Johnston et al. 2010).

Nematode chitin is synthesized by chitin synthases. Initial molecular studies in PPNs *Meloidogyne artiellia* (Veronico et al. 2001) and APNs *Brugia malayi* (Harris et al. 2000) and *Dirofilaria immitis* (Harris and Fuhrman 2002) revealed the presence of a single chitin synthase gene but more recently, two chitin synthase genes have been identified in filarial nematodes *Brugia malayi* and *Onchocerca volvulus* (Foster et al. 2005). However, the exact number of chitin synthases in PPNs is unknown. In fact, five putative chitin synthase genes (Minc3s01800g26401, Minc3s00767g17084, Minc3s00218g07846, Minc3s02226g28968 and Minc3s03844g34963) have been identified in BLAST searches of the WormBase genome of *Meloidogyne incognita* using the *Caenorhabditis elegans* chitin synthase gene (F48A11.1) as the template. The specific functions of these chitin synthase genes should be further studied.

Two putative chitin synthases (encoded by T25G3.2 and F48A11.1) exist in the whole genome of the free-living nematode *Caenorhabditis elegans*(Veronico et al. 2001; Harris et al. 2000). F48A11.1 is expressed in the pharynx and its RNA transcription is correlated with larval moulting, while T25G3.2 transcripts were found in late larvae and adult hermaphrodites (Veronico et al. 2001). The requirement of chitin for *Caenorhabditis elegans* development by revealing its distribution pattern in whole worms and studying the roles of the two predicted chitin synthases were demonstrated. This results showed that chitin is an indispensable component of the eggshell and the pharynx (Zhang et al. 2005). The two chitin synthases have non-overlapping functions. Using RNAi-mediated knock out the *Caenorhabditis elegans* chitin synthase gene function, Hanazawa et al. (2001) had confirmed that the essential role of chitin synthase in producing chitin for the eggshell, which resulted in sterile hermaphrodites that lay defective eggs (Hanazawa et al. 2001; Curtis et al. 2011).

In *Meloidogyne*, maturing oocytes increase in size and become arranged in single file within the oviduct. Movement of the maturing oocytes down the female reproductive tract causes them to move into the spermatheca, where fertilization occurs. The eggs within the nematode are therefore at a sequence of developmental stages and it has been reported that, in *Meloidogyne javanica*, synthesis of the chitinous layer continues until the embryonated eggs achieve their final shape. RNAi silencing of a chitin synthase gene expressed in the eggs of the RKN *Meloidogyne artiella* was achieved by soaking alive eggs matrix in a dsRNA solution and led to a reduction in stainable chitin in eggshells, and delayed hatching of second stage juveniles (Fanelli et al. 2005). The potential for using bacterial and fungal chitinases and chitin synthase inhibitors to control RKNs has been demonstrated (Spiegel and Chet 1985; Jung et al. 2002; van Nguyen et al. 2007).

In addition to its presence in the eggshells of nematodes, chitin has also been detected in the pharynx of *Caenorhabditis elegans* (Zhang et al. 2005). A chitin synthase gene is expressed in cells that form the pharynx of *Caenorhabditis elegans* before the moult. It was suggested that this gene might be involved in the synthesis of the feeding apparatus, which is replaced during each moult (Veronico et al. 2001). The APN *Oesophagostomum dentatum* also possessed pharyngeal chitin (Neuhaus et al. 1997), suggesting that enzymatic remodelling of pharyngeal chitin occurs during moulting.

10.5 Chitinases

Chitin is the main component of the fungal cell wall. The fungivorous nematode *Bursaphelenchus xylophilus* has several chitinase genes in its genome. Six of these genes possess the GH18 catalytic domain and a conserved DxxDxDxE catalytic motif supports their enzymatic activity. Knockdown of a chitinase gene of *Bursaphelenchus xylophilus*, Bx-chi-2, by RNAi decreased the speed of clearance of the fungal mycelium support this suggestion (Ju et al. 2016).

Obligate phytoparasitic nematodes that do not feed on fungi, also have chitinase genes. A chitinase full-length cDNA (designated Hg-chi-1) was isolated from a *Heterodera glycines* oesophageal gland cell-specific long-range PCR cDNA library. Gao et al. (2002) cloned and identified *Hg-chi-1* gene, it contains an open reading frame encoding 350 amino acids with the first 23 amino acids being a putative signal peptide for secretion. *Hg-chi-1* contains a GH18 family catalytic domain. Hg-CHI-1 is expressed specifically in the subventral glands and is likely to be secreted from the stylet (Gao et al. 2002). it was detected that this gene expression only in parasitic stages, implying that it has a function in the parasitic process. However, plants do not contain chitin and the role of the chitinase in *Heterodera glycines*, therefore, remains to be determined (Gao et al. 2002). Recent studies of the transcriptional responses of wheat and the cereal cyst nematodes demonstrated that expression of an effector gene *c72543*, for a protein homologous to Hg-CHI-1, is up-regulated at the early contact stage and functions to suppress the plant's defences by inhibiting BAX-triggered programmed cell death (Chen et al. 2017).

It is possible that chitinases in nematodes serve as antifungal defences for free-living species or are involved in eggshell degradation, as well as being effectors for fungivorous nematodes. The genome sequence of *Meloidogyne incognita* displays a radical reduction in chitinases and chitin-binding proteins in this species. *M incognita* possesses only 15 enzymes potentially involved in chitin degradation and binding, whereas *Caenorhabditis elegans* has 96 such enzymes (Abad et al. 2008). So, it is believed that the reason for this reduction is due to sedentary plant-parasitic nematodes like *M.incognita* spend most of their life cycles within the host plant roots, they may benefit from plant barriers for protection against fungi (Abad et al. 2008).

A secreted chitinase has been identified in the perivitelline fluid surrounding the infective larva of *Ascaris suum* prior to hatching, indicating that this enzyme might be responsible for the digestion of the eggshell during hatching of this nematode (Geng et al. 2002; Curtis et al. 2011). Suppressing the expression of the *Caenorhabditis elegans* gene *cht-1* by RNAi (WormBase ID: WBRNAi00000785) led to an embryonic lethal phenotype (Maeda et al. 2001), possibly caused by the failure of enzymatic hydrolysis of eggshell chitin by this chitinase. Two chitinases, Bx-chi-1 and Bx-chi-7, are highly expressed in eggs of *Bursaphelenchus xylophilus*. Egg hatching was delayed by RNAi silencing of either of these chitinases; 9.02% of eggs failed to hatch when treated with Bx-chi-7 dsRNA solution. Interestingly, Bx-chi-1 expression was significantly higher in female nematodes than in males and

RNAi knockout of Bx-chi-1 in female nematodes led to a significant decrease in reproduction (Ju et al. 2016), supporting an essential function for chitinases in nematode reproduction. A 57–68% decrease in the release of microfilariae, and a 42–58% inhibition of hatching of microfilariae after RNAi treatment of *Acanthocheilonema viteae* females targeting the GH18 chitinase gene *Av-cht-I*, also suggests that chitinase is necessary for nematode reproduction (Tachu et al. 2008).

Av-cht-I also functions in moulting of the filarial, and 93% suppression of expression of this chitinase caused 87% moulting inhibition (Tachu et al. 2008). The potential role of the chitinase Ov-Cht-1 in the moulting process was also inferred in the filarial nematode *Onchocerca volvulus*. The expression of Ov-Cht-1 initiated in late L2 larvae and increases markedly in infective L3 larvae (Wu et al. 2001), and the Ov-Cht-1 inhibitor closantel or its analogues showed promise in abrogating the L3 moulting (Gloeckner et al. 2010; Garner et al. 2011; Gooyit et al. 2015).

As chitinases play essential roles in egg hatching and moulting of parasitic nematodes, chitinase is a promising drug target for nematode infection control.

10.6 Applications of Chitin in Nematode Control

The applications of the chitin and chitosan in agriculture were reviewed by Zargar et al. (2015). They pointed out that it has four main objectives such as (1) plant protection against pre- and post-harvest diseases and epidemics; (2) enhancing antagonist microorganisms and biological control; (3) supporting beneficial plant microorganisms and symbiotic relationships; and (4) plant growth, regulation and development. Chitin and its derivatives have been employed extensively for boosting defensive mechanisms in plants (Zargar et al. 2015). The chitooligosaccharides of a specific size as potent elicitor signals in plants were evaluated to protect them against many vegetable diseases (Spiegel et al. 1986). Chitin and chitosan have fungicidal activity against many pathogen fungi. Antiviral and antibacterial activities of chitosan and its derivatives have also been demonstrated. These polysaccharides have been used successfully to control parasitic nematodes in soils. The addition of chitin to the soil raises the population of chitinolytic microorganisms (Zargar et al. 2015), which break down the polysaccharide chitin to the disaccharide chitobiose, disrupting the eggs and cuticles of young nematodes that contain chitin. Chitinase and chitosanase activities have also been reported in seeds protected by films of chitin and its derivatives. The antimicrobial properties of chitosan and its outstanding film-creating capacity have been exploited in the post-harvest preservation of fruits and vegetables. Covering fruits and vegetables with a chitosan film confers antimicrobial protection and enhanced shelf life. Addition of chitin and chitosan to the soil increases beneficial plant–microorganism symbiotic interactions (http://portal.ufrrj.br/), e.g. in the case of mycorrhiza. These compounds also improve the action of plague-controlling biological organisms such as *Trichoderma* sp., and *Bacillus* sp., and can be used for encapsulation of biocides. Therefore, their efficiency in the control against pathogenic microorganisms and plagues is enhanced. Chitosan and its derivatives also cause desirable changes in the metabolism of plants and fruits, improving germination and increasing crop yields.

Application of chitin to the soil results in increasing populations of chitinolytic microorganisms able to degrade chitin in nematode eggshells, and chitosan enhances plant growth, activating defences and increasing the activity of enzymes, such as chitinase. In the specific case of chitinolytic fungi, pressure within the egg cuticle, together with the activity of chitinase, causes disruption and subsequently, penetration and degradation of the egg (Mota and dos Santos 2016).

Chitin and chitosan products, such as HYT-C and HYT-D, are being implicated as alternative control measures for management of plant-parasitic nematodes. The chitin product HYT-C is based on micronized chitin extracted from shrimps. HYT-D is the deacetylated form, chitosan. Both products strengthen root formation and cell structure, and stimulate the natural ability of plants to resist soil-borne plant pathogens such as nematodes. The cattle manure vermicompost enriched with chitin in soil infested with *Meloidogyne incognita* were studied and a significant reduction in nematode reproduction in tomatoes grown in chitin-enriched soils compared with the control (Castro et al. 2011). An increase of bacteria, fungi and soil actinomycetes was also detected. When application chitin to soil 2 weeks before infesting with *Meloidogyne incognita* and planting tomato seedings, an increase in the dry weight of plant was observed. With 100 and 200 g of chitin in 150 cm^3 of soil, the numbers of galls and nematode eggs were significantly reduced. It concluded that chitin's nematicidal effect could be attributed partly to direct toxicity to the nematode, and partly to the stimulation of fungal chitinolytic activity against the microflora (Mota and dos Santos 2016). To evaluate the effect of chitosan on resistance against *Meloidogyne incognita*, sprayed chitin onto tomato leaves 5, 10 and 15 days prior to inoculation with the pathogen were carried out by Melo et al (2012). A reduction in the number of eggs and in nematode reproduction was observed, at all application times. Mota and dos Santos (2016) evaluated the efficacy of chitin and chitosan on the control of *Meloidogyune javanica*, on the increase of chitinolytic microorganisms in the soil, on chitinase activity on tomato leaves and on plant development. Leaf-sprayed chitosan increased dry matter of tomato shoots. Soil-applied chitin, associated with leaf-sprayed chitosan, reduced *Meloidogyune javanica* reproduction and its population associated with an increase in chitinolytic microorganisms in the soil, and in the activity of chitinase in tomato leaves 4 days after application (Mota and dos Santos 2016).

10.7 Further Perspectives

Parasitic nematode infection is a worldwide threat to human health, livestock production and agricultural production. However, the number of nematicides currently available in the market is small and reducing by prohibition for environmental impact or losing effectiveness by nematode resistant. Thus, novel nematicides are emergency in need. Chitin metabolism is a complicated and ordered process in nematode and essential for nematode survival. This process interfered by chitinolytic microorganisms resulted in nematode death. Thus, the key enzymes in chitin synthesis and hydrolysis processes are promising targets for nematicides development.

Though series of potential insecticides or fungicides are targeting to chitin synthesis or hydrolysis process, there is pitifully few inhibitors of a chitin metabolic enzyme of nematodes. This might be due to the parasitic nematodes are difficult experimental organisms and the structural and biochemical information of chitin metabolic enzymes of nematodes are unknown. In this context, the main challenge of nematicides development targeting to chitin metabolic enzymes is to obtain the purified enzymes, solve the crystal structures and establish a high throughput nematicides evaluation method.

References

Abad P et al (2008) Genome sequence of the metazoan plant-parasitic nematode Meloidogyne incognita. Nat Biotech 26:909–915

Adam R, Kaltmann B, Rudin W, Friedrich T, Marti T, Lucius R (1996) Identification of chitinase as the immunodominant filarial antigen recognized by sera of vaccinated rodents. J Biol Chem 271(3):1441–1447

Atkinson HJ, Taylor JD, Fowler M (1987) Changes in the 2nd stage juveniles of Globodera rostochiensis prior to hatching in response to potato root diffusate. Ann Appl Biol. 110:105–114

Bird AF, Mcclure MA (1976) The tylenchid (Nematoda) egg shell: structure, composition and permeability. Parasitology 72:19–28

Burgwyn B, Nagel B, Ryerse J, Bolla RI (2003) Heterodera glycines: eggshell ultrastructure and histochemical localization of chitinous components. Exp Parasitol 104:47–53

Bird AF, Bird J (1991) The egg. The Structure of Nematodes (2nd edn). Academic Press Inc., New York, pp 7–43

Bird AF, Self PG (1995) Chitin in Meloidogyne javanica. Fundam Appl Nematol. 18:235–239

Castro L, Flores L, Uribe L (2011) Efecto del vermicompost y quitina sobre el control de Meloidogyne incognita en tomate a nível de invernadero. Agronomía Costarricense 35:21–32

Chen C, Cui L, Chen Y, Zhang H, Liu P, Wu P, Qiu D, Zou J, Yang D, Yang L, Liu H, Zhou Y, Li H (2017) Transcriptional responses of wheat and the cereal cyst nematode Heterodera avenae during their early contact stage. Sci Rep. 7(1):14471

Clarke AJ, Cox PM, Shepherd AM (1967) Chemical composition of egg shells of potato cyst nematode Heterodera rostochiensis woll. Biochem J 104:1056–1060

Curtis RHC, Jones JT, Davies KG, Sharon E, Spiegel Y (2011) Chapter 5, Plant nematode surfaces. pp 115–144. In: Davies K, Spiegel Y (eds.), Biological control of plant-parasitic nematodes: building coherence between microbial ecology and molecular mechanisms. Prog Biol Control 11. https://doi.org/10.1007/978-1-4020-9648-8_5,

Dubinský P, Rybos M, Turceková L (1986) Properties and localization of chitin synthase in Ascaris suum eggs. Parasitology 92(Pt 1):219–225

Dusenbery DB (1997) Prospects for exploring sensory stimuli in nematode control. In: Veech JA, Dickson DW (eds) Vistas on nematology. Society of Nematologists, Hyattsville, MD, pp 131–135

Fanelli E, Di Vito M, Jones JT, De Giorgi C (2005) Analysis of chitin synthase function in a plant parasitic nematode, Meloidogyne artiellia, using RNAi. Gene 349:87–95

Foster JM, Zhang Y, Kumar S, Carlow CK (2005) Parasitic nematodes have two distinct chitin synthases. Mol Biochem Parasitol 142(1):126–132

Fuhrman JA, Lane WS, Smith RF, Piessens WF, Perler FB (1992) Transmission-blocking antibodies recognize microfilarial chitinase in brugian lymphatic filariasis. Proc Natl Acad Sci USA. 89(5):1548–1552

Gao B, Allen R, Maier T, McDermott JP, Davis EL, Baum TJ, Hussey RS (2002) Characterisation and developmental expression of a chitinase gene in Heterodera glycines. Int J Parasitol 32(10):1293–1300

Garner AL, Gloeckner C, Tricoche N, Zakhari JS, Samje M, Cho-Ngwa F, Lustigman S, Janda KD (2011) Design, synthesis, and biological activities of closantel analogues: structural promiscuity and its impact on Onchocerca volvulus. J Med Chem 54(11):3963–3972

Geng JM, Plenefisch J, Komuniecki PR et al (2002) Secretion of a novel developmentally regulated chitinase (family 19 glycosyl hydrolase) into the perivitelline fluid of the parasitic nematode, Ascaris suum. Mol Biochem Parasitol 124:11–21

Gloeckner C, Garner AL, Mersha F, Oksov Y, Tricoche N, Eubanks LM, Lustigman S, Kaufmann GF, Janda KD (2010) Repositioning of an existing drug for the neglected tropical disease Onchocerciasis. Proc Natl Acad Sci USA. 107(8):3424–3429

Gooyit M, Harris TL, Tricoche N, Javor S, Lustigman S, Janda KD (2015) Onchocerca volvulus Molting Inhibitors Identified through Scaffold Hopping. ACS Infect Dis. 1(5):198–202

Hanazawa M, Mochii M, Ueno N, Kohara Y, Iino Y (2001) Use of cDNA subtraction and RNA interference screens in combination reveals genes required for germ-line development in Caenorhabditis elegans. Proc Natl Acad Sci USA. 98(15):8686–8691

Harris MT, Lai K, Arnold K, Martinez HF, Specht CA, Fuhrman JA (2000) Chitin synthase in the filarial parasite *Brugia malayi.* Mol Biochem Parasitol. 111(2):351–362

Harris MT, Fuhrman JA (2002) Structure and expression of chitin synthase in the parasitic nematode Dirofilaria immitis. Mol Biochem Parasitol 122(2):231–234

Johnston WL, Krizus A, Dennis JW (2010) Eggshell chitin and chitin-interacting proteins prevent polyspermy in C. elegans. Curr Biol 20(21):1932–1937

Ju Y, Wang X, Guan T, Peng D, Li H (2016). Versatile glycoside hydrolase family 18 chitinases for fungi ingestion and reproduction in the pinewood nematode Bursaphelenchus xylophilus. Int J Parasitol. 46(12):819–828

Jung WJ, Jung SJ, An KN, Jin YL, Park RD, Kim KY, Shon BK, Kim TH (2002) Effect of chitinase-producing Paenibacillus illinoisensis KJA-424 on egg hatching of root-knot nematode (Meloidogyne incognita). J Microbiol Biotechnol 12:865–871

Maeda I, Kohara Y, Yamamoto M, Sugimoto A (2001) Large-scale analysis of gene function in Caenorhabditis elegans by high-throughput RNAi. Curr Biol 11(3):171–176

Melo TA, Serra IMRS, Silva GSS, Sousa RMS (2012) Produtos naturais aplicados para manejo de Meloidogyne incognita em tomateiros. Summa Phytopathol 3:223–227

Mota L, dos Santos MA (2016) Chitin and chitosan on *Meloidogyne javanica* management and on chitinase activity in tomato plants. Trop Plant Pathol. 2:84–90

Neuhaus B, Bresciani J, Peters W (1997) Ultrastructure of the pharyngeal cuticle and lectin labelling with wheat germ agglutinin-gold conjugate indicating chitin in the pharyngeal cuticle of Oesophagostomum dentatum (Strongylida Nematoda). Acta Zool. 78:205–213

Nicol JM, Turner SJ, Coyne DL, Nijs Ld, Hockland S, Maafi ZT. (2011) Current nematode threats to world agriculture. In: Jones J, Gheysen G, Fenoll C(Eds.) Genomics and Molecular Genetics of Plant-Nematode Interactions, Netherlands: Springer, pp 21–43

Perry RN, Clarke AJ (1981) Hatching mechanisms of nematodes. Parasitology 83:435–449

Premachandran D, Von Mende N, Hussey RS, McClure MA (1988) A method for staining nematode secretions and structures. J Nematol. 20:70–78

Sharon E, Chet I, Bar-Eyal M, Spiegel Y (2009) Biocontrol of root-knot nematodes by Trichoderma-modes of action. In: Steinberg C, Edel-Hermann V, Friberg H, Alabouvette C, Tronsmo A (eds) Multitrophic interactions in soil. IOBC/WPRS Bull, vol 42, pp 159–163

Siddique S, Grundler FM (2018) Parasitic nematodes manipulate plant development to establish feeding sites. Curr Opin Microbiol 46:102–108

Spiegel Y, Chet I (1985) Chitin synthetase inhibitors and their potential to control the root-knot nematode, Meloidogyne javanica. Nematologica. 31:480–482

Spiegel Y, Cohn E (1985) Chitin is present in gelatinous matrix of Meloidogyne. Revue Nematol. 8:184–186

Tachu B, Pillai S, Lucius R, Pogonka T (2008) Essential role of chitinase in the development of the filarial nematode Acanthocheilonema viteae. Infect Immun 76(1):221–228

Twomey U, Warrior P, Kerry BR, Perry RN (2000) Effects of the biological nematicide, DiTera®, on hatching of Globodera rostochiensis and G. pallida. Nematology. 2:355–362

van Megen H, van den Elsen S, Holterman M, Karssen G, Mooyman P, Bongers T, Holovachov O, Bakker J, Helder J (2009) A phylogenetic tree of nematodes based on about 1200 full-length small subunit ribosomal DNA sequences. Nematology. 11:927–950

van Nguyen N, Kim Oh KT, Jung W et al (2007) The role of chitinase from Lecanicillium antillanum B-3 in parasitism to root-knot nematode Meloidogyne incognita eggs. Biocontrol Sci Technol 17:1047–1058

Veronico P, Gray LJ, Jones JT, Bazzicalupo P, Arbucci S, Cortese MR, Di Vito M, De Giorgi C (2001) Nematode chitin synthases: gene structure, expression and function in Caenorhabditis elegans and the plant parasitic nematode Meloidogyne artiellia. Mol Genet Genom 266:28–34

Wu Y, Egerton G, Underwood AP, Sakuda S, Bianco AE (2001) Expression and secretion of a larval-specific chitinase (Family 18 glycosyl hydrolase) by the infective stages of the parasitic nematode. Onchocerca volvulus. J Biol Chem. 276(45):42557–42564

Zhang Y, Foster JM, Nelson LS, Ma D, Carlow CK (2005) The chitin synthase genes chs-1 and chs-2 are essential for C. elegans development and responsible for chitin deposition in the eggshell and pharynx, respectively. Dev Biol 285(2):330–339

Zargar V, Asghari M, Dashti A (2015) A review on chitin and chitosan polymers: structure, chemistry, solubility, derivatives, and applications. Chem Bio Eng Rev 2(3):204–226

Chapter 11
Human Chitinases: Structure, Function, and Inhibitor Discovery

Ashutosh Kumar and Kam Y. J. Zhang

Abstract Chitinases are glycosyl hydrolases that hydrolyze the β-(1-4)-linkage of N-acetyl-D-glucosamine units present in chitin polymers. Chitinases are widely distributed enzymes and are present in a wide range of organisms including insects, plants, bacteria, fungi, and mammals. These enzymes play key roles in immunity, nutrition, pathogenicity, and arthropod molting. Humans express two chitinases, chitotriosidase 1 (CHIT1) and acid mammalian chitinase (AMCase) along with several chitinase-like proteins (CLPs). Human chitinases are reported to play a protective role against chitin-containing pathogens through their capability to degrade chitin present in the cell wall of pathogens. Now, human chitinases are gaining attention as the key players in innate immune response. Although the exact mechanism of their role in immune response is not known, studies in recent years begin to relate chitin recognition and degradation with the activation of signaling pathways involved in inflammation. The roles of both CHIT1 and AMCase in the development of various diseases have been revealed and several classes of inhibitors have been developed. However, a clear understanding could not be established due to complexities in the design of the right experiment for studying the role of human chitinase in various diseases. In this chapter, we will first outline the structural features of CHIT1 and AMcase. We will then review the progress in understanding the role of human chitinases in the development of various diseases. Finally, we will summarize the inhibitor discovery efforts targeting both CHIT1 and AMCase.

Keywords Chitin · Chitinase · Chitotriosidase 1 · Acid mammalian chitinase · Inflammation · Inhibitors

---

A. Kumar · K. Y. J. Zhang (✉)
Laboratory for Structural Bioinformatics, Center for Biosystems Dynamics Research, RIKEN, 1-7-22 Suehiro, Tsurumi, Yokohama, Kanagawa 230-0045, Japan
e-mail: kamzhang@riken.jp

Q. Yang and T. Fukamizo (eds.), *Targeting Chitin-containing Organisms*, Advances in Experimental Medicine and Biology 1142,
https://doi.org/10.1007/978-981-13-7318-3_11

11.1 Introduction

Chitin is a polymer found in nature where β-(1-4)-linked N-acetyl-D-glucosamine (GlcNAc) units are linearly bonded together to form long chains. In chitin, N,N′-diacetylchitobiose $(GlcNAc)_2$ forms the structural subunit (Fig. 11.1). Chitin, unlike starch and glycogen that are storage polysaccharide, is a structural polysaccharide. It is the second most abundant naturally occurring polymer and is the main structural component of arthropod exoskeleton (Neville et al. 1976), crustacean shells, house dust mites (HDM), fungal cell walls (Lenardon et al. 2010) and microfilarial sheath of parasitic nematodes (Veronico et al. 2001; Foster et al. 2005). In fungal cell walls, chitin is the key structural polymer and is equivalent to peptidoglycan in bacteria. Chitin provides rigidity and structural integrity to cells, tissues and body surfaces.

X-ray diffraction analysis showed that chitin exists in three different crystalline forms, namely, α-chitin, β-chitin, and γ-chitin based on the arrangement of chitin polysaccharide chains in crystalline fibers (Rudall and Kenchington 1973; Kramer and Koga 1986; Imai et al. 2003; Rinaudo 2006) (Fig. 11.1). In addition, a non-crystalline form has also been reported in fungi (Vermeulen and Wessels 1986). Among these three forms, α-chitin is the most abundant and compact form where chitin chains are arranged in an antiparallel manner (Fig. 11.1). This type of arrangement ensures strong hydrogen bonding (Kramer and Koga 1986). The α-chitin is a major structural component of arthropod exoskeleton and fungal cell

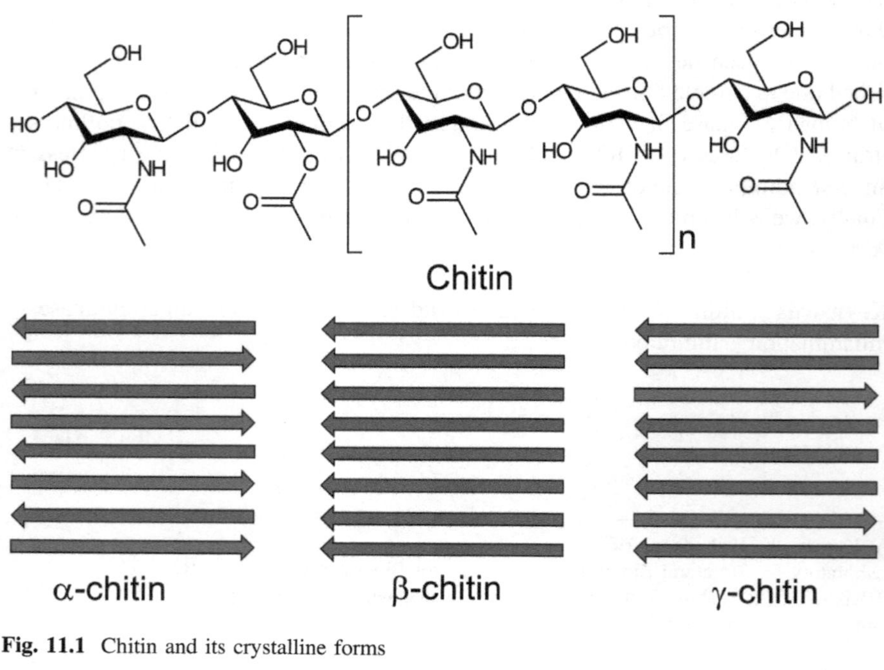

Fig. 11.1 Chitin and its crystalline forms

walls. It is also found in shells of crustaceans like shrimps and crabs. Antiparallel arrangement of chitin chains in α-chitin ensures mechanical strength and stability (Giraud-Guille and Bouligand 1986). Contrary to α-chitin, chitin chains are arranged in a parallel manner in β-chitin resulting in a less compact form due to weak intermolecular forces (Fig. 11.1). Inter-chain hydrogen bonding contacts are reduced which results in a more flexible and soft structure due to a high degree of hydration and reduced packing. β-chitin is commonly found in squid pens, peritrophic matrices and cocoons. Another chitin form, γ-chitin is the mixture of both α-chitin and β-chitin chains and is commonly found in mushrooms (Fig. 11.1). Interestingly, some animals utilize multiple forms of chitin for different biological functions, for example, squid possess α-chitin in its beak, β-chitin in its pen and γ-chitin in the stomach lining (Gooday 1990).

In chitin-containing organism, chitin is produced by a highly conserved enzyme, chitin synthase (EC 2.4.1.16). Chitin synthases belong to GT2 family of polymerizing glycosyltransferases including synthases for polymers such as cellulose, mannan, hyaluronate, etc. Chitin synthases add GlcNAc units to the nonreducing end of chitin chains. It utilizes UDP-N-acetylglucosamine (UDP-GlcNAc) as sugar donor to form a polymer of chitin (Glaser and Brown 1957). Linear polymers are first produced which are then assembled into microfibrils of various length and diameter (Imai et al. 2003; Merzendorfer 2006). These microfibrils are then extruded from cell walls in a similar manner as cellulose (Morgan et al. 2012; Gonçalves et al. 2016). Chitin biosynthesis is extensively studied; however, the detailed description is beyond the scope of this book chapter. Some reviews may be referred for chitin synthases and biosynthesis (Merzendorfer and Zimoch 2003; Merzendorfer 2006, 2011). The focus of this book chapter will be another group of enzymes which is responsible for chitin degradation.

11.2 Chitinase and Their Classification

Despite being one of the most abundant polymers, chitin doesn't accumulate in environment due to the presence of chitinolytic enzymes known as chitinases which hydrolyze β-(1-4)-linkage of GlcNAc present in chitin chains. Chitinases are widely distributed enzymes and are present in many organisms including insects, plants, bacteria, fungi, and mammals. These enzymes play roles in immunity and defense, nutrient acquisition, digestion, pathogenicity, and arthropod molting. Chitinases belong to the glycosyl hydrolase (GH) family containing a wide diversity of proteins. Based on sequence similarity, chitinases are classified into five classes (class I–V) (Hamid et al. 2013). Class I chitinases have most conserved catalytic domain with a cysteine-rich domain at N-terminus. Class II chitinases possess high similarity with catalytic domain of class I chitinases but lack cysteine-rich domain at N-terminus. Class III chitinases do not share any identity with class I and II. Class IV chitinases are similar to class I but are much smaller compared to class I due to four deletions in catalytic domain. Class V chitinases lack sequence identity with any other class.

Based on sequence identity, structural homology and biochemical studies, chitinases are also classified into family GH18 chitinase, family GH19 chitinase (mostly plant chitinase) and family GH20 N-acetyl glucosaminidase (Fukamizo 2000). Chitinases of family GH18 and GH19 have very low sequence similarity with each other and have completely different structures. Chitinases of class III and V form family GH18 while class I, II, and IV belong to family GH19. Family GH18 chitinases hydrolyze chitin into shorter $(GlcNAc)_2$ and $(GlcNAc)_3$ fragments while β-N-acetyl-D-hexosaminidases hydrolyze these oligosaccharides into GlcNAc. Family GH18 can be further divided into chitinase and enzymatically inactive chitinase-like-lectins or chitinase-like proteins. The domain architectures and structural features of family GH18 chitinases are highly conserved (Huang et al. 2012) and consist of a catalytic domain and a cysteine-rich carbohydrate or chitin binding domain. A hinge connects the catalytic domain with carbohydrate binding domain. The catalytic domain is mostly located at the N-terminus while chitin binding domain is located at C-terminus. The catalytic domain of members of GH family is known to adopt triosephosphate isomerase (TIM) fold. TIM fold consists of a highly conserved $(\beta/\alpha)_8$-barrel structure in which β-sheets are surrounded by α-helices (Henrissat 1999; Stam et al. 2005). The active site is highly conserved and composed of aromatic amino acid residues that help in substrate binding (Perrakis et al. 1994; van Aalten et al. 2000; Boot et al. 2001; Fusetti et al. 2002; Olland et al. 2009; Fadel et al. 2015). Substrate binding occurs at a cleft in the center of $(\beta/\alpha)_8$-barrel. Substrate binding sites in GH18 chitinases employ –n to +n nomenclature where n represents GlcNAc binding subsites (Davies et al. 1997). Subsites –n are located toward the nonreducing end while +n subsites are located at the reducing end of the substrate relative to substrate cleavage site. The cleavage of chitin substrate molecule occurs between −1 and +1 GlcNAc binding subsites (Davies et al. 1997). The catalytic triad lies at the bottom of active site at −1 subsite. A conserved sequence motif in the β_4 strand (DXXDXDXE, where D stands for aspartic acid, E stands for glutamic acid, and X represents any amino acid) form the active site pocket. Glutamic acid and central aspartic acid in the conserved sequence motif are the key amino acid residues in catalysis. Glutamic acid donates a proton required for hydrolyzing the β-(1-4) glycosidic bond in chitin (Chou et al. 2006) while the second aspartate in the motif contributes to stabilization of substrate molecule during catalysis (McCarter and Stephen Withers 1994; Perrakis et al. 1994; van Aalten et al. 2000, 2001; Fusetti et al. 2002). Family GH18 chitinases make use of substrate-assisted reaction mechanism (Terwisscha van Scheltinga et al. 1995; van Aalten et al. 2001) whereas family GH19 chitinases employ fold-and-reaction mechanism (Monzingo et al. 1996) suggesting independent evolution of these two families. Family GH18 includes chitin hydrolyzing enzymes from bacteria, fungi, viruses, animals, and some chitinases from plants (class III and V) (Ohno et al. 1996). Family GH19 contains chitinases mostly from plants (class I, II, and IV) and nematodes.

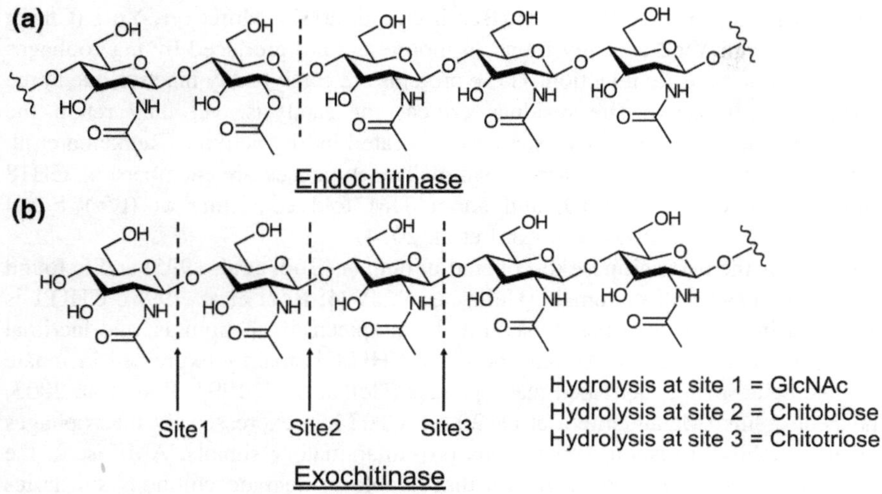

Fig. 11.2 Chitin hydrolysis by chitinase types **a** Endochitinase and **b** Exochitinase

Chitinase can be further classified based on their mode of cleavage into two classes: (a) endochitinase and (b) exochitinase. Endochitinases (EC 3.2.1.14) catalyze the hydrolysis of internal chitin chains from inner random position generating low oligomeric units of GlcNAc such as chitotriose, chitobiose, and diacetylchitobiose (Fig. 11.2a). Exochitinases (EC 3.2.1.52) cleave from the nonreducing end of chitin chains (Nikolov et al. 2010) (Fig. 11.2b). Exochitinases (EC 3.2.1.52) have been further classified into two categories, namely, chitobiosidases and β-N-acetyl-D-hexosaminidases. Chitobiosidases (EC 3.2.1.29) catalyze the release of diacetylchitobiose from the nonreducing end of chitin chains while β-N-acetyl-D-hexosaminidases cleave oligomeric GlcNAc units produced by endochitinases into GlcNAc (Fig. 11.2b).

11.3 Human Chitinases

Although chitin is neither present nor utilized as a nutrient source in humans, yet they express two chitinases: acidic mammalian chitinase (AMCase or CHIA) and chitotriosidase 1 (CHIT1) (Hollak et al. 1994; Boot et al. 1995, 2001). Both of these enzymes function as endochitinases and are members of GH18 chitinase family. These two proteins are recognized as true chitinases as they are the only ones that possess chitin hydrolyzing activity (Boot et al. 2001). In addition, structurally related chitinase-like proteins (CLPs) or chitinase-like-lectins (Chi-lectins) are expressed such as chitinase 3-like-1 (CHI3L1 or YKL-40), chitinase 3-like-2 (CHI3L2 or YKL-39), chitinase domain–containing 1 (CHID1), stabilin-1-interacting chitinase-like protein (SI-CLP) (Kzhyshkowska et al. 2006) and oviductal glycoprotein 1 (OVGP1) (Bussink et al. 2007). Moreover, Ym1 (Chang et al. 2001) and Ym2 are only found in mouse and are produced by macrophages after fungal or parasitic infection. These proteins are catalytically inactive due to the substitutions in active site residues critical for catalysis, yet they retain the TIM-barrel structure and active site carbohydrate binding activity (Renkema et al. 1998; Houston et al. 2003). Most mammalian chitinases are members of GH18 family (Li and Greene 2010) and adopt TIM fold consisting of $(\beta/\alpha)_8$-barrel structure (Stam et al. 2005; Lombard et al. 2014).

CHIT1 is the first chitinase discovered in human (Boot et al. 1995) and is found in the genomes of all mammals (Hollak et al. 1994; Boot et al. 1995). CHIT1 is expressed in a variety of tissues such as lung, spleen, liver, thymus, and lacrimal gland (Ohno et al. 2013). Among the cells, CHIT1 is mainly expressed in innate immune cells such as activated macrophages (Hollak et al. 1994; Rao et al. 2003) and neutrophils (Malaguarnera et al. 2006). CHIT1 is expressed by macrophages and neutrophils in response to various pro-inflammatory signals. AMCase is the second chitinase identified in human that is able to degrade chitinous substrates (Boot et al. 2001, 2005). AMCase is mainly expressed in the stomach and gastrointestinal tract. AMCase has been also detected in the lung where it is secreted into the airway lumen by epithelial cells, club cells, and type 2 alveolar cells. AMCase has been reported as the major endochitinase in airway fluid (Fitz et al. 2012; Van Dyken et al. 2017). The presence of AMCase in the stomach and gastrointestinal tract suggests the role of AMCase in digestion (Boot et al. 2001, 2005; Chou et al. 2006; Ohno et al. 2012) for breaking chitin-containing food while the expression in lung implicates its role in defense against chitin-containing pathogens.

AMCase gene is located on 1q13.1-21.3 chromosome and consists of 12 exons (Boot et al. 2001). CHIT1, on the other hand, is located on 1q31-32 chromosome and also contains 12 exons. The location of both genes in chromosome 1, sequence and structural similarities and intron–exon boundary conservation suggest that these genes might have arisen due to gene duplication (Boot et al. 2001) which is believed to have taken place during the emergence of jawed and jawless fish (Hussain and Wilson 2013). Further loss of function mutations and duplications might have given rise to CLPs. Some of these CLPs are species-specific while others are found in all mammals (Hussain and Wilson 2013). Both AMCase and CHIT1 are secretory proteins with approximately 50 kD molecular weight (Boot et al. 1995, 2001). AMCase has about 51% sequence identity and 66% sequence similarity with CHIT1. AMCase like other GH18 family members contains both chitin hydrolysis and chitin binding domain. These two domains are connected to each other by a short hinge region (Renkema et al. 1997). Despite both sequence and structural similarity, these two proteins differ in their enzymatic behavior at acidic pH. AMCase is extremely stable at acidic pH and displays an optimum activity at pH 2 (Boot et al. 2001). The name AMCase is derived from this property of the enzyme. CHIT1, on the other hand, shows an optimum around pH 5 (Renkema et al. 1995; Zheng et al. 2005) and is inactive at low pH (Boot et al. 2001, 2005).

CHIT1 is found to exist in two isoforms: a 39 kDa isoform containing the catalytic domain only and a 50 kDa isoform containing the full-length CHIT1. The full-length CHIT1 is composed of a large catalytic domain connected to a comparatively small chitin binding domain (Fig. 11.3a). The two domains are linked by a proline-rich hinge. Initially, several crystal structures of the catalytic domain alone were solved both in apo form as well in complex with CHIT1 inhibitors (Fusetti et al. 2002; Rao et al. 2005a; Fadel et al. 2015) including a 1.0 Å apo structure of CHIT1 (Fadel et al. 2015). The catalytic domain structure of CHIT1 consists of a core $(\beta/\alpha)_8$-barrel domain which is similar to AMCase and other family GH18 chitinases. CHIT1 structure contains a DXXDXDXE motif at the end of fourth β sheet with Glu140 as catalytic residue. An additional α/β domain is present in between the seventh α-helix and the seventh β-strand. This additional domain consists of six β-sheets and one α-helix and gives a groove like character to the

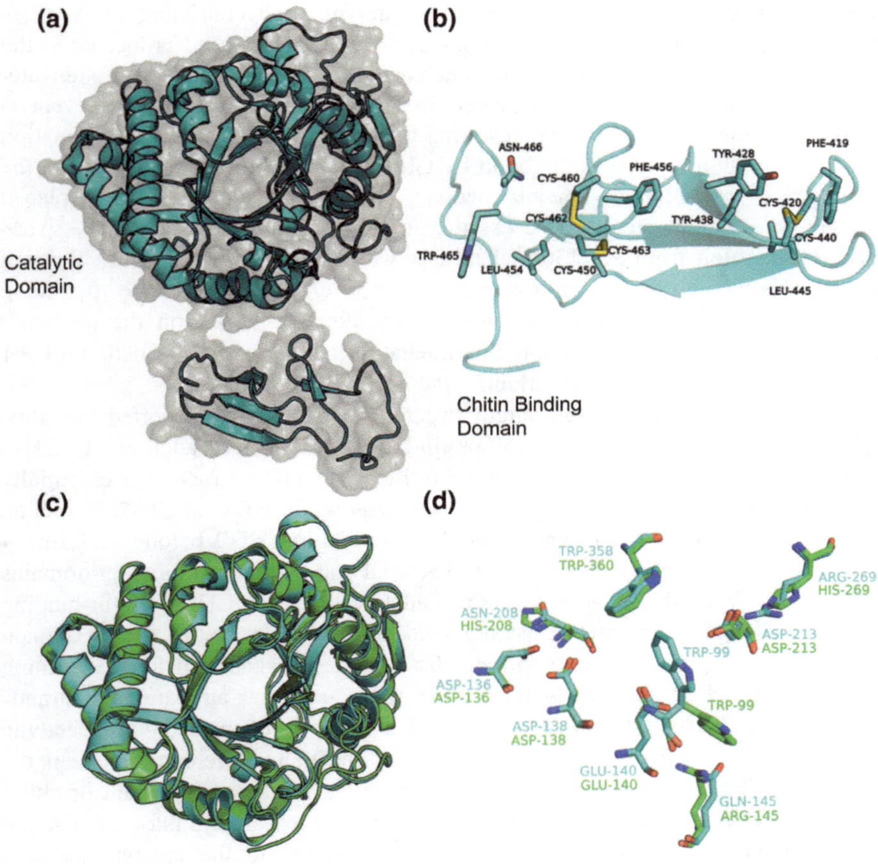

Fig. 11.3 **a** A cartoon representation of full-length CHIT1 showing catalytic and chitin binding domain. **b** CHIT1 chitin binding domain. **c** Superposition of AMCase (green) with CHIT1 (cyan). **d** Active site amino acid residues of AMCase (green) and CHIT1 (cyan)

active site (Fusetti et al. 2002). CHIT1 active site is lined by several aromatic amino acids that stack against the hydrophobic face of sugar units. The first stretch of aromatic amino acid residues lies along the −6 to −1 subsites and includes Trp31, Tyr34, Trp71 and Trp358. The second stretch of aromatic residues is located at +1 and +2 subsites and includes Trp99 and Trp218. These residues are not only identical in both CHIT1 and AMCase but also highly conserved in other family GH18 chitinases. Trp358 is especially important as it is stacked with −1 subsite GlcNAc and is responsible for accommodating −1 subsite GlcNAc in boat conformation critical for catalysis. Several studies revealed that the mutation of Trp358 abolished chitinase activity (van Aalten et al. 2000; Fusetti et al. 2002; Songsiriritthigul et al. 2008; Yang et al. 2010).

The mechanism of substrate hydrolysis in CHIT1 is similar to other GH18 family enzymes and has been recently elucidated utilizing X-ray crystallography and QM/MM studies (van Aalten et al. 2001; Fadel et al. 2015). The cleavage occurs between −1 and +1 subsites. The catalytic triad consisting of Asp136, Asp138 and Glu140 amino acid residues of DXXDXDXE motif is located at the bottom of −1 subsite. Binding of substrate causes displacement of active site water molecules and the transfer of a proton from Asp138 to Glu140. This event is followed by the distortion of pyranose ring to a boat conformation and protonation of scissile O atom of glycosidic bond by Glu140 which subsequently leads to the formation of oxazolinium ion intermediate. An aglycon sugar is next displaced allowing a water molecule to access the active site. At the same time, Asp138 receives a proton from Asp136 and turns toward Glu140 which in turn forms a hydrogen bond with N-acetyl nitrogen of −1 GlcNAc. A water molecule after being activated by Asp213 probably carries out nucleophilic attack on the anomeric carbon atom of oxazolinium ion intermediate leading to the formation of −1 GlcNAc in its original configuration (Fadel et al. 2015).

About two years ago, a full-length structure of CHIT1 was reported that shed light on the structural features of chitin-binding domain (Fadel et al. 2016) (Fig. 11.3a). The catalytic domain in the full-length CHIT1 structure is essentially the same as other CHIT1 catalytic domain structures (Fusetti et al. 2002; Rao et al. 2005a; Fadel et al. 2015). The chitin binding domain in CHIT1 belongs to CBM14 family and is structurally different from bacterial and plant chitin-binding domains (Ikegami et al. 2000; Akagi et al. 2006; Fadel et al. 2016). CHIT1 chitin binding domain is elongated in structure as compared to the globular chitin-binding domain of bacteria and plants. It is composed of a distorted β-sandwich fold containing three N-terminal antiparallel β-strands and two C-terminal antiparallel β-strands (Fadel et al. 2016). CHIT1 chitin binding domain is connected with catalytic domain by a stretch of proline-rich residues. The main difference between the full-length CHIT1 and other bacterial chitinases lies in the arrangement of chitin binding domain relative to catalytic domain. In S. marcescens chitinases, ChiA and ChiB, the chitin-binding domain is rigidly connected to the catalytic domain resulting in a completely different orientation of chitin binding domain relative to catalytic domain (Perrakis et al. 1994; van Aalten et al. 2000). Chitin substrate interacts with chitin binding domain via a stretch of aromatic amino acid residues (Fig. 11.3b). Most of these residues are surface-exposed, especially a conserved residue Trp465 enabling strong binding with chitin substrates (Fadel et al. 2016). The active site cleft in CHIT1 contains nine GlcNAc binding sites (from −6 to +3 subsites). CHIT1 was initially believed to be an exochitinase as it has been shown to hydrolyze chitotriose, however, structural and biochemical studies demonstrated CHIT1 as an endochitinase (Fusetti et al. 2002; Kuusk et al. 2017).

The AMCase catalytic domain crystal structure was solved in apo form as well as in complex with several inhibitors including allosamidin derivative methylallosamidin (Olland et al. 2009), bisdionin C and F (Sutherland et al. 2011) and many hits from in silico, high-throughput and fragment screening (Cole et al. 2010). No structure of full-length AMCase is available till date. Catalytic domain of AMCase possesses about 57% identity with CHIT1 and it very similar to CHIT1 catalytic domain crystal structure (Fusetti et al. 2002; Rao et al. 2003, 2005b; Fadel et al. 2015, 2016) consisting of a $(\beta/\alpha)_8$-barrel core (Fig. 11.3c). The most notable difference between the structures of AMCase and CHIT1 is the three residues near the active site (Arg145, His208, and His269). Two of these residues His208 and His269 are different in CHIT1 but are conserved in human AMCase and AMcase of other species (Bussink et al. 2007) (Fig. 11.3d). These three residues make contacts with all the conserved residues within the active site and are proposed to change the pH optimum of AMCase by influencing the pK_a of the catalytic residues Asp138 and Glu140 (Olland et al. 2009). Amino acid residue His269 is Arg269 in CHIT1 and it may be responsible for higher negative charge of AMCase active site lowering down its pH optimum (Fig. 11.3d).

CLPs, such as YKL-40, YKL-39, CHID1, SI-CLP, OVGP1, Ym1, Ym2, are a diverse set of proteins expressed in species-specific manner. In humans and mice, they are expressed in immune and structural cells such as macrophages, neutrophils, epithelial cells, dendritic cells and chondrocytes (Sutherland 2018). CLPs are structurally related to GH18 family chitinases including both human chitinases, AMCase and CHIT1. CLPs structure consists of a 39 kDa TIM-barrel like fold similar to the catalytic domain of active chitinases (Boot et al. 2001; Fusetti et al. 2002). However, most CLPs lack chitin binding domain. Crystal structures of several human or mouse CLPs are solved (Tsai et al. 2004; Meng et al. 2010; Schimpl et al. 2012; Ranok et al. 2015) which clearly illustrates that CLPs bind to chitin oligosaccharides. CLPs retain their ability to bind chitin oligosaccharides with high affinity, however, lack chitinase activity due to the substitution of a key catalytic glutamate in DXXDXDXE motif with either leucine, isoleucine or tryptophan (Lee et al. 2011). Furthermore, it has been shown that substitution of the catalytic residue back to the one in active chitinases recovered the chitinase activity of CLPs (Schimpl et al. 2012).

11.4 Role of Human Chitinases in Diseases

Chitinases are widely distributed in many organisms and play a myriad of biological roles. In bacteria and archaea, chitinases degrade chitin for nitrogen and carbon source (Gao et al. 2003; Bhattacharya et al. 2007). In fungi, chitinases are responsible for ensuring plasticity by remodeling chitin components of cell wall (Langner and Göhre 2016). In insects, chitinases are essential for molting and the growth of insects and are required for the breakdown of chitin components in cell walls (Arakane and Muthukrishnan 2010). Plant chitinases play a major role in dealing with environmental stress such as cold, drought, or salinity (Grover 2012). Mammalian chitinases serve as a protective mechanism against chitin-containing pathogens through their capability to degrade both colloidal chitin as well as chitin present in cell walls of pathogens (Boot et al. 2001). Apart from their protective role against pathogens through chitin degradation, mammalian chitinases are now gaining attention as the key players in innate immune response against fungi, bacteria, and other pathogens.

Several studies pointed out the involvement of two human chitinases, CHIT1 and AMCase, in inflammation-related diseases. Among these two active chitinases, AMCase was widely studied and reported to express at higher levels during type II inflammatory responses in several diseases in both allergic patients and murine models (Zhu et al. 2004; Shen et al. 2015). Chronic respiratory diseases such as asthma is the inflammation of airways. Asthma is characterized by the influx of eosinophils into the lung tissue, mucus metaplasia, hyper-responsiveness, and airflow obstruction. There is compelling evidence for the involvement of T-helper 2 (Th2) cells in asthma via the induction and maintenance of inflammation. Most evidences for the role of chitinases in asthma came from biomarker studies which suggested the involvement of AMCase. AMCase is highly expressed in the lung tissue of asthmatic patients and in animal models of asthma (Zhu et al. 2004; Bierbaum et al. 2005; Yang et al. 2009; Shen et al. 2015). Additionally, higher expression of AMCase has been found in the lung epithelium and alveolar macrophages of ovalbumin(OVA)-sensitized mice (Yang et al. 2009). Administration of AMCase antisera or chitinase inhibitor allosamidin reduced inflammatory cell in BAL fluid of OVA-sensitized mice and thereby alleviated asthma symptoms (Zhu et al. 2004). Moreover, it has been observed that the treatment of allergen-challenged mice with chitinase inhibitors, allosamidin or demethylallosamidin, significantly reduced eosinophilia, a hallmark of allergic inflammation (Matsumoto et al. 2009). Furthermore, the treatment of allergen-challenged mice with AMCase selective inhibitor Bisdionin F alleviated the primary features of allergic inflammation including eosinophilia (Sutherland et al. 2011). Additionally, the administration of compound **3** (Table 11.1), a highly potent AMCase inhibitor showed significant anti-inflammatory efficacy in HDM-induced allergic airway inflammation in mice (Mazur et al. 2018b). These studies suggest that inhibition of AMCase is a good strategy for the development of therapeutics against allergic airway inflammation-related diseases.

Table 11.1 A few chemical classes displaying inhibitory activities against human chitinases

Chemical class	Compound	Structure	Activity		Reference
			AMCase	CHIT1	
Allosamidine and its derivatives	Allosamidin	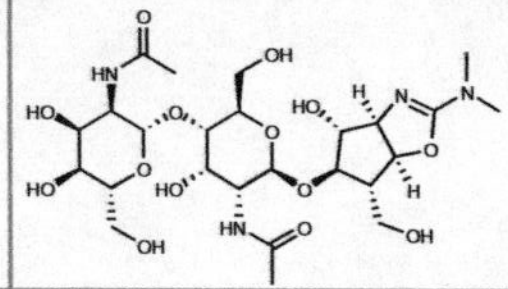	mAMCase IC_{50} = 400 nM	IC_{50} = 40 nM	Rao et al. (2003)
	Demethylallosamidin	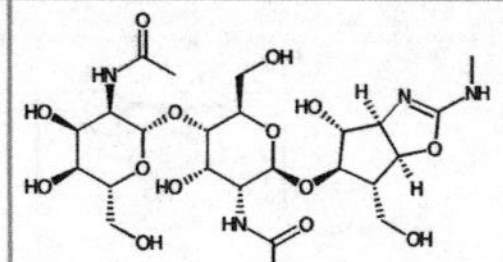	NA	IC_{50} = 1.9 nM	
Cyclopentapeptides	Argadin	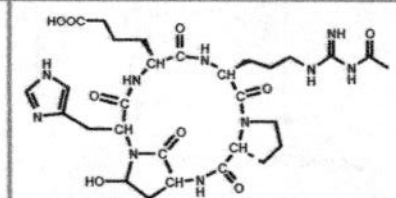	IC_{50} = 1.2 μM	IC_{50} = 0.013 μM	Rao et al. (2005b)
	Argifin	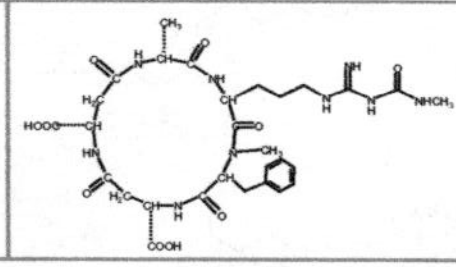	IC_{50} = 0.2 μM	IC_{50} = 4.5 μM	

(continued)

Table 11.1 (continued)

Chemical class	Compound	Structure	Activity		Reference
			AMCase	CHIT1	
Other natural products	Berberine	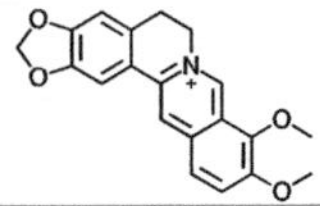	$K_i = 65$ µM	$K_i = 19$ µM	Duan et al. (2018)
	Thalifendine	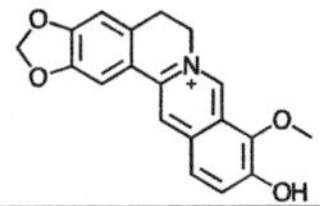	$K_i = 55$ µM	$K_i = 15$ µM	
	Palmatine	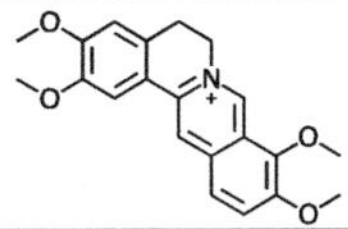	$K_i = 70$ µM	$K_i = 15$ µM	
Deacetylated chitooligosaccharides	$(GlcN)_6$	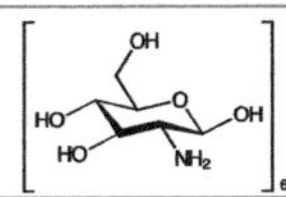	NA	$IC_{50} = 69.5 \pm 10.1$ µM	Chen et al. (2014)
	$(GlcN)_7$	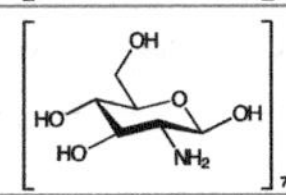	NA	$IC_{50} = 37.8 \pm 8.6$ µM	

(continued)

Table 11.1 (continued)

Chemical class	Compound	Structure	Activity		Reference
			AMCase	CHIT1	
Methyl xanthine derivatives	Thiophylline	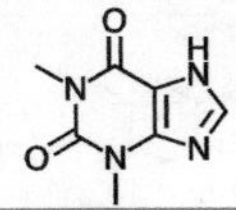	36% inhibition at 1 mM	IC_{50} = > 500 µM	Rao et al. (2005a)
	Caffeine	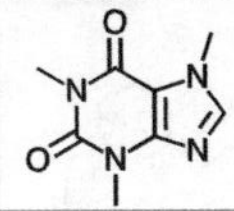	36% inhibition at 1 mM	$IC_{50} = 257 \pm 8$ µM	
	Pentoxiphylline	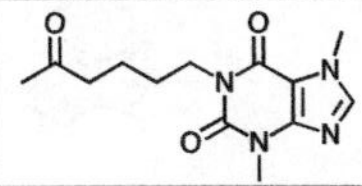	49% inhibition at 1 mM	$IC_{50} = 98 \pm 8$ µM	
Bisdionins	1 (Bisdionin B)	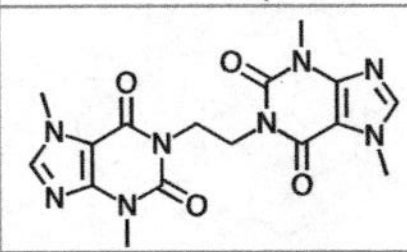	$IC_{50} = 90 \pm 4$ µM	$IC_{50} = 110 \pm 10$ µM	Schüttelkopf et al. (2006)
	Bisdionin C	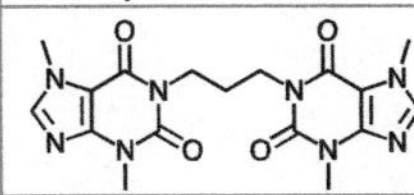	$IC_{50} = 3.4 \pm 0.2$ µM	$IC_{50} = 8.3 \pm 0.7$ µM	Schüttelkopf et al. (2011)
	Bisdionin F	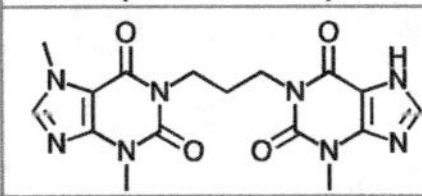	$IC_{50} = 0.92 \pm 0.04$ µM	$IC_{50} = 17.1 \pm 1$ µM	Sutherland et al. (2011)

(continued)

Table 11.1 (continued)

Chemical class	Compound	Structure	Activity		Reference
			AMCase	CHIT1	
Aminotriazoles	2	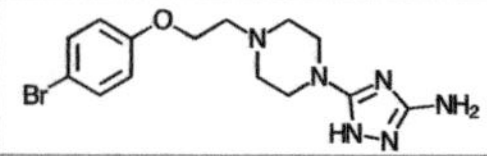	IC_{50} = 0.21 μM	IC_{50} = 4.23 μM	Cole et al. (2010)
	3	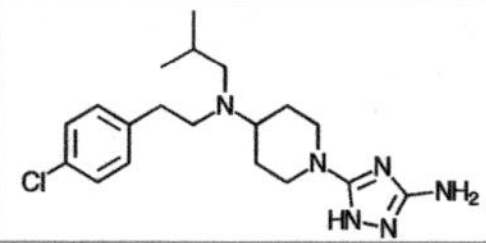	IC_{50} = 14.2 ± 1.0 nM	IC_{50} = 232 ± 48 nM	Mazur et al. (2018b)
	4	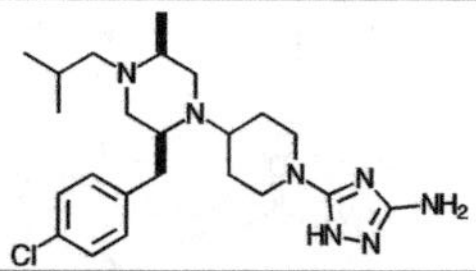	mAMCase IC_{50} = 4170 ± 42 nM	mCHIT1 IC_{50} = 29 ± 4 nM	Mazur et al. (2018a)
Other classes	5	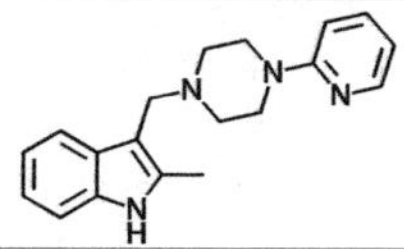	IC_{50} = 0.7 μM	IC_{50} = 1.34 μM	Cole et al. (2010)
	6	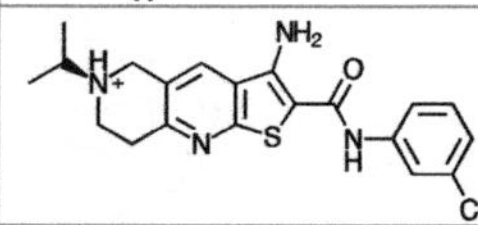	NA	IC_{50} = 54.6 ± 7.2 μM	Jiang et al. (2016)
	7	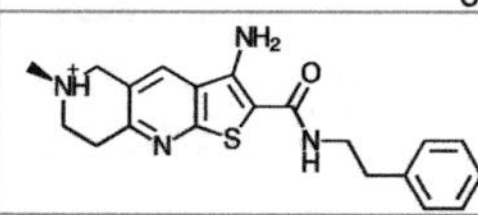	NA	IC_{50} = 67.6 ± 8.0 μM	

However, several other studies reported contrasting results. A transgenic mice overexpressing AMCase showed normal lung function and no signs of inflammation (Reese et al. 2007). In fact, the type II inflammation resulted after the chitin challenge was ameliorated in mice with overexpressing AMCase. Moreover, AMCase deficient mice revealed no role for this enzyme in mouse model of HDM or OVA-induced allergy in the lung (Fitz et al. 2012). Another study reported that in human lungs contrary to mice lungs, the expressed AMCase is mostly inactive (Seibold et al. 2008). Furthermore, AMCase deficient mice exhibit mortality with accumulation of chitin and expression of pro-fibrotic cytokines. These mice develop pulmonary fibrosis, which was ameliorated with the restoration of AMCase activity (Van Dyken et al. 2017). These recent studies suggested that AMCase has protective role and may not be a good drug target against type II inflammation related pathologies such as asthma.

Numerous studies similarly found higher levels of another human chitinase, CHIT1 during type II inflammatory responses in several diseases in both allergic patients and murine models. Most evidences for the role of CHIT1 in various diseases came from the biomarker studies. CHIT1 has also been reported to be involved in asthma and airway hyper-responsiveness (Gavala et al. 2013). Elevated CHIT1 activity and levels were observed after allergen challenge and these levels correlated with the levels of inflammatory cells, T cell chemokines, and other pro-fibrotic factors. CHIT1 was also found to be responsible for chitinase activity in human lung (Seibold et al. 2008). Another study reported the prevalence of CHIT1 24-base pair duplication allele in patients with severe asthma (Livnat et al. 2014). This allele is reported to have reduced CHIT1 activity. However, numerous studies also showed no association between CHIT1 and airway diseases (Shuhui et al. 2009; Létuvé et al. 2010). Mice with CHIT1 null mutant showed significantly higher type II inflammatory responses to HDM or OVA challenge. Furthermore, this study suggests a protective role of CHIT1 in allergic airway responses via regulation of TGF expression (Hong et al. 2018).

As described here, various studies tried to elucidate the role of chitinases in inflammatory diseases. However, whether chitinase functions in a protective or adverse role in inflammation is not very clear. Studies in recent years begin to relate differences in inflammatory responses with chitin degradation and recognition. As chitin is not synthesized by mammals, it is considered as a target of mammalian immune system (Elieh Ali Komi et al. 2018). One study showed the direct involvement of chitin in allergic responses (Reese et al. 2007). Administration of chitin bead in the lungs of mice expressing GFP-enhanced IL-4 led to the recruitment of GFP positive basophils and eosinophils. Further, it was shown that macrophage activation was a crucial step in the recruitment of these cells. Moreover, it has been shown that intranasal administration of chitin particles activates alveolar macrophages to express cytokines including IL-12, tumor necrosis factor-α (TNF-α) and IL-18 (Shibata et al. 1997). Several receptor proteins have been identified that are believed to recognize and bind chitin. These include FIBCD1, NKR-P1, and RegIIIc (Semeňuk et al. 2001; Cash et al. 2006; Thomsen et al. 2011; Bueter et al. 2013). In addition, Toll-like receptor 2, dectin-1 and mannose receptor are also known to be involved in immune response to chitin (Bueter et al. 2013; Elieh Ali Komi et al. 2018; Fuchs et al. 2018). These chitin receptors are present on the surface of macrophages. These receptors upon interaction with chitin stimulate the production of cytokines and mediators such as IL-17, IL-18, IL-23, and TNF-α which in turn stimulate the production of chitinases and CLPs (Zhu et al. 2004;Da Silva et al. 2008, 2009; Amarsaikhan and Templeton, 2015). It has been also reported that chitin exposure increased the expression of IL-25, IL-33, TLSP, and CCL2 in lung epithelial cells. These factors induce type II innate lymphoid cells to secrete IL-5 and IL-13 cytokines which are critical for the accumulation of eosinophils and macrophages (Van Dyken et al. 2014). Chitinases modulate the local and/or circulating concentration of chitin in the body and thereby regulating the immune response to chitin. However, the mechanism of immune response stimulation is not well understood and two alternative hypotheses have been suggested previously. According to one hypothesis, chitinases degrade exogenous chitin from sources such as fungi or HDM which consequently prevent chitin from stimulating immune responses. In the absence of chitinases, chitin may accumulate in tissues, activate innate immune cells, thereby triggering an excessive inflammatory response (Alvarez 2014). Interestingly, a contrasting hypothesis suggests that the size of chitin fragment is important, as large fragments are generally inert while smaller fragments produced due to chitinase activity trigger inflammatory response (Da Silva et al. 2009; Kogiso et al. 2011). Increasing number of evidences suggest that immune response and inflammatory cell recruitment to chitin exposure is influenced by the size of chitin particles, shape of particles, tissue of exposure, exposure duration, etc. (Alvarez 2014; Amarsaikhan and Templeton 2015). In one study, when chitin was intranasally and intraperitoneally delivered to mice, macrophage activation and eosinophil migration were observed in both routes of administration. However, a transient neutrophilic response was only observed in the case of intraperitoneal challenge (Reese et al. 2007). Similarly, highly purified chitin did not stimulate, while chitin particles of 40–70 μm in size were able to stimulate the production of TNF-α and anti-inflammatory response (Da Silva et al. 2009; Mora-Montes et al. 2011). Chitin particles of 1–10 μm in size induced both anti-inflammatory and Th1 protective responses (Da Silva et al. 2009). Chitin particles of about 0.2 μm in size were not immunogenic (Alvarez 2014). One recent study identified six GlcNAc unit long chitin chain as the smallest immunogenic unit (Fuchs et al. 2018). They further demonstrated that Toll-like receptor 2 is a primary fungal chitin sensor on human and murine immune cells and chitin oligomers bind to Toll-like receptor 2 with affinity in nanomolar range. The extent of chitin degradation is determined by chitinase activity in airways, which further influence immune responses. Differences in inflammatory responses (Roy et al. 2012) by different particle or fragment sizes might be due to the deployment of distinct chitin receptors that activate different signaling pathways. In fact, several studies have reported evidences for the stimulation of multiple signaling pathways upon chitin challenge (Reese et al. 2007; Van Dyken et al. 2014).

AMCase and CHIT1 were implicated in several other diseases such as Gaucher disease, idiopathic pulmonary fibrosis, sarcoidosis, chronic obstructive pulmonary disease (COPD), and Alzheimer's disease. However, their role is not clearly understood. Most of these evidences came from biomarker studies where either high expression or high activity of AMCase or CHIT1 was observed in patients or animal models. Gaucher disease is a lysosomal storage disorder caused by the accumulation of glucocerebroside in macrophages due to the deficiency of beta-glucosidase caused by autosomal recessive inheritance (Grabowski 2012). These affected macrophages known as Gaucher cells than displace normal cells in bone marrow and visceral organs causing organ dysfunction, skeletal manifestations, thrombopenia, etc. (Grabowski et al. 2015). Gaucher cells secrete biomarkers in the blood and CHIT1 is one such biomarkers used for the diagnosis of Gaucher disease (Hollak et al. 1994; Wajner et al. 2004; van Dussen et al. 2014), and Niemann–Pick diseases (Wajner et al. 2004). It has been observed that circulating levels of CHIT1 increased by 1000 folds in patients with Gaucher disease when compared with healthy persons. CHIT1 concentrations in serum were higher in patients with sarcoidosis than that of healthy persons (Bargagli et al. 2013). Increased CHIT1 activity in airways was observed in patients with chronic obstructive pulmonary disease (COPD) as compared to control subjects (Seibold et al. 2008; Létuvé et al. 2010; James et al. 2016). Numerous evidences suggest that inflammation is involved in the pathogenesis of Alzheimer's disease (Stefano et al. 2007; Heppner et al. 2015). CHIT1 activity was found to be significantly increased in cerebrospinal fluid of patients with Alzheimer's disease as compared to control (Mattsson et al. 2011; von Arnim et al. 2011; Watabe-Rudolph et al. 2012). However, the exact mechanism of CHIT1 involvement in Alzheimer's disease is not known. CHIT1 has been shown to play a protective role by enhancing TGFβ1 mediated clearance of amyloid β (Wang et al. 2018).

Apart from AMCase and CHIT1, increased expression of CLPs such as YKL-40 was also observed in Th2 type inflammation (Chupp et al. 2007; Komi et al. 2016). Additionally, increased levels of YKL-40 in serum and lungs of asthma patients were observed as compared to control (Chupp et al. 2007). Furthermore, it was shown that YKL-40 is the central component of Th2 inflammatory responses. In BRP-39 (homolog of YKL-40 in mice) deleted mice, type II immune response declined after OVA exposure. However, these immune responses were rescued by YKL-40 overexpression (Lee et al. 2009; Lee et al. 2011). It was further shown that YKL-40 deficiency alleviates IL-13 dependent fibrosis suggesting its critical role in Th2 inflammation (Lee et al. 2009; Kang et al. 2015). High YKL-40 levels were also found in lungs of patients with idiopathic pulmonary fibrosis (Furuhashi et al. 2010; Zhou et al. 2014). Other CLPs, Ym1, and Ym2 were also identified as allergy-associated protein in mice allergy model (Zhao et al. 2005; Song et al. 2008). Higher number of epithelial cells and macrophages expressing YKL-40 in lung samples of idiopathic pulmonary fibrosis patients was observed (Zhou et al. 2014). YKL-40 expression was also found to be elevated in various cancer cells compared with normal cells (Johansen et al. 2009; Choi et al. 2010; Lee et al. 2011). Several studies also reported the association of YKL-40 and cancer metastasis (Jensen et al. 2003; Ma et al. 2015). Knock-down of YKL-40 resulted in the decrease of metastases in mice lung tissues and human cancer cell lines (Kim et al. 2018). Increased levels of YKL-40 in cerebrospinal fluid were also found in Alzheimer's disease patients (Hellwig et al. 2015; Janelidze et al. 2016; Zhang et al. 2018).

Although the studies described above provide clues about the role of chitin, human chitinases, and CLPs in inflammation and development of various inflammatory diseases, their specific role needs to be elucidated. Moreover, there is no clear mechanistic understanding of how they regulate inflammation and immune responses. Whether they function in a protective or adverse role in inflammation have been found to be controversial. As inflammation is a very complex process with multiple triggers, effectors, and mechanisms, and the observed outcome (pro- or anti-inflammatory) may depend on how the experiment was conducted and what "markers" were monitored. Studies have shown that size and shape of chitin particles, dose of administration, tissue of exposure and exposure duration could affect inflammatory response in different manner. Chitin particles with different sizes produced as a result of chitinase activity could interact with different cell surface receptors and may stimulate macrophages to express different effector molecules activating distinct signaling cascades. Activated macrophages may secrete pro-inflammatory cytokines, such as TNF-α, IL-12, IL-18, which recruit eosinophils, neutrophils, and basophils. These cells produce type II inflammatory response by secreting Th2 cytokines such as IL-4, IL-5, and IL-13. On the other hand, chitin can also activate macrophages to produce type I cytokines to suppress type II inflammatory response. Furthermore, studies have shown that chitinases and CLPs may not only be playing direct protective role by degrading chitin but also may be involved in augmenting the immune response against chitin and other allergens.

11.5 Current Status of CHIT1 and AMCase Inhibitor Development

Allosamidin was the first chitinase inhibitor discovered about three decades ago from the mycelia of Streptomyces species (Sakuda et al. 1986, 1987b). Allosamidin has a pseudotrisaccharide structure consisting of one allosamizoline and two N-acetyl-D-allosamine moieties (Sakuda et al. 1987a, 1988) (Table 11.1). It mimics a transition state intermediate in chitin hydrolysis (Sakuda et al. 2001). Allosamidin is a potent inhibitor against all family GH18 chitinases (Berecibar et al. 1999). However, it does not inhibit family GH19 chitinases (Sakuda et al. 1993). Family GH18 chitinases hydrolyze chitin via a substrate-assisted mechanism which involves the production of an oxazolium ion intermediate. Allosamizoline moiety in allosamidin mimics this transition state intermediate (Tews et al. 1997). Family GH19 chitinases, on the other hand, employ a fold-and-reaction mechanism (Monzingo et al. 1996) which is different from the substrate-assisted mechanism of family GH18 chitinases. Up to now, seven naturally occurring allosamidins (allosamidin, methylallosamidin, demethylallosamidin, glucoallosamidin A, glucoallosamidin B, methyl-N-demethylallosamidin, and didemethylallosamidin) have been identified (Nishimoto et al. 1991). Allosamidin is the most widely studied among them. The inhibitory potency of allosamidin against both AMCase (Boot et al. 2001) and CHIT1 (Boot et al. 2001; Rao et al. 2003) is listed in Table 11.1. Allosamidin was utilized as a chemical probe to demonstrate the association between AMCase and asthma. It was shown that AMCase expression elevated upon exposure to allergen or IL-13 induced inflammation in lung (Zhu et al. 2004) and the administration of AMCase inhibitor allosamidin or demethylallosamidin (Table 11.1) suppressed allergen-induced eosinophilia in murine asthma model (Zhu et al. 2004; Matsumoto et al. 2009). Both allosamidin and demethylallosamidin inhibited chitinase activity in vivo, however, reduction in allergen or IL-13-induced airway hyper-responsiveness was only observed in case of demethylallosamidin.

Another class of reported family GH18 chitinase inhibitors include peptide-like compounds that can mimic protein–carbohydrate interactions. These include cyclic proline-containing dipeptides (Izumida et al. 1996; Houston et al. 2002) and cyclic pentapeptides argadin (Arai et al. 2000b) and argifin (Arai et al. 2000a; Omura et al. 2000) (Table 11.1). Cyclic proline-containing dipeptides were isolated from the broth of a marine bacterium while the cyclopeptides were isolated from Clonostachys sp. FO-7314 and Gliocladium sp. FTD-0668. Argadin and argifin were reported to inhibit both human chitinases (Rao et al. 2005b; Goedken et al. 2011). AMCase inhibitory activity of argadin and argifin were evaluated using human and murine AMCases (Goedken et al. 2011). This study reported slight differences in the AMCase inhibitory activity of argadin and argifin which might be due to differences in the purification or method of expression for human and murine AMCases (Goedken et al. 2011). In this study, Argifin was found to be much more potent than argadin for both human and murine AMCases (Table 11.1). Argadin and argifin also inhibited chitinase activity in BAL fluid obtained from OVA-challenged mice with similar potency as recombinant enzyme (Goedken et al. 2011). Although both argadin and argifin displayed inhibitory activity against human CHIT1, Argadin was found to be much more potent than Argifin (Rao et al. 2005b). Berberine and its analogs constitute another class of natural product chitinase inhibitors (Duan et al. 2018) (Table 11.1). These compounds were competitive inhibitors of family GH18 chitinases and GH20 β-*N*-acetyl-D-hexosaminidase. Barberine and two of its analogs inhibited AMCase and CHIT1 with moderate potency and are nonselective against either enzyme (Duan et al. 2018). Deacetylated chitooligosaccharides represent another class of human chitinase inhibitors. A series of fully deacetylated chitooligosaccharides $(GlcN)_{2-7}$ was reported to inhibit CHIT1 where the potency was observed to increase with the addition of GlcN unit. $(GlcN)_7$ was found to be the most potent whereas $(GlcN)_2$ was the least potent (Chen et al. 2014) (Table 11.1).

Some of the aforementioned natural products were potent AMCase and CHIT1 inhibitors, but their use as lead molecules in drug discovery was significantly impeded by their high molecular weight, presence of several stereocenters, limited availability, chemical complexity, and difficulty in synthesis. Furthermore, these properties also hampered their use in in vivo studies to investigate the role of AMCase and CHIT1 in the development of various diseases. With a goal to develop drug-like chitinase inhibitors, several groups reported different chemical classes of drug-like compounds. Xanthine derivatives were among the first reported drug-like inhibitors of human chitinases identified by screening a commercially available library of drug molecules (Rao et al. 2005a). Screening resulted in the identification of three hits (theophylline, caffeine, and pentoxifylline) with 1,3-dimethylxanthine substructure. All three hits, however, were found to be only weak inhibitors of CHIT1 and AMCase (Rao et al. 2005a) (Table 11.1). Crystallographic analyses of theophylline, caffeine, and pentoxifylline with AfChiB1 revealed interactions mimicking allosamidin (Rao et al. 2005a). Although the inhibitory values of these methylxanthine derivatives were not very high, they represent a chemical class which is easily available, low molecular weight, low cost and generally considered safe. Additionally, their binding mode with AfChiB1 stimulated the identification of other drug-like inhibitors. In one such study, a library of 5.1 million commercially available compounds was filtered for 3-methylxanthine substructure (Schüttelkopf et al. 2006). Further prioritization of hits utilizing molecular docking and visual inspection followed by evaluation of inhibitory activity against AfChiB1 resulted in the identification of compound 1 (1-(2-(theobromine-1-yl)ethyl)-theobromine) which is actually two linked caffeine molecules. Compound 1 showed improvement of two orders of magnitude when compared with caffeine, its parent compound (Schüttelkopf et al. 2006). Compound 1 also displayed moderate inhibition of CHIT1 and murine AMCase. Crystal structure of Compound 1 in complex with AfChiB1 revealed strained geometry of this compound within the binding pocket (Schüttelkopf et al. 2006) when compared to its parent compound (Rao et al. 2005a). Syntheses of several dicaffeine scaffold derivatives (named as Bisdionins) with variable linker length to alleviate ligand strain resulted in compounds with improved human chitinase inhibitory potency (Table 11.1) (Schüttelkopf et al. 2011). Especially, Bisdionin C displayed low micromolar inhibition of both AMCase and CHIT1. Although Bisdionin C displayed reasonable potency with excellent drug-like properties, it is nonselective and inhibits both human chitinases with more or less equal potency. As selective inhibitors are important to understand functional differences in AMCase and CHIT1, Bisdionin F was synthesized taking hints from the co-crystal structure of Bisdionin C in complex with AMCase (Sutherland et al. 2011). Crystal structure of Bisdionin C in complex with AMCase revealed that Bisdionin C occupies −1, −2 and −3 GlcNAc binding subsites of AMCase and a methyl group at N7 position of xanthine ring imposes an unfavorable conformation of AMCase pocket residue Asp138 (Fig. 11.4a). Bisdionin F was synthesized by removing this methyl group from the xanthine scaffold. This structure-guided optimization resulted in the improvement of AMCase inhibitory activity by one order of magnitude for Bisdionin F when compared with Bisdionin C (Sutherland et al. 2011) (Table 11.1). Co-crystal structure of Bisdionin F with AMCase revealed that Asp138 adopted a favorable conformation interacting with the compound via an additional hydrogen bond at N7 position (Sutherland et al. 2011) (Fig. 11.4b). Furthermore, as improvement in potency was

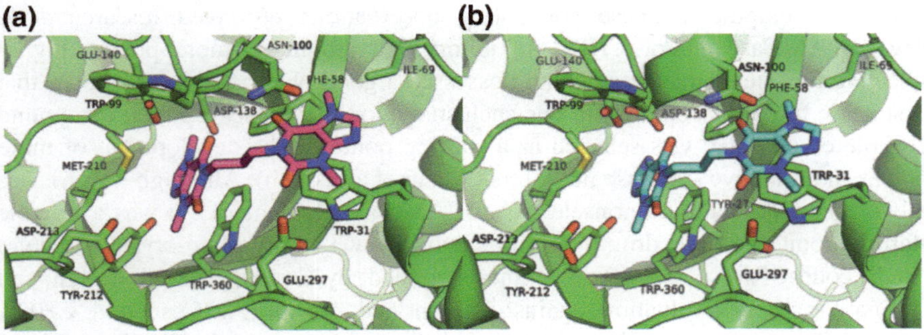

Fig. 11.4 Co-crystal structures of CHIT1 with **a** Bisdionin C and **b** Bisdionin F

observed only for AMCase but not for CHIT1, a 20-fold selectivity for AMCase was obtained (Sutherland et al. 2011). Bisdionin F also displayed similar effects when evaluated in vivo and its treatment attenuated chitinase activity while alleviating some hallmarks of allergic inflammation including eosinophilia.

Another study employed a combination of high-throughput screening, fragment-based drug design and in silico screening to identify several drug-like inhibitors of AMCase and CHIT1 (Cole et al. 2010). Although some of these reported compounds were highly potent AMCase and reasonably active CHIT1 inhibitors (compound 2 and 5), most of them lack selectivity toward either enzyme (Table 11.1). The most potent of the reported AMCase inhibitors (compound 2) was orally active and reduced the chitinase activity in BAL fluid of mice challenged with the combination of HDM and cockroach allergens (Cole et al. 2010). Co-crystal structures of many of the reported compounds were obtained that stimulated research toward the development of more potent and selective human chitinase inhibitors. Another class of drug-like inhibitors were derived from the natural product chitinase inhibitor argifin (Arai et al. 2000a; Omura et al. 2000). The amino-(3-methylureido)-methaniminium functional group of argifin was used as a query to identify structurally similar compounds. Resulting hits were further prioritized with molecular docking and selected compounds were evaluated for inhibitory activities against AMCase. Seven of the tested compounds showed IC_{50} values of ≤ 100 µM (Wakasugi et al. 2013). No CHIT1 inhibitory activities were reported for these compounds. In pursuit of the development of agrochemicals, our group has also identified two new chemical classes of compounds demonstrating inhibitory activity against chitinase of destructive crop pest Asian corn borer, *Ostrinia furnacalis*. These two chemical scaffolds were identified following a hierarchical virtual screening approach (Kumar and Zhang 2015) where a combination of shape similarity and molecular docking was employed to filter a library of about four million commercially available compounds. One of these classes (compound 6 and 7) exhibited broad-spectrum activity against various chitinases with moderate inhibitory activity against CHIT1 (Jiang et al. 2016).

As previously described in this manuscript, contrasting observations were obtained in different experiments meant to study the role of AMCase/CHIT1 in disease development. These contrasting observations prompted researchers to develop AMCase and/or CHIT1 inhibitors that are much more potent, highly selective for one of the enzyme and possessed a good pharmacokinetic profile that is suitable for animal studies. In one such study, the previously reported compound 2 (Cole et al. 2010) was selected as a starting point for the development of more potent and selective AMCase inhibitors (Mazur et al. 2018b). Although wyeth1 was relatively less potent and nonselective for AMCase (Table 11.1), it was a suitable starting point due to its drug-like profile and the availability of co-crystal structure (PDB code 3RM4). Chemical synthesis guided by structure-based design and chitinase activity evaluation against human and murine AMCase and CHIT1 resulted in a series of aminotriazoles with high AMCase and CHIT1 inhibitory activity (Mazur et al. 2018b). Several compounds were found to be potent against AMCase with IC_{50} in low nanomolar range and reasonable selectivity against CHIT1. Compound 3 was reported to be especially interesting as AMCase inhibitor due to its high potency, specificity and good pharmacokinetic properties (Table 11.1). Compound 3 also demonstrated significant reduction in inflammation in HDM-induced allergic airway inflammation model where reduction in AMCase activity highly correlated with inflammatory cell influx in BAL fluid (Mazur et al. 2018b). However, compound 3 was not pursued further due to dopamine receptor off-target activity and potential concerns about safety. Utilizing compound 2 scaffold as a starting point the same research group also reported compound 4 that was highly potent and possessed 143-fold selectivity for mouse CHIT1 over mouse AMCase (Mazur et al. 2018a) (Table 11.1).

11.6 Conclusion

Humans express two chitinases, AMCase and CHIT1 to degrade chitin from environmental sources. Recent evidences suggest that chitinases may also play a key role in innate immune response against pathogens. Several studies have shown the involvement of both AMCase and CHIT1 in the development of various diseases such as asthma, COPD, idiopathic pulmonary fibrosis, Alzheimer disease, and cancer. However, the mechanism of their involvement is not well understood. Most of the information about their role came from biomarker studies, where either human chitinases were highly expressed or elevated chitinase activity was observed. Moreover, their role in disease development is not devoid of controversies. There is no clear understanding about the mechanism of their immune regulation and whether they play a protective or adverse role in inflammation is still controversial. As inflammation is a very complex process with multiple triggers, effectors, and mechanisms, and the observed outcome (pro- or anti-inflammatory) may depend on how the experiment was conducted, what inducer was used, dose and method of administration, exposure duration and what "markers" were monitored. Recent studies also pointed out the role of chitin recognition and degradation in the activation of pro- and anti-inflammatory responses. Chitin fragments of different sizes may interact with different receptors on the surface of macrophages triggering a distinct immune response. These studies suggest that chitinases may play a direct protective role by degrading chitin as well as may be involved in augmenting the immune response against chitin and other allergens. However, further characterization of interactions between chitin and chitin receptors and associated pathways will be required to understand the biological role of chitinases in humans. Recent reports also highlighted the involvement of various CLPs and some of the CLPs have been shown to play a central role in human responses to pathogens or disease conditions. Further work is required to study their role in inflammation. Lack of understanding of chitinase role in disease development did not prevent researchers from developing inhibitors of AMCase and CHIT1. Though initial efforts were focused toward natural product inhibitors, drug-like inhibitors with high potency and better pharmacokinetic properties were later developed. These inhibitors played a significant role in studying both AMCase and CHIT1 and implications of inhibiting these proteins. However, studying the role of these proteins is much more challenging and requires the specific inhibition of one of these proteins. Identifying selective inhibitors against these proteins is challenging due to the similar fold, functions and very conserved active site. This may be the reason why almost all of the inhibitors developed for these two proteins are nonselective or only weakly selective. The level of selectivity of current generation of inhibitors may not be adequate for in vivo studies as much higher dose is utilized and with that dose it is very difficult to avoid inhibition of both proteins. Therefore, the discovery of highly selective inhibitors for AMCase and CHIT1 is necessary in order to better understand the roles of chitin and chitinases in inflammation and development of inflammatory diseases.

References

Akagi K-I, Watanabe J, Hara M, Kezuka Y, Chikaishi E, Yamaguchi T, Akutsu H, Nonaka T, Watanabe T, Ikegami T (2006) Identification of the substrate interaction region of the chitin-binding domain of streptomyces griseus chitinase C. J Biochem 139:483–493

Alvarez F (2014) The effect of chitin size, shape, source and purification method on immune recognition. Molecules 19:4433

Amarsaikhan N, Templeton SP (2015) Co-recognition of β-glucan and chitin and programming of adaptive immunity to Aspergillus fumigatus. Front Microbiol 6:344–344

Arai N, Shiomi K, Iwai Y, Omura S (2000a) Argifin, a new chitinase inhibitor, produced by Gliocladium sp. FTD-0668. II. Isolation, physico-chemical properties, and structure elucidation. J Antibiot (Tokyo) 53:609–614

Arai N, Shiomi K, Yamaguchi Y, Masuma R, Iwai Y, Turberg A, Kolbl H, Omura S (2000b) Argadin, a new chitinase inhibitor, produced by clonostachys sp.FO-7314. Chem Pharm Bull 48:1442–1446

Arakane Y, Muthukrishnan S (2010) Insect chitinase and chitinase-like proteins. Cell Mol Life Sci 67:201–216

Bargagli E, Bennett D, Maggiorelli C, Di Sipio P, Margollicci M, Bianchi N, Rottoli P (2013) Human chitotriosidase: a sensitive biomarker of sarcoidosis. J Clin Immunol 33:264–270

Berecibar A, Grandjean C, Siriwardena A (1999) Synthesis and biological activity of natural aminocyclopentitol glycosidase inhibitors: mannostatins, trehazolin, allosamidins, and their analogues. Chem Rev 99:779–844

Bhattacharya D, Nagpure A, Gupta RK (2007) Bacterial chitinases: properties and potential. Crit Rev Biotechnol 27:21–28

Bierbaum S, Nickel R, Koch A, Lau S, Deichmann KA, Wahn U, Superti-Furga A, Heinzmann A (2005) Polymorphisms and haplotypes of acid mammalian chitinase are associated with bronchial asthma. Am J Respir Crit Care Med 172:1505–1509

Boot RG, Blommaart EFC, Swart E, Ghauharali-Van Der Vlugt K, Bijl N, Moe C, Place A, Aerts JMFG (2001) Identification of a novel acidic mammalian chitinase distinct from chitotriosidase. J Biol Chem 276:6770–6778

Boot RG, Bussink AP, Verhoek M, de Boer PA, Moorman AF, Aerts JMFG (2005) Marked differences in tissue-specific expression of chitinases in mouse and man. J Histochem Cytochem 53:1283–1292

Boot RG, Renkema GH, Strijland A, Van Zonneveld AJ, Aerts JMFG (1995) Cloning of a cDNA encoding chitotriosidase, a human chitinase produced by macrophages. J Biol Chem 270:26252–26256

Bueter CL, Specht CA, Levitz SM (2013) Innate sensing of chitin and chitosan. PLoS Pathog 9: e1003080

Bussink AP, Speijer D, Aerts JMFG, Boot RG (2007) Evolution of mammalian chitinase(-Like) members of family 18 glycosyl hydrolases. Genetics 177:959–970

Cash HL, Whitham CV, Behrendt CL, Hooper LV (2006) Symbiotic bacteria direct expression of an intestinal bactericidal lectin. Science 313:1126–1130

Chang N-CA, Hung S-I, Hwa K-Y, Kato I, Chen J-E, Liu C-H, Chang AC (2001) A macrophage protein, Ym1, transiently expressed during inflammation is a novel mammalian lectin. J Biol Chem 276:17497–17506

Chen L, Zhou Y, Qu M, Zhao Y, Yang Q (2014) Fully deacetylated chitooligosaccharides act as efficient glycoside hydrolase family 18 chitinase inhibitors. J Biol Chem 289:17932–17940

Choi IK, Kim YH, Kim JS, Seo JH (2010) High serum YKL-40 is a poor prognostic marker in patients with advanced non-small cell lung cancer. Acta Oncol 49:861–864

Chou Y-T, Yao S, Czerwinski R, Fleming M, Krykbaev R, Xuan D, Zhou H, Brooks J, Fitz L, Strand J, Presman E, Lin L, Aulabaugh A, Huang X (2006) Kinetic characterization of recombinant human acidic mammalian chitinase. Biochemistry 45:4444–4454

Chupp GL, Lee CG, Jarjour N, Shim YM, Holm CT, He S, Dziura JD, Reed J, Coyle AJ, Kiener P, Cullen M, Grandsaigne M, Dombret M-C, Aubier M, Pretolani M, Elias JA (2007) A chitinase-like protein in the lung and circulation of patients with severe asthma. N Engl J Med 357:2016–2027

Cole DC, Olland AM, Jacob J, Brooks J, Bursavich MG, Czerwinski R, Declercq C, Johnson M, Joseph-Mccarthy D, Ellingboe JW, Lin L, Nowak P, Presman E, Strand J, Tam A, Williams CMM, Yao S, Tsao DHH, Fitz LJ (2010) Identification and characterization of acidic mammalian chitinase inhibitors. J Med Chem 53:6122–6128

Da Silva CA, Chalouni C, Williams A, Hartl D, Lee CG, Elias JA (2009) Chitin is a size-dependent regulator of macrophage TNF and IL-10 production. J Immunol 182:3573–3582

Da Silva CA, Hartl D, Liu W, Lee CG, Elias JA (2008) TLR-2 and IL-17A in Chitin-Induced macrophage activation and acute inflammation. J Immunol 181:4279–4286

Davies GJ, Wilson KS, Henrissat B (1997) Nomenclature for sugar-binding subsites in glycosyl hydrolases. Biochem J 321:557–559

Duan Y, Liu T, Zhou Y, Dou T, Yang Q (2018) Glycoside hydrolase family 18 and 20 enzymes are novel targets of the traditional medicine berberine. J Biol Chem

Elieh Ali Komi D, Sharma L, Dela Cruz CS (2018) Chitin and its effects on inflammatory and immune responses. Clin Rev Allergy Immunol 54:213–223

Fadel F, Zhao Y, Cachau R, Cousido-Siah A, Ruiz FX, Harlos K, Howard E, Mitschler A, Podjarny A (2015) New insights into the enzymatic mechanism of human chitotriosidase (CHIT1) catalytic domain by atomic resolution X-ray diffraction and hybrid QM/MM. Acta Crystallograph Sect D 71:1455–1470

Fadel F, Zhao Y, Cousido-Siah A, Ruiz FX, Mitschler A, Podjarny A (2016) X-Ray crystal structure of the full length human chitotriosidase (CHIT1) reveals features of its chitin binding domain. PLoS One 11:e0154190

Fitz LJ, Declercq C, Brooks J, Kuang W, Bates B, Demers D, Winkler A, Nocka K, Jiao A, Greco RM, Mason LE, Fleming M, Quazi A, Wright J, Goldman S, Hubeau C, Williams CMM (2012) Acidic mammalian chitinase is not a critical target for allergic airway disease. Am J Respir Cell Mol Biol 46:71–79

Foster JM, Zhang Y, Kumar S, Carlow CKS (2005) Parasitic nematodes have two distinct chitin synthases. Mol Biochem Parasitol 142:126–132

Fuchs K, Gloria YC, Wolz OO, Herster F, Sharma L, Dillen CA, Täumer C, Dickhöfer S, Bittner Z, Dang TM, Singh A, Haischer D, Schlöffel MA, Koymans KJ, Sanmuganantham T, Krach M, Roger T, Le Roy D, Schilling NA, Frauhammer F, Miller LS, Nürnberger T, Leibundgut-Landmann S, Gust AA, Macek B, Frank M, Gouttefangeas C, Dela Cruz CS, Hartl D, Weber AN (2018) The fungal ligand chitin directly binds TLR2 and triggers inflammation dependent on oligomer size. EMBO Rep

Fukamizo F (2000) Chitinolytic enzymes: catalysis, substrate binding, and their application. Curr Protein Pept Sci 1:105–124

Furuhashi K, Suda T, Nakamura Y, Inui N, Hashimoto D, Miwa S, Hayakawa H, Kusagaya H, Nakano Y, Nakamura H, Chida K (2010) Increased expression of YKL-40, a chitinase-like protein, in serum and lung of patients with idiopathic pulmonary fibrosis. Respir Med 104:1204–1210

Fusetti F, Von Moeller H, Houston D, Rozeboom HJ, Dijkstra BW, Boot RG, Aerts JMFG, Van Aalten DMF (2002) Structure of human chitotriosidase: implications for specific inhibitor design and function of mammalian chitinase-like lectins. J Biol Chem 277:25537–25544

Gao J, Bauer MW, Shockley KR, Pysz MA, Kelly RM (2003) Growth of hyperthermophilic archaeon *Pyrococcus furiosus* on chitin involves two family 18 chitinases. Appl Environ Microbiol 69:3119–3128

Gavala ML, Kelly EA, Esnault S, Kukreja S, Evans MD, Bertics PJ, Chupp GL, Jarjour NN (2013) Segmental allergen challenge enhances chitinase activity and levels of CCL18 in mild atopic asthma. Clin Experiment Allergy 43:187–197

Giraud-Guille MM, Bouligand Y (1986) Chitin-protein molecular organization in arthropod. In: Muzzarelli R, Jeuniaux C, Gooday GW (eds.) Chitin in nature and technology, Boston, MA, Springer US, pp 29–35

Glaser L, Brown DH (1957) The synthesis of chitin in cell-free extracts of neurospora crassa. J Biol Chem 228:729–742

Goedken ER, O'Brien RF, Xiang T, Banach DL, Marchie SC, Barlow EH, Hubbard S, Mankovich JA, Jiang J, Richardson PL, Cuff CA, Cherniack AD (2011) Functional comparison of recombinant acidic mammalian chitinase with enzyme from murine bronchoalveolar lavage. Protein Expr Purificat 75:55–62

Gonçalves IR, Brouillet S, Soulié M-C, Gribaldo S, Sirven C, Charron N, Boccara M, Choquer M (2016) Genome-wide analyses of chitin synthases identify horizontal gene transfers towards bacteria and allow a robust and unifying classification into fungi. BMC Evol Biol 16:252

Gooday GW (1990) Physiology of microbial degradation of chitin and chitosan. Biodegradation 1:177–190

Grabowski GA (2012) Gaucher disease and other storage disorders. ASH Educat Prog Book 2012:13–18

Grabowski GA, Zimran A, Ida H (2015) Gaucher disease types 1 and 3: Phenotypic characterization of large populations from the ICGG gaucher registry. Am J Hematol 90: S12–S18

Grover A (2012) Plant chitinases: genetic diversity and physiological roles. Crit Rev Plant Sci 31:57–73

Hamid R, Khan MA, Ahmad M, Ahmad MM, Abdin MZ, Musarrat J, Javed S (2013) Chitinases: an update. J Pharm Bioallied Sci 5:21–29

Hellwig K, Kvartsberg H, Portelius E, Andreasson U, Oberstein TJ, Lewczuk P, Blennow K, Kornhuber J, Maler JM, Zetterberg H, Spitzer P (2015) Neurogranin and YKL-40: independent markers of synaptic degeneration and neuroinflammation in Alzheimer's disease. Alzheimers Res Ther 7:74–74

Henrissat B (1999) Classification of chitinases modules

Heppner FL, Ransohoff RM, Becher B (2015) Immune attack: the role of inflammation in Alzheimer disease. Nat Rev Neurosci 16:358

Hollak CE, Van Weely S, Van Oers MH, Aerts JM (1994) Marked elevation of plasma chitotriosidase activity. A novel hallmark of Gaucher disease. J Clin Investig 93:1288–1292

Hong JY, Kim M, Sol IS, Kim KW, Lee C-M, Elias JA, Sohn MH, Lee CG (2018) Chitotriosidase inhibits allergic asthmatic airways via regulation of TGF-β expression and Foxp3 + Treg cells. Allergy 73:1686–1699

Houston DR, Eggleston I, Synstad BR, Eijsink VGH, Aalten DMFV (2002) The cyclic dipeptide CI-4 [cyclo-(l-Arg-d-Pro)] inhibits family 18 chitinases by structural mimicry of a reaction intermediate. Biochem J 368:23–27

Houston DR, Recklies AD, Krupa JC, Van Aalten DMF (2003) Structure and ligand-induced conformational change of the 39-kDa glycoprotein from human articular chondrocytes. J Biol Chem 278:30206–30212

Huang Q-S, Xie X-L, Liang G, Gong F, Wang Y, Wei X-Q, Wang Q, Ji Z-L, Chen Q-X (2012) The GH18 family of chitinases: Their domain architectures, functions and evolutions. Glycobiology 22:23–34

Hussain M, Wilson JB (2013) New paralogues and revised time line in the expansion of the vertebrate GH18 family. J Mol Evol 76:240–260

Ikegami T, Okada T, Hashimoto M, Seino S, Watanabe T, Shirakawa M (2000) Solution structure of the chitin-binding domain of bacillus circulans WL-12 chitinase A1. J Biol Chem 275:13654–13661

Imai T, Watanabe T, Yui T, Sugiyama J (2003) The directionality of chitin biosynthesis: a revisit. Biochem J 374:755–760

Izumida H, Imamura N, Sano H (1996) A novel chitinase inhibitor from a marine bacterium, Pseudomonas sp. J Antibiot (Tokyo) 49:76–80

James AJ, Reinius LE, Verhoek M, Gomes A, Kupczyk M, Hammar U, Ono J, Ohta S, Izuhara K, Bel E, Kere J, Söderhäll C, Dahlén B, Boot RG, Dahlén S-E (2016) Increased YKL-40 and chitotriosidase in asthma and chronic obstructive pulmonary disease. Am J Respir Crit Care Med 193:131–142

Janelidze S, Hertze J, Zetterberg H, Landqvist Waldö M, Santillo A, Blennow K, Hansson O (2016) Cerebrospinal fluid neurogranin and YKL-40 as biomarkers of Alzheimer's disease. Ann Clinic Transl Neurol 3:12–20

Jensen BV, Johansen JS, Price PA (2003) High levels of serum HER-2/neu and YKL-40 independently reflect aggressiveness of metastatic breast cancer. Clin Cancer Res 9:4423–4434

Jiang X, Kumar A, Liu T, Zhang KYJ, Yang Q (2016) A novel scaffold for developing specific or broad-spectrum chitinase inhibitors. J Chem Inf Model 56:2413–2420

Johansen JS, Schultz NA, Jensen BV (2009) Plasma YKL-40: a potential new cancer biomarker? Fut Oncol 5:1065–1082

Kang M-J, Yoon CM, Nam M, Kim D-H, Choi J-M, Lee CG, Elias JA (2015) Role of chitinase 3–Like-1 in interleukin-18–induced pulmonary type 1, type 2, and type 17 inflammation; alveolar destruction; and airway fibrosis in the murine lung. Am J Respir Cell Mol Biol 53:863–871

Kim KC, Yun J, Son DJ, Kim JY, Jung JK, Choi JS, Kim YR, Song JK, Kim SY, Kang SK, Shin DH, Roh YS, Han SB, Hong JT (2018) Suppression of metastasis through inhibition of chitinase 3-like 1 expression by miR-125a-3p-mediated up-regulation of USF1. Theranostics 8:4409–4428

Kogiso M, Nishiyama A, Shinohara T, Nakamura M, Mizoguchi E, Misawa Y, Guinet E, Nouri-Shirazi M, Dorey CK, Henriksen RA, Shibata Y (2011) Chitin particles induce size-dependent but carbohydrate-independent innate eosinophilia. J Leukoc Biol 90:167–176

Komi DEA, Kazemi T, Bussink AP (2016) New insights into the relationship between chitinase-3-like-1 and asthma. Curr Allerg Asthma Rep 16:57

Kramer KJ, Koga D (1986) Insect chitin: Physical state, synthesis, degradation and metabolic regulation. Ins Biochem 16:851–877

Kumar A, Zhang KYJ (2015) Hierarchical virtual screening approaches in small molecule drug discovery. Methods 71:26–37

Kuusk S, Sørlie M, Väljamäe P (2017) Human chitotriosidase is an endo-processive enzyme. PLoS One 12:e0171042–e0171042

Kzhyshkowska J, Mamidi S, Gratchev A, Kremmer E, Schmuttermaier C, Krusell L, Haus G, Utikal J, Schledzewski K, Scholtze J, Goerdt S (2006) Novel stabilin-1 interacting chitinase-like protein (SI-CLP) is up-regulated in alternatively activated macrophages and secreted via lysosomal pathway. Blood 107:3221–3228

Langner T, Göhre V (2016) Fungal chitinases: function, regulation, and potential roles in plant/pathogen interactions. Curr Genet 62:243–254

Lee CG, Hartl D, Lee GR, Koller B, Matsuura H, Da Silva CA, Sohn MH, Cohn L, Homer RJ, Kozhich AA, Humbles A, Kearley J, Coyle A, Chupp G, Reed J, Flavell RA, Elias JA (2009) Role of breast regression protein 39 (BRP-39)/chitinase 3-like-1 in Th2 and IL-13–induced tissue responses and apoptosis. J Exp Med 206:1149–1166

Lee CG, Da Silva CA, Dela Cruz CS, Ahangari F, Ma B, Kang MJ, He CH, Takyar S, Elias JA (2011) Role of chitin and chitinase/chitinase-like proteins in inflammation, tissue remodeling, and injury. Annu Rev Physiol 73:479–501

Lenardon MD, Munro CA, Gow NAR (2010) Chitin synthesis and fungal pathogenesis. Curr Opin Microbiol 13:416–423

Létuvé S, Kozhich A, Humbles A, Brewah Y, Dombret M-C, Grandsaigne M, Adle H, Kolbeck R, Aubier M, Coyle AJ, Pretolani M (2010) Lung chitinolytic activity and chitotriosidase are elevated in chronic obstructive pulmonary disease and contribute to lung inflammation. Am J Pathol 176:638–649

Li H, Greene LH (2010) Sequence and structural analysis of the chitinase insertion domain reveals two conserved motifs involved in chitin-binding. PLoS ONE 5:e8654

Livnat G, Bar-Yoseph R, Mory A, Dagan E, Elias N, Gershoni R, Bentur L (2014) Duplication in CHIT1 gene and the risk for aspergillus lung disease in CF patients. Pediatr Pulmonol 49:21–27

Lombard V, Golaconda Ramulu H, Drula E, Coutinho PM, Henrissat B (2014) The carbohydrate-active enzymes database (CAZy) in 2013. Nucleic Acids Res 42:D490–D495

Ma B, Herzog EL, Lee CG, Peng X, Lee C-M, Chen X, Rockwell S, Koo JS, Kluger H, Herbst RS, Sznol M, Elias JA (2015) Role of chitinase 3–like-1 and semaphorin 7a in pulmonary melanoma metastasis. Can Res 75:487–496

Malaguarnera L, Di Rosa M, Zambito AM, Dell'ombra N, Nicoletti F, Malaguarnera M (2006) Chitotriosidase gene expression in Kupffer cells from patients with non-alcoholic fatty liver disease. Gut 55:1313–1320

Matsumoto T, Inoue H, Sato Y, Kita Y, Nakano T, Noda N, Eguchi-Tsuda M, Moriwaki A, Kan-O K, Matsumoto K, Shimizu T, Nagasawa H, Sakuda S, Nakanishi Y (2009) Demethylallosamidin, a chitinase inhibitor, suppresses airway inflammation and hyperresponsiveness. Biochem Biophys Res Commun 390:103–108

Mattsson N, Tabatabaei S, Johansson P, Hansson O, Andreasson U, Månsson J-E, Johansson J-O, Olsson B, Wallin A, Svensson J, Blennow K, Zetterberg H (2011) Cerebrospinal fluid microglial markers in Alzheimer's disease: elevated chitotriosidase activity but lack of diagnostic utility. NeuroMol Med 13:151–159

Mazur M, Bartoszewicz A, Dymek B, Salamon M, Andryianau G, Kowalski M, Olejniczak S, Matyszewski K, Pluta E, Borek B, Stefaniak F, Zagozdzon A, Mazurkiewicz M, Koralewski R, Czestkowski W, Piotrowicz M, Niedziejko P, Gruza MM, Dzwonek K, Golebiowski A,

Golab J, Olczak J (2018a) Discovery of selective, orally bioavailable inhibitor of mouse chitotriosidase. Bioorg Med Chem Lett 28:310–314

Mazur M, Olczak J, Olejniczak S, Koralewski R, Czestkowski W, Jedrzejczak A, Golab J, Dzwonek K, Dymek B, Sklepkiewicz PL, Zagozdzon A, Noonan T, Mahboubi K, Conway B, Sheeler R, Beckett P, Hungerford WM, Podjarny A, Mitschler A, Cousido-Siah A, Fadel F, Golebiowski A (2018b) Targeting acidic mammalian chitinase is effective in animal model of asthma. J Med Chem 61:695–710

Mccarter JD, Stephen Withers G (1994) Mechanisms of enzymatic glycoside hydrolysis. Curr Opin Struct Biol 4:885–892

Meng G, Zhao Y, Bai X, Liu Y, Green TJ, Luo M, Zheng X (2010) Structure of human Stabilin-1 interacting chitinase-like protein (SI-CLP) reveals a saccharide-binding cleft with lower sugar-binding selectivity. J Biol Chem 285:39898–39904

Merzendorfer H (2006) Insect chitin synthases: a review. J Comp Physiol B 176:1–15

Merzendorfer H (2011) The cellular basis of chitin synthesis in fungi and insects: Common principles and differences. Eur J Cell Biol 90:759–769

Merzendorfer H, Zimoch L (2003) Chitin metabolism in insects: structure, function and regulation of chitin synthases and chitinases. J Exp Biol 206:4393–4412

Monzingo AF, Marcotte EM, Hart PJ, Robertas JD (1996) Chitinases, chitosanases, and lysozymes can be divided into procaryotic and eucaryotic families sharing a conserved core. Nat Struct Biol 3:133

Mora-Montes HM, Netea MG, Ferwerda G, Lenardon MD, Brown GD, Mistry AR, Kullberg BJ, O'Callaghan CA, Sheth CC, Odds FC, Brown AJP, Munro CA, Gow NaR (2011) Recognition and blocking of innate immunity cells by candida albicans chitin. Infect Immun 79:1961–1970

Morgan JLW, Strumillo J, Zimmer J (2012) Crystallographic snapshot of cellulose synthesis and membrane translocation. Nature 493:181

Neville AC, Parry DA, Woodhead-Galloway J (1976) The chitin crystallite in arthropod cuticle. J Cell Sci 21:73

Nikolov S, Petrov M, Lymperakis L, Friák M, Sachs C, Fabritius H-O, Raabe D, Neugebauer J (2010) Revealing the design principles of high-performance biological composites Using Ab initio and multiscale simulations: the example of lobster cuticle. Adv Mater 22:519–526

Nishimoto Y, Sakuda S, Takayama S, Yamada Y (1991) Isolation and characterization of new allosamidins. J Antibiot (Tokyo) 44:716–722

Ohno M, Togashi Y, Tsuda K, Okawa K, Kamaya M, Sakaguchi M, Sugahara Y, Oyama F (2013) Quantification of chitinase mRNA levels in human and mouse tissues by real-time PCR: species-specific expression of acidic mammalian chitinase in stomach tissues. PLoS ONE 8: e67399

Ohno M, Tsuda K, Sakaguchi M, Sugahara Y, Oyama F (2012) Chitinase mRNA levels by quantitative PCR using the single standard DNA: acidic mammalian chitinase is a major transcript in the mouse stomach. PLoS ONE 7:e50381

Ohno T, Armand S, Hata T, Nikaidou N, Henrissat B, Mitsutomi M, Watanabe T (1996) A modular family 19 chitinase found in the prokaryotic organism Streptomyces griseus HUT 6037. J Bacteriol 178:5065–5070

Olland AM, Strand J, Presman E, Czerwinski R, Joseph-Mccarthy D, Krykbaev R, Schlingmann G, Chopra R, Lin L, Fleming M, Kriz R, Stahl M, Somers W, Fitz L, Mosyak L (2009) Triad of polar residues implicated in pH specificity of acidic mammalian chitinase. Protein Sci 18:569–578

Omura S, Arai N, Yamaguchi Y, Masuma R, Iwai Y, Namikoshi M, Turberg A, Kolbl H, Shiomi K (2000) Argifin, a new chitinase inhibitor, produced by Gliocladium sp. FTD-0668. I. Taxonomy, fermentation, and biological activities. J Antibiot (Tokyo) 53:603–608

Perrakis A, Tews I, Dauter Z, Oppenheim AB, Chet I, Wilson KS, Vorgias CE (1994) Crystal structure of a bacterial chitinase at 2.3 Å resolution. Structure 2:1169–1180

Ranok A, Wongsantichon J, Robinson RC, Suginta W (2015) Structural and thermodynamic insights into chitooligosaccharide binding to human cartilage chitinase 3-like protein 2 (CHI3L2 or YKL-39). J Biol Chem 290:2617–2629

Rao FV, Andersen OA, Vora KA, Demartino JA, Van Aalten DMF (2005a) Methylxanthine drugs are chitinase inhibitors: investigation of inhibition and binding modes. Chem Biol 12:973–980
Rao FV, Houston DR, Boot RG, Aerts JM, Hodkinson M, Adams DJ, Shiomi K, van Aalten DMF (2005b) Specificity and affinity of natural product cyclopentapeptide inhibitors against A. fumigatus, human, and bacterial chitinases. Chem Biol 12:65–76
Rao FV, Houston DR, Boot RG, Aerts JMFG, Sakuda S, Van Aalten DMF (2003) Crystal structures of allosamidin derivatives in complex with human macrophage chitinase. J Biol Chem 278:20110–20116
Reese TA, Liang H-E, Tager AM, Luster AD, Van Rooijen N, Voehringer D, Locksley RM (2007) Chitin induces accumulation in tissue of innate immune cells associated with allergy. Nature 447:92
Renkema GH, Boot RG, Au FL, Donker-Koopman WE, Strijland A, Muijsers AO, Hrebicek M, Aerts JMFG (1998) Chitotriosidase, a chitinase, and the 39-kDa human cartilage glycoprotein, a chitin-binding lectin, are homologues of family 18 glycosyl hydrolases secreted by human macrophages. Eur J Biochem 251:504–509
Renkema GH, Boot RG, Muijsers AO, Donker-Koopman WE, Aerts JMFG (1995) Purification and characterization of human chitotriosidase, a novel member of the chitinase family of proteins. J Biol Chem 270:2198–2202
Renkema GH, Boot RG, Strijland A, Donker-Koopman WE, Berg M, Muijsers AO, Aerts JMFG (1997) Synthesis, sorting, and processing into distinct isoforms of human macrophage chitotriosidase. Eur J Biochem 244:279–285
Rinaudo M (2006) Chitin and chitosan: properties and applications. Prog Polym Sci 31:603–632
Roy RM, Wüthrich M, Klein BS (2012) Chitin elicits CCL2 from airway epithelial cells and induces CCR112-dependent innate allergic inflammation in the lung. J Immunol 189:2545–2552
Rudall KM, Kenchington W (1973) The chitin system. Biol Rev 48:597–633
Sakuda S, Isogai A, Makita T, Matsumoto S, Koseki K, Kodama H, Suzuki A (1987a) Structures of allosamidins, novel insect chitinase inhibitors, produced by actinomycetes. Agric Biol Chem 51:3251–3259
Sakuda S, Isogai A, Matsumoto S, Suzuki A (1987b) Search for microbial insect growth-regulators. 2. Allosamidin, a novel insect chitinase inhibitor. J Antibiot 40:296–300
Sakuda S, Isogai A, Matsumoto S, Suzuki A, Koseki K (1986) The structure of allosamidin, a novel insect chitinase inhibitor, produced by streptomyces Sp. Tetrahedron Lett 27:2475–2478
Sakuda S, Isogai A, Matsumoto S, Suzuki A, Koseki K, Kodama H, Yamada Y (1988) Absolute configuration of allosamizoline, an aminocyclitol derivative of the chitinase inhibitor allosamidin. Agric Biol Chem 52:1615–1617
Sakuda S, Isogai A, Suzuki A, Yamada Y (1993) Chemistry and biochemistry of the chitinase inhibitors, allosamidins. Actinomycetologica 7:50–57
Sakuda S, Sugiyama Y, Zhou Z-Y, Takao H, Ikeda H, Kakinuma K, Yamada Y, Nagasawa H (2001) biosynthetic studies on the cyclopentane ring formation of allosamizoline, an aminocyclitol component of the chitinase inhibitor allosamidin. J Organ Chem 66:3356–3361
Schimpl M, Rush CL, Betou M, Eggleston IM, Recklies AD, Van Aalten DM (2012) Human YKL-39 is a pseudo-chitinase with retained chitooligosaccharide-binding properties. Biochem J 446:149–157
Schüttelkopf AW, Andersen OA, Rao FV, Allwood M, Lloyd C, Eggleston IM, Van Aalten DMF (2006) Screening-based discovery and structural dissection of a novel family 18 chitinase inhibitor. J Biol Chem 281:27278–27285
Schüttelkopf AW, Andersen OA, Rao FV, Allwood M, Rush CL, Eggleston IM, Van Aalten DMF (2011) Bisdionin C—A rationally designed, submicromolar inhibitor of family 18 chitinases. ACS Med Chem Lett 2:428–432
Seibold MA, Donnelly S, Solon M, Innes A, Woodruff PG, Boot RG, Burchard EG, Fahy JV (2008) Chitotriosidase is the primary active chitinase in the human lung and is modulated by genotype and smoking habit. J Allergy Clin Immunol 122:944–950.e943

Semeňuk T, Krist P, Pavlíček J, Bezouška K, Kuzma M, Novák P, Křen V (2001) Synthesis of chitooligomer-based glycoconjugates and their binding to the rat natural killer cell activation receptor NKR-P1. Glycoconj J 18:817–826

Shen C-R, Juang H-H, Chen H-S, Yang C-J, Wu C-J, Lee M-H, Hwang Y-S, Kuo M-L, Chen Y-S, Chen J-K, Liu C-L (2015) The correlation between chitin and acidic mammalian chitinase in animal models of allergic asthma. Int J Mol Sci 16:27371–27377

Shibata Y, Metzger WJ, Myrvik QN (1997) Chitin particle-induced cell-mediated immunity is inhibited by soluble mannan: mannose receptor-mediated phagocytosis initiates IL-12 production. J Immunol 159:2462–2467

Shuhui L, Mok YK, Wong WSF (2009) Role of mammalian chitinases in asthma. Int Arch Allergy Immunol 149:369–377

Song H-M, Jang A-S, Ahn M-H, Takizawa H, Lee S-H, Kwon J-H, Lee Y-M, Rhim T, Park C-S (2008) Ym1 and Ym2 expression in a mouse model exposed to diesel exhaust particles. Environ Toxicol 23:110–116

Songsiriritthigul C, Pantoom S, Aguda AH, Robinson RC, Suginta W (2008) Crystal structures of Vibrio harveyi chitinase A complexed with chitooligosaccharides: Implications for the catalytic mechanism. J Struct Biol 162:491–499

Stam MR, Blanc E, Coutinho PM, Henrissat B (2005) Evolutionary and mechanistic relationships between glycosidases acting on α- and β-bonds. Carbohyd Res 340:2728–2734

Stefano S, Piras MR, Rita B, Giannina A, Fois ML, Giulio R, Salvatore M (2007) Chitotriosidase and Alzheimers disease. Curr Alzheimer Res 4:295–296

Sutherland TE (2018) Chitinase-like proteins as regulators of innate immunity and tissue repair: helpful lessons for asthma? Biochem Soc Trans 46:141–151

Sutherland TE, Andersen OA, Betou M, Eggleston IM, Maizels RM, van Aalten D, Allen JE (2011) Analyzing airway inflammation with chemical biology: dissection of acidic mammalian chitinase function with a selective drug-like inhibitor. Chem Biol 18:569–579

Terwisscha Van Scheltinga AC, Armand S, Kalk KH, Isogai A, Henrissat B, Dijkstra BW (1995) Stereochemistry of chitin hydrolysis by a plant chitinase/lysozyme and x-ray structure of a complex with allosamidin evidence for substrate assisted catalysis. Biochemistry 34:15619–15623

Tews I, Terwisscha Van Scheltinga AC, Perrakis A, Wilson KS, Dijkstra BW (1997) Substrate-assisted catalysis unifies two families of chitinolytic enzymes. J Am Chem Soc 119:7954–7959

Thomsen T, Schlosser A, Holmskov U, Sorensen GL (2011) Ficolins and FIBCD1: Soluble and membrane bound pattern recognition molecules with acetyl group selectivity. Mol Immunol 48:369–381

Tsai M-L, Liaw S-H, Chang N-C (2004) The crystal structure of Ym1 at 1.31Å resolution. J Struct Biol 148:290–296

Van Aalten DMF, Komander D, Synstad B, Gåseidnes S, Peter MG, Eijsink VGH (2001) Structural insights into the catalytic mechanism of a family 18 exo-chitinase. Proc Natl Acad Sci 98:8979–8984

Van Aalten DMF, Synstad B, Brurberg MB, Hough E, Riise BW, Eijsink VGH, Wierenga RK (2000) Structure of a two-domain chitotriosidase from Serratia marcescens at 1.9-Å resolution. Proc Natl Acad Sci 97:5842–5847

Van Dussen L, Hendriks EJ, Groener JEM, Boot RG, Hollak CEM, Aerts JMFG (2014) Value of plasma chitotriosidase to assess non-neuronopathic Gaucher disease severity and progression in the era of enzyme replacement therapy. J Inherit Metab Dis 37:991–1001

Van Dyken SJ, Liang H-E, Naikawadi RP, Woodruff PG, Wolters PJ, Erle DJ, Locksley RM (2017) Spontaneous chitin accumulation in airways and age-related fibrotic lung disease. Cell 169:497–509.e413

Van Dyken SJ, Mohapatra A, Nussbaum JC, Molofsky AB, Thornton EE, Ziegler SF, McKenzie AN, Krummel MF, Liang HE, Locksley RM (2014) Chitin activates parallel immune modules that direct distinct inflammatory responses via innate lymphoid type 2 and γδ T cells. Immunity 40:414–424

Vermeulen CA, Wessels JGH (1986) Chitin biosynthesis by a fungal membrane preparation. Eur J Biochem 158:411–415

Veronico P, Gray L, Jones J, Bazzicalupo P, Arbucci S, Cortese M, Di Vito M, De Giorgi C (2001) Nematode chitin synthases: gene structure, expression and function in Caenorhabditis elegans and the plant parasitic nematode Meloidogyne artiellia. Mol Genet Genom 266:28–34

Von Arnim C, Watabe-Rudolph M, Song Z, Schnack C, Begus-Nahrmann Y, Rolyan H, Otto M, Tumani H, Thal D, Attems J, Jellinger K, Kestler H, Rudolph K (2011) Chitinase enzyme activity in cerebral spinal fluid is strongly associated with Alzheimer's Disease. Alzheimer's Dementia 7:S288

Wajner A, Michelin K, Burin MG, Pires RF, Pereira MLS, Giugliani R, Coelho JC (2004) Biochemical characterization of chitotriosidase enzyme: comparison between normal individuals and patients with Gaucher and with Niemann-Pick diseases. Clin Biochem 37:893–897

Wakasugi M, Gouda H, Hirose T, Sugawara A, Yamamoto T, Shiomi K, Sunazuka T, Ōmura S, Hirono S (2013) Human acidic mammalian chitinase as a novel target for anti-asthma drug design using in silico screening. Bioorg Med Chem 21:3214–3220

Wang X, Yu W, Fu X, Ke M, Xiao Q, Lü Y (2018) Chitotriosidase enhances TGFβ-Smad signaling and uptake of β-amyloid in N9 microglia. Neurosci Lett 687:99–103

Watabe-Rudolph M, Song Z, Lausser L, Schnack C, Begus-Nahrmann Y, Scheithauer M-O, Rettinger G, Otto M, Tumani H, Thal DR, Attems J, Jellinger KA, Kestler HA, Von Arnim CaF, Rudolph KL (2012) Chitinase enzyme activity in CSF is a powerful biomarker of Alzheimer disease. Neurology 78:569–577

Yang C-J, Liu Y-K, Liu C-L, Shen C-N, Kuo M-L, Su C-C, Tseng C-P, Yen T-C, Shen C-R (2009) Inhibition of acidic mammalian chitinase by RNA interference suppresses ovalbumin-sensitized allergic asthma. Hum Gene Ther 20:1597–1606

Yang J, Gan Z, Lou Z, Tao N, Mi Q, Liang L, Sun Y, Guo Y, Huang X, Zou C, Rao Z, Meng Z, Zhang K-Q (2010) Crystal structure and mutagenesis analysis of chitinase CrChi1 from the nematophagous fungus Clonostachys rosea in complex with the inhibitor caffeine. Microbiology 156:3566–3574

Zhang H, Ng KP, Therriault J, Kang MS, Pascoal TA, Rosa-Neto P, Gauthier S, Initiative TaSDN, Aging NIO, Imaging NIOB, Bioengineering, Abbvie ASA, Foundation ASDD, Biotech A, Bioclinica I, Biogen Company B-MS, Cerespir I, Inc., CE Elan Pharmaceuticals, I Lilly, E, Company, Euroimmun, Hoffmann-La, F., Ltd., R., Its Affiliated Company Genentech, I., Fujirebio, Healthcare G, Ltd., I, Research, JaI Development, L, Johnson, Research JP, Llc., D, Lumosity Lundbeck Merck Co., I, Meso Scale Diagnostics L, Research N, Technologies N, Corporation, NP, Inc. P, Imaging P, Servier, Company TP, Therapeutics T, Research, CIOH, Health, FFTNIO (2018) Cerebrospinal fluid phosphorylated tau, visinin-like protein-1, and chitinase-3-like protein 1 in mild cognitive impairment and Alzheimer's disease. Transl Neurodegenerat 7, 23

Zhao J, Zhu H, Wong CH, Leung KY, Wong WSF (2005) Increased lungkine and chitinase levels in allergic airway inflammation: a proteomics approach. Proteomics 5:2799–2807

Zheng T, Rabach M, Chen NY, Rabach L, Hu X, Elias JA, Zhu Z (2005) Molecular cloning and functional characterization of mouse chitotriosidase. Gene 357:37–46

Zhou Y, Peng H, Sun H, Peng X, Tang C, Gan Y, Chen X, Mathur A, Hu B, Slade MD, Montgomery RR, Shaw AC, Homer RJ, White ES, Lee C-M, Moore MW, Gulati M, Geun Lee C, Elias JA, Herzog EL (2014) Chitinase 3–Like 1 suppresses injury and promotes fibroproliferative responses in mammalian lung fibrosis. Sci Transl Med 6:240ra276-240ra276

Zhu Z, Zheng T, Homer RJ, Kim Y-K, Chen NY, Cohn L, Hamid Q, Elias JA (2004) Acidic mammalian chitinase in asthmatic Th2 inflammation and IL-13 pathway activation. Science 304:1678–1682

Chapter 12
Chitin/Chitosan-Active Enzymes Involved in Plant–Microbe Interactions

Tamo Fukamizo and S. Shinya

Abstract Plant chitinase hydrolyzing β-1,4-glycosidic linkages of chitin are major enzymes acting in plant–microbe interactions and are involved in self-defense against fungal pathogens. Chitosanases from soil bacteria are also involved in plant defense by hydrolyzing chitosan components of the fungal cell wall. The crystal structures of these enzymes in complex with their substrates have been solved, and the mechanisms of substrate binding were elucidated at the atomic level. These findings enabled us to speculate on the enzyme targets under physiological conditions, leading us to define the physiological roles of the enzymes. The structures and functions of chitin/chitosan-binding modules appended to modular chitinases/chitosanases were analyzed by NMR and isothermal titration calorimetry (ITC), and the enzymes were found to form an appropriate modular organization to fulfill their roles in plant–microbe interactions.

Keywords Chitinase · Chitosanase · Crystal structure · Nuclear magnetic resonance · Isothermal titration calorimetry

Abbreviations

GlcNAc	2-acetamido-2-deoxy-D-glucopyranose
$(GlcNAc)_n$	β-1,4-linked oligosaccharide of GlcNAc with a polymerization degree of n
GlcN	2-amino-2-deoxy-D-glucopyranose

T. Fukamizo (✉)
Biochemistry-Electrochemistry Research Unit, School of Chemistry, Suranaree University of Technology, 111, University Avenue, Muang District, Nakhon Ratchasima 30000, Thailand
e-mail: tamo0111fuka@gmail.com

S. Shinya
Laboratory of Molecular Biophysics, Institute of Protein Research, Osaka University, 3-2 Yamadaoka, Suita-Shi, Osaka 565-0871, Japan
e-mail: shinya-s@protein.osaka-u.ac.jp

Q. Yang and T. Fukamizo (eds.), *Targeting Chitin-containing Organisms*, Advances in Experimental Medicine and Biology 1142,
https://doi.org/10.1007/978-981-13-7318-3_12

$(GlcN)_n$	β-1,4-linked oligosaccharide of GlcN with a polymerization degree of n
ITC	isothermal titration calorimetry
NMR	nuclear magnetic resonance

12.1 Introduction

Plants utilized in agricultural productions interact with other living organisms, including insects and microbes, under field conditions throughout the entire life cycle. Since these interactions directly affect the physiological conditions, plant productivity is often affected, resulting in a significant reduction in food supply. Controlling the interactions between these organisms is one of the greatest concerns in agricultural food production. Since chitin and chitin-related enzymes in insects are fully described in other chapters, this chapter deals with chitin/chitosan-active enzymes involved in plant–microbe interactions: chitinases from plants and chitosanases from soil bacteria antagonizing with fungal pathogens.

Plant chitinases (E.C. 3.2.1.14) have been recognized as playing important roles in plant–microbe interactions. The enzymes themselves inhibit fungal growth by adhesion to the fungal cell wall and hydrolysis of the β-1,4-glycosidic linkages of chitinous components (Schlumbaum et al. 1986; Taira et al. 2002; Arakane et al. 2012). Furthermore, when plants are infected by fungal pathogens, chitinase products from the cell wall, chitooligosaccharides, are released and act as signals for triggering the plant immune system (Desaki et al. 2018). Thus, chitin and its derivatives can be regarded as a critical interface in the interactions between plants and fungi, in spite of plants not containing chitin (El Hadrami et al. 2010). However, it was reported that multiple chitinase genes with different sequences are constitutively expressed in plants, and the physiological roles of individual chitinases are still unclear (Kasprzewska 2003). Since chitin polysaccharide chains are de-*N*-acetylated to some extent in the fungal cell wall, chitosanases produced by soil bacteria, which hydrolyze β-1,4 glycosidic linkages of chitosan, may be involved in interactions between plants and pathogenic fungi by inhibiting the fungal growth (Saito et al. 2009). Thus, it is desirable to understand the molecular basis of interactions of chitinases/chitosanases with their substrates for efficiently controlling plant–microbe interactions. Some chitinases/chitosanases are known to have subdomains (chitin/chitosan-binding modules) assisting the catalytic action of the enzymes (Armenta et al. 2017). The structure and function of these enzymes and their subdomains have been investigated by crystallography and other physicochemical methods, including NMR and isothermal titration calorimetry (ITC), and the molecular basis of the enzyme–substrate interactions has been progressively better understood in recent years.

The authors herein review recent advances in the structural enzymology of chitinases from plants and chitosanases from soil bacteria. The crystal structures of the enzymes in complex with their substrates revealed the mechanism of substrate recognition, and NMR and ITC analysis of the substrate binding provided the quantitative details of the interactions. The findings from these analyses may lead to a comprehensive understanding of the physiological functions of these enzymes.

12.2 Plant GH18 Chitinases

In the well-known CAZy database (http://www.cazy.org), chitinases are classified into the GH18 and GH19 families based on the amino acid sequences. Among the GH18 chitinases, the first X-ray crystal structure was solved using a chitinase present in the lutoid bodies of latex from the rubber tree *Hevea brasiliensis*, hevamine (Terwisscha van Scheltinga et al. 1994). The overall structure of hevamine was shown to be an $(\alpha/\beta)_8$-barrel (TIM-barrel) fold (Fig. 12.1a). No additional subdomains are present in its structure. Plant GH18 chitinases of this type are designated as class III chitinases (Neuhaus et al. 1996). Based on the structural and biochemical data, the catalytic reaction of the enzyme was suggested to take place through a double displacement reaction with anomer retention. This reaction is involved in an oxazolinium ion intermediate formation and is called the substrate-assisted mechanism (Terwisscha van Scheltinga et al. 1995). Mutational analysis has shown that several aromatic amino acid residues contribute to the reaction by participating in the catalytic intermediate stabilization (Bokma et al. 2002) and binding of the substrate (Suzukawa et al. 2003). Sasaki et al. (2006) examined the mode of action of class III chitinases from rice and reported that enzymes hydrolyze the substrate in an endo-splitting/non-processive manner and that only subsite −1 was specific to GlcNAc residue.

On the other hand, GH18 chitinases with an insertion subdomain are present in plants and are designated as class V chitinases (Neuhaus et al. 1996). The first crystal structure of class V chitinase was reported by Ohnuma et al. (2011a) using the enzyme from *Nicotiana tabacum* (NtChiV) and is shown in Fig. 12.1b. The insertion subdomain is composed of one α-helix (α7) and five β-strands (β10-β14) and is inserted between β9 and α8 of the core $(\alpha/\beta)_8$ domain. The DxDxE motif, that is crucial for the catalytic reaction of GH18 chitinases, is present inside the central hollow portion of the $(\alpha/\beta)_8$ barrel fold. The insertion subdomain is positioned opposite the DxDxE motif, forming a deep binding groove. Both ends of the binding groove are opened outward, possibly accepting a long chitin polysaccharide chain. Indeed, the class V chitinase catalyzes the hydrolysis in an endo-splitting/non-processive manner (Ohnuma et al. 2011a). The crystal structure of a class V chitinase from *Arabidopsis thaliana* (AtChiC), of which the amino acid sequence is 57% homologous to that of NtChiV, was also reported by the same research group (Ohnuma et al. 2011b), and NtChiV was found to be almost identical to AtChiC in their three-dimensional structures, as shown in Fig. 12.1b, c. Both structures are also very similar to that of the catalytic domain of *Serratia marcescens* chitinase B

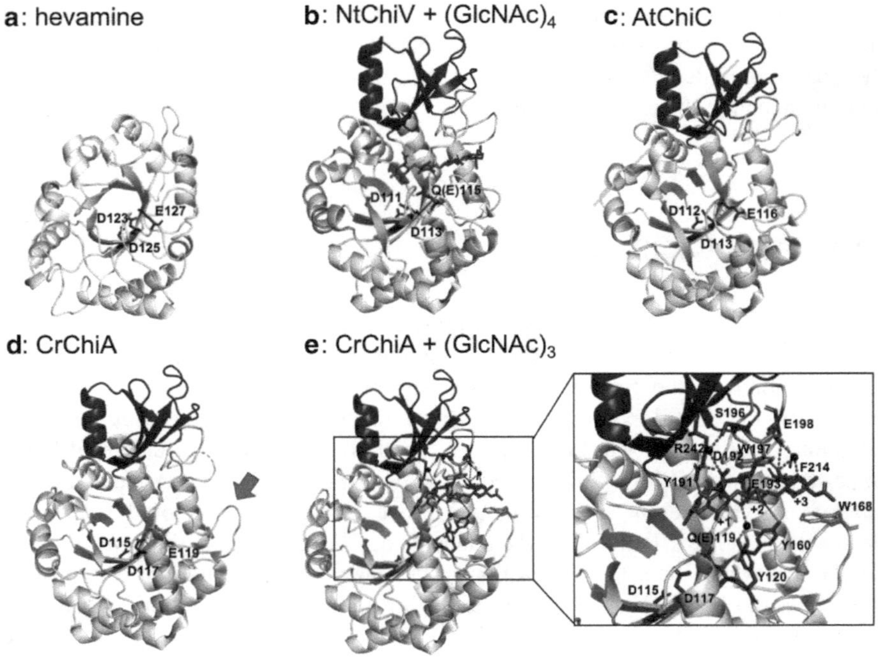

Fig. 12.1 Crystal structures of plant GH18 chitinases: **a** Hevamine, a class III chitinase from *Hevea brasiliensis* (2HVM Terwisscha van Scheltinga et al. 1994). **b** NtChiV, a class V chitinase from *Nicotiana tabacum* in complex with $(GlcNAc)_4$ (3ALG, Ohnuma et al. 2011a). **c** AtChiC, a class V chitinase from *Arabidopsis thaliana* (3AQU, Ohnuma et al. 2011b). **d** CrChiA, a class V chitinase from *Cycas revolute* (4MNJ, Umemoto et al. 2015a). **e** CrChiA in complex with $(GlcNAc)_3$ (3WIJ, Umemoto et al. 2015a). $(GlcNAc)_3$ binds to subsites +1, +2, and +3, acceptor-binding site. Three contiguous acidic amino acids highlighted in the central hollow are the DxDxE motif essential for catalysis. The light-gray region is an $(\alpha/\beta)_8$-barrel fold, while the dark-gray region is an insertion domain specific to class V chitinases. Broken lines are possible hydrogen bonds between the enzyme and $(GlcNAc)_3$

(van Aalten et al. 2001), indicating that the catalytic mechanism of the class V chitinases is similar to that of *Serratia* chitinase B.

Taira et al. (2009) isolated a novel type of class V chitinase from cycad *Cycas revolta* (CrChiA). The crystal structure of CrChiA is similar to those of NtChiV and AtChiC, as shown in Fig. 12.1d (Umemoto et al. 2015a). However, CrChiA was found to catalyze the transglycosylation reaction in addition to hydrolysis (Taira et al. 2010), whereas NtChiV and AtChiC have no significant transglycosylation activity. To identify the structure conferring transglycosylation activity to the enzyme, a closer examination of the crystal structures of the three class V chitinases were conducted. Although the three class V chitinases resemble each other in their crystal structures, the authors noticed an additional loop structure in CrChiA, which was not found in NtChiV or AtChiC, at the rightmost site of the binding groove (designated by an arrow, Fig. 12.1d). Umemoto et al. (2015a) successfully solved the CrChiA structure in complex with the transglycosylation acceptor, chitin trimer $(GlcNAc)_3$, bound to +1, +2, and +3 (Fig. 12.1e). Interestingly, the additional loop structure in CrChiA has a tryptophan residue (Trp168), which makes a stacking CH-π interaction with the +3 pyranose ring. A stacking interaction was also found in Trp197, and several hydrogen bonds are likely to have stabilized the bound acceptor molecule, which is used for the following transglycosylation reaction. The strong interaction of the acceptor molecule confers the efficient transglycosylation activity to CrChiA. Indeed, mutation of Trp168 of CrChiA significantly reduced the transglycosylation activity (Umemoto et al. 2015b). It is now desirable to examine whether the transglycosylation reaction is essential for normal growth of the cycad plant itself.

Under normal growth conditions, an *Arabidopsis* class III chitinase (AtChiA) gene (At5g24090) was not expressed at all, whereas the gene was expressed exclusively when the plants were exposed to environmental stress, especially salt and wounding. The class III chitinase appears to participate in tolerances to environmental stress (Takenaka et al. 2009). Some physiological studies suggested that class III chitinases are important for nodulation process in plant–microbe symbiosis as well as for protein storage (Goormachtig et al. 1998; Yang et al. 2011). The first class V chitinase (NtChiV) gene was isolated from tobacco leaves inoculated with tobacco mosaic virus (Melchers et al. 1994). In contrast to the class III chitinase, AtChiA, which is only responsive to abiotic stress, the expression level of the gene encoding AtChiC, *Arabidopsis* class V chitinase gene (At4g19810), was elevated by abiotic as well as biotic stresses (Ohnuma et al. 2011b). Taken together, it appears that plant GH18 chitinases have multiple roles, and at least participate in biotic and abiotic stress tolerance through an action toward a GlcNAc-containing-sugar molecule. This enzyme target is still unknown but may be a precursor of the signal triggering the genes for stress tolerance. Indeed, antifungal activities of GH18 chitinases are much lower than those of the GH19 chitinases described below (Arakane et al. 2012).

12.3 Plant GH19 Chitinases

The first crystal structure of GH19 chitinase was reported by Hart et al. (1993) for the enzyme from barley, *Hordeum vulgare*, seeds (BSC-c). As shown in Fig. 12.2a, the enzyme consists of two lobes, both of which are rich in α-helices, and the substrate-binding groove was suggested to lie in between the two lobes from the modeling study. The plant GH19 chitinases of this type are designated as class II chitinases, while the GH19 enzymes, in which CBM18 chitin-binding module is appended to the N-terminus of the class II enzyme, are designated as class I chitinases (Neuhaus et al. 1996). In spite of numerous trials conducted for solving the structures of the class II enzymes in complex with their substrate, the complexed structure was not elucidated until Ohnuma et al. (2012) reported the crystal structure of a class II chitinase from rye seeds (RSC-c) in complex with $(GlcNAc)_4$.

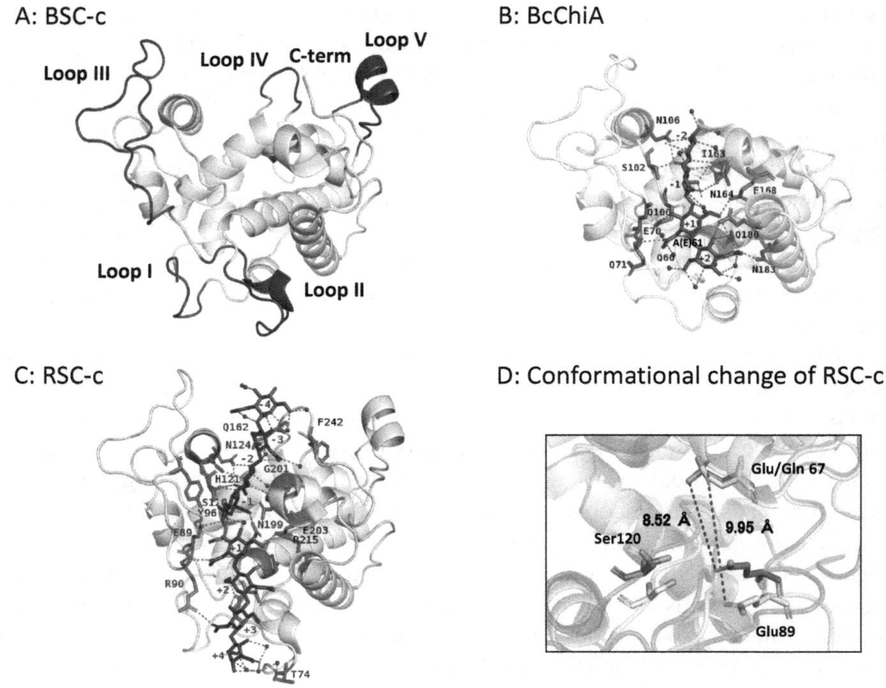

Fig. 12.2 Crystal structures of plant GH19 chitinases. **a** BSC-c, a class II chitinase from barley, *Hordeum vulgare*, seeds (2BAA, Hart et al. 1993). Six loop structures are highlighted by dark gray. Individual loop structures are numbered from N-terminus. **b** and **c** BcChiA, a class II chitinase from *Bryum coronatum* (3WH1, Ohnuma et al. 2014), and RSC-c, a class II chitinase from rye, *Secale cereal*, seeds (4J0L, Ohnuma et al. 2013), respectively. Dark gray, amino acid residues involved in chitin oligosaccharide binding; Black, bound chitin oligosaccharides. Small black spheres are bound to water molecules. Broken lines are possible hydrogen bonds between the enzyme and ligands. **d** Conformational change induced by chitin oligosaccharide binding in RSC-c (Ohnuma et al. 2013). The catalytic center is enlarged. Two broken lines are distances between the catalytic acid and base. The longer one is for the free state (4DWX), while the shorter one for the bound state (4J0L). Light gray represents the free state, while dark gray the bound state

The amino acid sequence of RSC-c is similar to that of BSC-c by 92%. In this complexed structure of RSC-c the bound $(GlcNAc)_4$ was only located on the aglycon-binding site (positively numbered subsites: +1, +2, +3, and +4; Davies et al. 1997), which could not provide any information on the catalytic mechanism. After this report, the same research group reported the crystal structure of a class II chitinase from *Bryum coronatum* (BcChiA) in complex with $(GlcNAc)_4$, which binds to subsites −2, −1, +1, and +2, as shown in Fig. 12.2b (Ohnuma et al. 2014). This is the first structure in which $(GlcNAc)_4$ binds to the binding groove spanning the catalytic center. This structure provided the experimental evidence that the enzymes catalyze the hydrolysis of glycosidic linkages through the single displacement reaction; that is, after donating a proton to the glycosidic oxygen by the catalytic acid (Glu61), a water molecule activated by the catalytic base (Glu70) attacks the C1 carbon of the −1 GlcNAc residue from the opposite side to complete hydrolysis. In the same report, the substrate binding to BcChiA was also investigated by NMR spectroscopy, providing information on the substrate binding in solution based on the chemical shift perturbations. Although most perturbations were found in the substrate-binding groove defined by the crystal structure (Fig. 12.2b), chemical shift perturbations were observed in the back of the binding groove as well, suggesting conformational changes induced by substrate binding. Since the conformational changes appear to be involved in the catalytic action, this finding may be useful for obtaining additional insights into the catalytic mechanism of GH19 chitinases (Ohnuma et al. 2013). This will be discussed later.

On the other hand, for GH19 chitinases from barley and rye seeds (BSC-c and RSC-c, respectively), researchers were unable to obtain the crystal structure of the enzyme–substrate complex, in which $(GlcNAc)_n$ binds to the enzyme spanning the catalytic center. Comparison in the crystal structures between BSC-c/RSC-c and BcChiA revealed that BSC-c/RSC-c have additional loop structures (I, II, IV, V, and C-term, labeled in Fig. 12.2a) at both ends of the binding groove, whereas BcChiA does not (Fig. 12.2b). These loop structures appeared to extend the binding groove outward. Indeed, Ohnuma et al. (2013) successfully obtained the crystal structure of RSC-c in complex with two molecules of $(GlcNAc)_4$ in which one $(GlcNAc)_4$ molecule binds to subsites −4 to −1 and the other binds +1 to +4, as shown in Fig. 12.2c. Thus, BcChiA was found to have a binding groove consisting of −2, −1, +1, and +2, while BSC-c/RSC-c a binding groove consisting of −4, −3, −2, −1, +1, +2, +3, and +4. As seen from these structures, four GlcNAc residues from −2 to +2 are strongly recognized by a number of direct hydrogen bonds with RSC-c, but the other GlcNAc residues (−4, −3, +3, and +4) are more loosely recognized by water-mediated hydrogen bonds. So the loop structures (I, II, IV, and C-term) found in BSC-c/RSC-c do not strongly contribute to substrate recognition; however, the core structure (subsites −2 to +2) of BSC-c/RSC-c (Fig. 12.2c), as well as the entire binding groove of BcChiA (Fig. 12.2b) strongly recognizes the chitin substrate by direct hydrogen bonding networks. The hydrogen bonds are less intensive in the interactions of GH18 chitinases (Ohnuma et al. 2011a, b). The GH19 chitinases with a full set of loop structures (I, II, III, IV, V, and C-term) are designated as "loopful" chitinases and those with only loop III as "loopless" chitinases. Comparison in the structures between free and bound states of RSC-c revealed a conformational change upon substrate binding, which narrows the catalytic site, as shown in Fig. 12.2d (Ohnuma et al. 2013). Conformational changes of this type are designated as "domain motion". The narrowing the catalytic site may probably result in an optimum arrangement of the catalytic acid (Glu61/67), the base (Glu70/89), and the catalytic water molecule so that the triad most efficiently catalyze the cleavage of the β-1,4-glycosidic bond.

In contrast to GH18 chitinases, GH19 chitinases were reported to be constitutively expressed in *Arabidopsis* plants (Takenaka et al. 2009), and the enzymes recognize three contiguous GlcNAc residues at subsites −2, −1, and +1 (Sasaki et al. 2006). The GH19 chitinases appeared to act directly on chitinous components of the fungal cell wall, and are likely involved in antifungal action. A number of papers have reported the antifungal activity of plant GH19 chitinases (Schlumbaum et al. 1986; Taira et al. 2002, 2005a, b; Roberts and Selitrennikoff 1986; Mauch et al. 1988a, b; Verburg and Huynh 1991; Leah et al. 1991; Brogue et al. 1991). Taira et al. (2005a) reported that a basic GH19 chitinase from pineapple leaves exhibited strong antifungal activity, whereas an acidic chitinase from the same origin exhibited very little antifungal activity. This suggests the importance of the net charge of the chitinase protein in the antifungal action. A basic class II chitinase tightly bound to a column packed with a cell wall fraction prepared from the mycelia of *Trichoderma* sp. at pH 6.0 and low ionic strength; however, the binding ability of chitinases to the same column was decreased by raising the pH or ionic strength. Since many antifungal peptides are highly basic, the basicity may be an important factor for antifungal activity (Broekaert et al. 1992); that is, the positive charges of the basic chitinases are likely to contribute to the electrostatic interactions with the negative charges of the anionic phospholipids of the fungal cell surface. Catalytic acid-deficient mutants of the GH19 chitinases (completely inactive) were produced, and the antifungal activities were determined to compare the activities with those of the wild-type chitinases. An inactive mutant from BSC-c (class II) had only 15% of the antifungal activity when compared with that of the wild type (Andersen et al. 1997). Ohnuma et al. (2004) showed that an inactive mutant of class I chitinase from rye seeds (RSC-a) did not significantly exhibit antifungal activity under any ionic strength condition, while the wild-type RSC-a considerably inhibited the fungal growth. The chitinolytic activity of the GH19 chitinases appears to be advantageous for antifungal activity, but not absolutely essential. It is most likely that a cooperative action of chitin-binding and chitin hydrolysis may bring about the strong antifungal activity.

12.4 GH46 Chitosanases

Chitosan is a β-1,4-linked heteropolysaccharide consisting of GlcNAc and D-glucosamine (GlcN), and is biosynthesized through a tandem action of chitin synthetase and chitin deacetylase in fungi (Davis and Bartnicki-Garcia 1984). The fungal cell wall has been recognized to contain chitosans with various degrees of *N*-acetylation, which are involved in maintaining cell integrity (Baker et al. 2007). Thus, chitosanases from soil bacteria can destabilize the cell wall of fungal pathogens, controlling the phytopathogenicity of soil fungi. Although chitosanases are widely distributed in various families of glycoside hydrolases, including GH3, GH5, GH7, GH8, GH46, GH75, and GH80, a GH46 chitosanase from *Streptomyces* sp. N174 (CsnN174) has been the most intensively studied with respect to the structure and function. The structure of CsnN174 (Fig. 12.3a) resemble those of GH19 chitinases, and the catalytic mechanisms are proposed to be similar to those of GH19 enzymes (Monzingo et al. 1996; Marcotte et al. 1996). Two carboxylic acids, Glu22 and Asp40 (catalytic acid and base, respectively), and the hydroxyl group of Thr45 (holding a catalytic water molecule) concertedly act in the catalytic

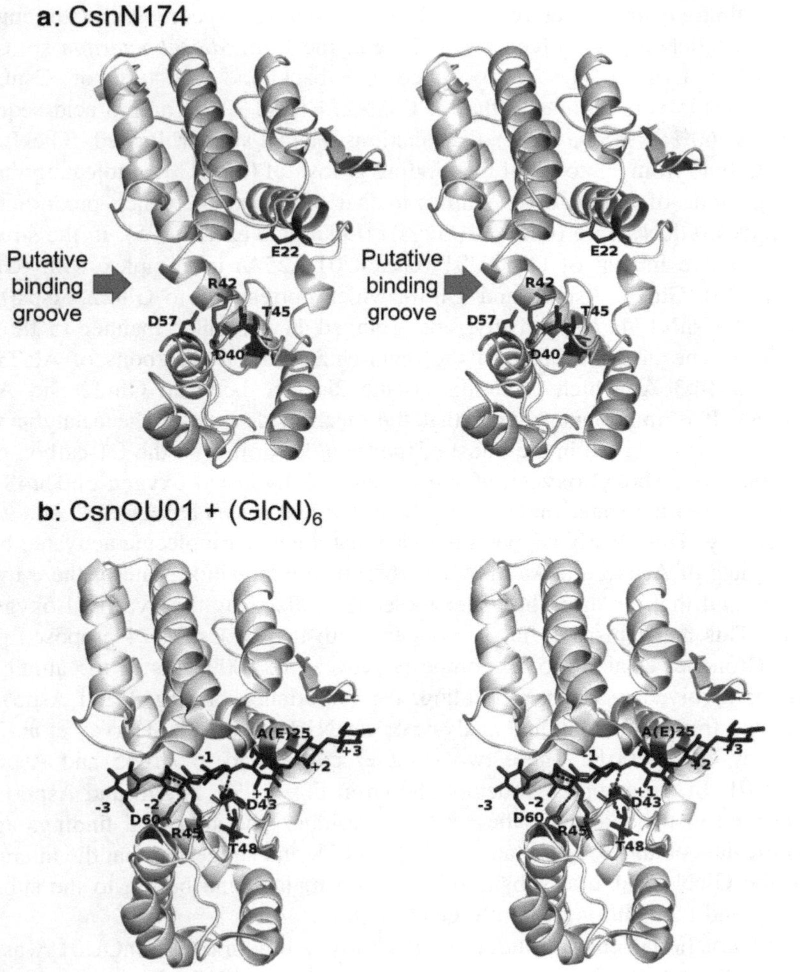

Fig. 12.3 Stereoviews of the crystal structures of CsnN174, a chitosanase from *Streptomyces* sp. N174 in the free state (1CHK, Marcotte et al. 1996) (**a**), and CsnOU01, a chitosanase from *Microbacterium* sp. OU01 in complex with $(GlcN)_6$ (4OLT, Lyu et al. 2014) (**b**). Black, the bound $(GlcN)_6$. Dark gray, important amino acid residues for catalysis and substrate binding

reaction (Boucher et al. 1995; Lacombe-Harvey et al. 2009), and the proton-donating potency of Glu22 is reported to be maintained by an electrostatic interaction network located behind the catalytic acid (Fukamizo et al. 2000). An acidic environment in the substrate-binding groove is most likely involved in the recognition of the positive charges of the polysaccharide substrate (Tremblay et al. 2001; Katsumi et al. 2005). Although, the X-ray crystal structure of CsnN174 in the free state (Fig. 12.3a) was solved by Marcotte et al. (1996), structure analysis of the enzyme in complex with the chitosan substrate had been unsuccessful until recently.

The first crystal structure of GH46 chitosanase in complex with chitosan hexamer, $(GlcN)_6$, was solved using the enzyme from *Microbacterium* sp. OU01 (CsnOU01; Lyu et al. 2014). Since the backbone structure of CsnOU01 (Fig. 12.3b) is very similar to that of CsnN174 (Fig. 12.3a; amino acid sequence similarity, 60%; RMSD, 1.4 Å), the functions may be similar. Indeed, $(GlcN)_6$ was found to bind from −3 to +3 of the binding groove of CsnOU01, indicating that the binding mode of CsnOU01 is similar to that of CsnN174, which predominantly hydrolyzes $(GlcN)_6$ into $(GlcN)_3$+$(GlcN)_3$ (Fukamizo et al. 1995). In the structure of an inactive mutant of CsnOU01 (CsnOU01-E25A) in complex with $(GlcN)_6$ (Fig. 12.3b), Glu25, Asp43, and Thr48, which correspond to Glu22, Asp40, and Thr45 of CsnN174, respectively, are arranged in a similar manner to those of CsnN174. The distance between the main-chain carbonyl carbons of Ala25 and Asp43 is 14.3 Å, which is similar to the distance between Glu22 and Asp40 (13.8 Å). It is important to note that the electron density of the catalytic water molecule was observed in the midst of the triangle formed by the C1-carbon of the −1 sugar, the carboxyl oxygen of Asp43, and the hydroxyl oxygen of Thr48. The distances from the water molecule to the individual atoms are 3.6, 2.7, and 2.7 Å, respectively. This clearly supports the idea that the water molecule activated by the carboxylate of Asp43 attacks the C1 carbon of the transition state of the −1 sugar residue, and that the catalytic water molecule is fixed by the hydroxyl oxygen of Thr48. This mechanism is fully consistent with that of CsnN174 proposed previously (Boucher et al. 1995; Lacombe-Harvey et al. 2009). As to the amino acid residues involved in chitosan binding, the importance of Arg42 and Asp57 was suggested from the mutational analysis of CsnN174 (Lacombe-Harvey et al. 2013; Tremblay et al. 2001). These two residues correspond to Arg45 and Asp60 in CsnOU01. In the complex structure shown in Fig. 12.3b, Arg45 and Asp60 form direct hydrogen bonds with the −2 GlcN residue. The structural findings agreed with the data obtained by mutations of CsnN174, and suggested that the interaction with the GlcN residue at subsite −2 makes a major contribution to the substrate binding and recognition of family GH46 chitosanases.

Conformational change induced by chitosan binding to CsnOU01 was also reported from the same research group (Lyu et al. 2015). They proposed that CsnOU01 undergoes open–closed–open conformational transition upon chitosan binding and the release of enzyme products. This type of conformational change has been observed for glycoside hydrolases with bilobal structures such as family GH19 chitinases (Ohnuma et al. 2013, 2014). These experimental data revealing the conformational changes were obtained from the snapshots of the crystal structures in the free and bound states; however, no experimental evidence on the conformational changes in solution was reported for these enzymes until recently. Shinya et al. (2017) conducted NMR titration experiments with $(GlcN)_6$ using stable isotope-labeled CsnN174 and observed unusual migration profiles of the chemical shifts of the protein resonances as the chitosan oligosaccharide concentration increased. The resonances in the free state were gradually shifting and disappearing upon titration progress while resonances in the ligand-bound state appeared at a different location without changing the chemical shifts. As an example, the chemical shift perturbation of the Trp28 main chain −NH resonance is shown in Fig. 12.4a. This migration profile may be caused by four possible molecular mechanisms (Kovrigin 2012): (A) preexisting dimerization equilibrium when the dimer is incapable of ligand binding, (B) dimerization of the bound state to form a dimer that cannot dissociate the ligands, (C) binding of two ligand molecules to different binding sites (1:2 enzyme–substrate complex formation), and (D) ligand binding followed by isomerization to a tightly bound complex (induced-fit model).

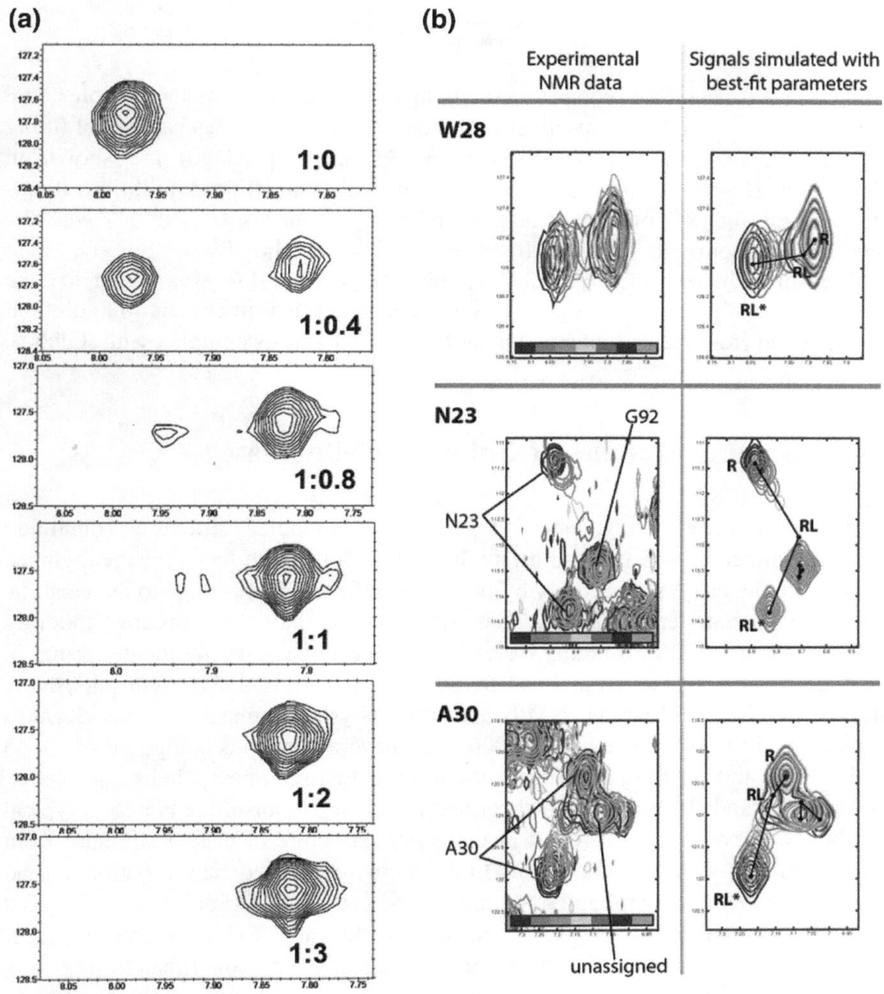

Fig. 12.4 NMR line-shape analysis of the interaction between CsnN174 and $(GlcNAc)_6$ (Shinya et al. 2017). **a** Chemical shift migration of the main-chain NH resonance of Trp28. The numerals in the figure represent molar ratios of $(GlcNAc)_6$ to the enzyme. **b** Experimental (left panels) and simulated (right panels) profiles of the chemical shift perturbations of Trp28, Asn23, and Ala30. R, RL, and RL* represent the free, bound, and induced-fit states of CsnN174

No experimental evidence was obtained for the protein dimerization of CsnN174. As described above, CsnN174 has six subsites in the substrate-binding groove, suggesting a very low probability of 1:2 complex formation between CsnN174 and $(GlcN)_6$. These facts clearly support the binding mechanism (D) as the most probable mechanism for the CsnN174-$(GlcN)_6$ interaction. Thus, we analyzed the NMR-line shapes based on the binding mechanism (D): the induced-fit model shown below,

$$\mathrm{P} + \mathrm{L} \underset{K_\mathrm{d}}{\overset{K_\mathrm{d}}{\longleftrightarrow}} PL \underset{K_\mathrm{if}}{\overset{K_\mathrm{if}}{\longleftrightarrow}} \mathrm{PL}*$$

where P, L, PL, and PL* represent protein, ligand, the enzyme–ligand complex, and the induced-fit complex, respectively. The induced-fit model was capable of fitting all datasets, and three examples (Trp28, Asn23, and Ala30) are shown in Fig. 12.4b. Based on this line-shape analysis, we obtained an equilibrium dissociation constant (K_d) of 42 μM and an induced-fit constant (K_if) of 3.9 with the reverse rate constants of 30,000 s^{-1} and 9 s^{-1} for the dissociation (k_d) and induced-fit (k_if), respectively. Substrate binding to CsnN174 was found to take place through fast binding followed by slow induced-fit isomerization in solution, as suggested from the crystal structures of CsnOU01 and its complex with $(GlcN)_6$.

12.5 Binding Modules of Chitinase/Chitosanase

Carbohydrate-active enzymes often exhibit a modular structure containing non-catalytic carbohydrate-binding modules (CBMs), which have been recognized to enhance the enzymatic activity by positioning the substrate closer to the catalytic domain (Armenta et al. 2017). It is well known that chitin-binding modules belonging to the CBM18 family, so-called hevein domain, are frequently found in plant chitinases. The structures and functions of hevein domains were intensively studied by X-ray diffraction, NMR, and other physicochemical methods (Kezuka et al. 2010; Jiménez-Barbero et al. 2006). The hevein domain is composed of 30–43 amino acids and has a common structural motif composed of a 3_{10} helix, an α-helix, and a two-stranded antiparallel β-sheet, with three or four disulfide bonds. A typical example of hevein domain found in the crystal structure of class I chitinase from *Oryza sativa* is shown in Fig. 12.5a. In the highly conserved central region, a triad of aromatic residues (Trp53, Trp55, and Phe62) and serine (Ser51) are conserved and are positioned on one side of the domain, forming a chitin-binding groove. According to the classification based on the state of the carbohydrate-binding site, hevein domains belong to Type C CBM (Boraston et al. 2004).

Ohnuma et al. (2008) first reported that lysin motifs (LysMs) appended to the N-terminus of a chitinase from the fern *Pteris ryukyuensis* (PrChi-A) act as chitin-binding modules, and the structure of one of the two LysMs from PrChiA was solved by the X-ray diffraction method (Ohnuma et al. 2017). LysMs are

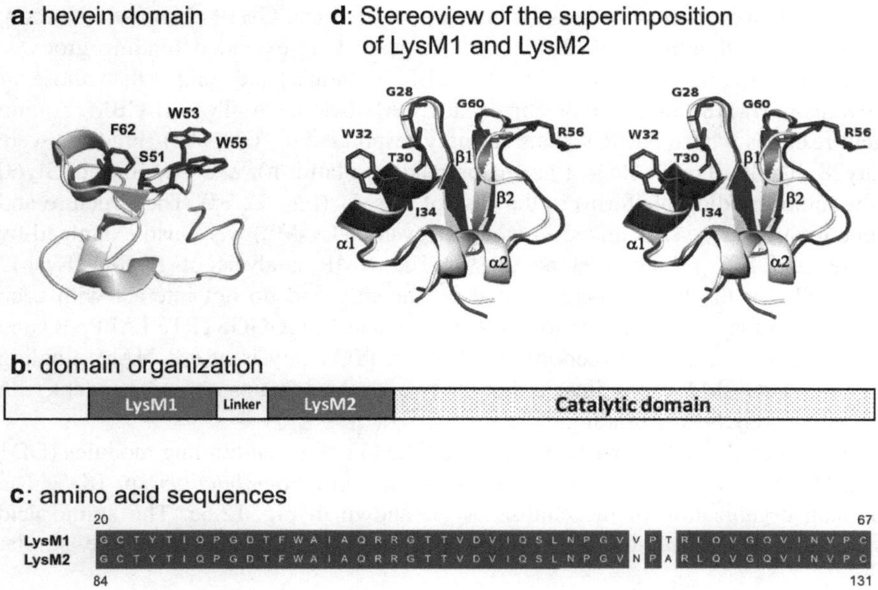

Fig. 12.5 **a** Crystal structure of a hevein domain found in class I chitinase from *Oryza sativa* (Kezuka et al. 2010). A triad of aromatic residues (Trp53, Trp55, and Phe62) and serine residue (Ser51) forming a chitin-binding groove are highlighted. **b** Domain organization of a chitinase from *Volvox carteri*; **c** Amino acid sequences of LysM1 and LysM2 domains (Kitaoku et al. 2017). White backgrounds are amino acid residues substituted between the two domains. **d** Stereoview of the superimposition between the solution structures of LysM1 and LysM2 domains (5YZ6 and 5YZK, respectively; Kitaoku et al. 2019). Black, LysM1; Light gray, LysM2. Amino acid residues, whose NMR signals responded to the chitin oligosaccharide binding, are highlighted

classified into the CBM50 family in the CAZy database and are ubiquitously distributed in living organisms from bacteria to humans. However, in plant chitinases, LysMs are only found in the enzymes from primitive plant species such as fern, horsetail, and a green alga (Onaga and Taira, 2008; Inamine et al. 2015; Kitaoku et al. 2017). The same research group also reported the structure and function of two LysMs found in the N-terminus of a chitinase from *Volvox carteri* (Kitaoku et al. 2019; Fig. 12.5b, c), of which the gene was triggered by a sex-inducing pheromone and wounding (Amon et al. 1998). As shown in Fig. 12.5d, LysMs adopt a βααβ fold comprising two antiparallel β-strands (β1 and β2) and two α-helices (α1 and α2), and the chitin-binding groove is present between two-binding platforms, a major platform is formed by a loop between β1 and α1 and the N-terminal part of α1 (Gln26/90-Trp32/96), while a minor platform is formed by a loop between α2 and β2 (Val53/Asn117-Gly60/124). ITC analysis of the chitin oligosaccharide binding to *Volvox* LysMs provided the thermodynamic parameters for the interaction between LysM and $(GlcNAc)_n$. The higher the polymerization degree of $(GlcNAc)_n$, the higher the binding affinities, the negative values of the Gibbs free energy changes of binding (ΔG°), toward LysM.

The results were similar to those obtained for GH18 and GH19 chitinases (Norberg et al. 2010; Ohnuma et al. 2011c), which have long-extended binding grooves. Thus, the binding grooves of LysMs (CBM50 family) are longer than those of heveins (CBM18 family), indicating that LysMs belongs to Type B CBM. Amino acid residues whose NMR signals strongly responded to $(GlcNAc)_6$ binding, were Gly28, Thr30, Trp32, Ile34 (the major binding platform), and Arg56 and Gly60 (the minor binding platform) in the case of LysM1 (Fig. 12.5d). The structure and function of two LysMs linked in tandem (LysM1+LysM2) were also examined by NMR and ITC (Kitaoku et al. 2019). The NMR analysis of LysM1+LysM2 revealed that the two domains fold independently and do not interact with each other. This may be due to the long (16 amino acids, SGGGGSTPTSTAPPAR) and flexible linker region between the two LysMs. ITC analysis of $(GlcNAc)_n$ binding to LysM1+LysM2 also indicated that the two LysM domains are independent with respect to $(GlcNAc)_n$ binding.

Shinya et al. (2013) first found the two CBM32 chitosan-binding modules (DD1 and DD2) at the C-terminus of a chitosanase from *Paenibacillus* sp. IK-5. The domain organization of the chitosanase is shown in Fig. 12.6a. The amino acid sequences and NMR-based solution structures of DD1 and DD2 are similar

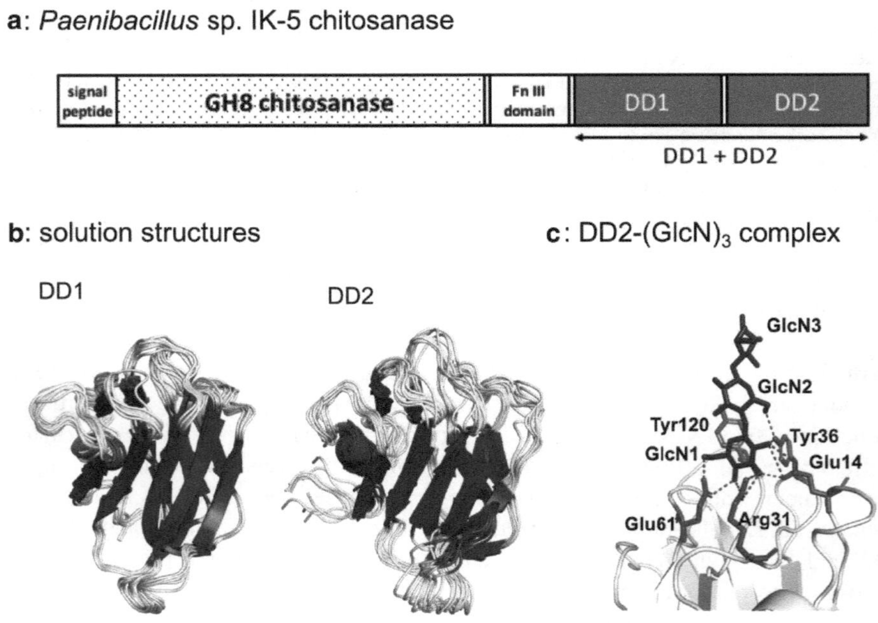

Fig. 12.6 **a** Domain organization of a chitosanase from *Paenibacillus* sp. IK-5 (Shinya et al. 2013); **b** NMR-based solution structures of DD1 (2RV9) and DD2 (2RVA) (Shinya et al. 2016). Ten structures with the lowest target function are superimposed. Dark gray, core β-sandwich domain; Light gray, loop structures. **c** Crystal structure of DD2 in complex with $(GlcN)_3$ (4ZZ8) (Shinya et al. 2016). Black, bound $(GlcN)_3$; Dark gray, amino acid residues involved in chitosan binding. Broken lines are possible hydrogen bonds between the protein and ligands

(sequence similarity, 74%; RMSD, 0.46 Å) as shown in Fig. 12.6b, however, the binding affinities considerably differed from each other (Shinya et al. 2016). The binding affinity of DD1 toward $(GlcN)_4$ was higher than that of DD2 by 2.6 kcal/mol of ΔG° (Shinya et al. 2013). The difference in binding affinity may be derived from amino acid substitutions at the binding site. The crystal structure of DD2 in complex with $(GlcN)_3$ shown in Fig. 12.6c revealed that Glu14, Arg31, Tyr36, Glu61, and Tyr120 are involved in the sugar residue binding. Among these amino acids, Tyr36 is substituted with glutamic acid in DD1, and the substitution at the 36th position may bring about the difference in binding affinity. Indeed, mutation of Tyr36 to glutamic acid in DD2 enhanced the binding affinity, while that of Glu36 to tyrosine in DD1 strongly reduced the affinity (Shinya et al. 2016). On the other hand, a two-modular protein composed of DD1 and DD2 linked in tandem (DD1+DD2) was also produced to analyze the interaction between the two modules. Comparison in NMR spectra among DD1, DD2, and DD1+DD2 revealed that NMR signals of DD1+DD2 were not overlapped with those of DD1 or DD2 alone. In contrast to LysMs in the *Volvox* chitinase, DD1 and DD2 modules interact with each other in DD1+DD2. This may be due to the short linker sequence (6 amino acids, GSTAPS). ITC analysis of $(GlcN)_n$ binding to DD1+DD2 also suggested cooperativity between the two modules; the binding affinities of the DD1 or DD2 module in DD1+DD2 were higher than those of DD1 or DD2 alone (Shinya et al. 2013). The function of multi-carbohydrate-binding modules may be strongly affected by the state of linker polypeptide chain.

12.6 Concluding Remarks

The authors herein reviewed the progress in understanding the structure and function of the enzymes involved in plant–microbe interactions: plant chitinases, chitosanases from soil bacteria, and the substrate-binding modules of these enzymes. For both chitinases and chitosanases, crystal structure analysis of the enzymes in complex with their substrates provided critical information on the mode of substrate binding. GH19 chitinases recognize chitin predominantly via the hydrogen bonding network (Ohnuma et al. 2012, 2014), while hydrogen bonds are less intensive in GH18 chitinases (Ohnuma et al. 2011a, b). This situation may result in a strong recognition of GlcNAc residues at three contiguous subsites −2, −1, and +1 in GH19 chitinases, and a weak recognition of GlcNAc only at subsite −1 in GH18 chitinases (Sasaki et al. 2006). These findings are also consistent with antifungal activities, which are much higher in GH19 chitinases than in GH18 chitinases (Arakane et al. 2012). These facts enabled us to speculate on the biological targets of the plant chitinases: GH19 chitinases may directly attack chitinous components of the fungal cell wall and release the products, chitin fragments, which act as the signal molecules for triggering plant immune responses. On the other hand, GH18 chitinases, of which the genes are expressed in response to both biotic and abiotic stress (Takenaka et al. 2009; Ohnuma et al. 2011b), may have more broad specificity toward various targets containing a GlcNAc sugar moiety, suggesting multiple roles in tolerances toward various types of stress under physiological conditions. Indeed, physiological functions appear to be much more diverse in GH18 chitinases than in GH19 chitinases.

Structures of chitin- and chitosan-binding modules, belonging to the CBM50 (LysM) and CBM32 families (DD1 and DD2), respectively, were successfully solved and found to be similar to those previously obtained for other binding modules belonging to the identical families (Kitaoku et al. 2017, 2019; Shinya et al. 2013, 2016). Two LysMs are appended to the N-terminus of a class III chitinase from *V. carteri*, and linked with a flexible polypeptide chain composed of 16 amino acid residues; hence, the two LysMs do not interact with each other, and are independent with respect to $(GlcNAc)_n$ binding. The flexible movements of the two LysMs may be advantageous for the class III chitinase, which may have multiple roles in stress tolerance rather than in chitin assimilation or aggressive action toward pathogens. On the other hand, DD1 and DD2 are appended to the C-terminus of a GH8 chitosanase from *Paenibacillus* sp. IK-5, and linked with a polypeptide chain of 6 amino acids. These two chitosan-binding modules interact with each other and cooperatively act toward $(GlcN)_n$. Since this chitosanase plays a major role in chitosan assimilation, the cooperative binding of the two chitosan-binding modules, DD1 and DD2, may be advantageous in the efficient assimilation of chitosan. Taken together, it is most likely that the modular chitinases/chitosanases were found to have appropriate domain organization, which is designed to fulfill their roles in plant–microbe interactions.

References

Amon P, Haas E, Sumper M (1998) The sex-inducing pheromone and wounding trigger the same set of genes in the multicellular green alga *Volvox*. Plant Cell 10:781–789

Andersen MD, Jensen A, Robertus JD, Leah R, Skriver K (1997) Heterologous expression and characterization of wild-type and mutant forms of a 26 kDa endochitinase from barley (*Hordeum vulgare* L.). Biochem J 322:815–822

Arakane Y, Taira T, Ohnuma T, Fukamizo T (2012) Chitin-related enzymes in agro-biosciences. Curr Drug Targets 13:442–470

Armenta S, Moreno-Mendieta S, Sánchez-Cuapio Z, Sánchez S, Rodríguez-Sanoja R (2017) Advances in molecular engineering of carbohydrate-binding modules. Proteins 85:1602–1617

Baker LG, Specht CA, Donlin MJ, Lodge JK (2007) Chitosan, the deacetylated form of chitin, is necessary for cell wall integrity in *Cryptococcus neoformans*. Eukaryot Cell 6:855–867

Bokma E, Rozeboom HJ, Sibbald M, Dijkstra BW, Beintema JJ (2002) Expression and characterization of active site mutants of hevamine, a chitinase from the rubber tree *Hevea brasiliensis*. Eur J Biochem 269:893–901

Boraston AB, Bolam DN, Gilbert HJ, Davies GJ (2004) Carbohydrate-binding modules: fine-tuning polysaccharide recognition. Biochem J 382:769–781

Boucher I, Fukamizo T, Honda Y, Willick GE, Neugebauer WA, Brzezinski R (1995) Site-directed mutagenesis of evolutionary conserved carboxylic amino acids in the chitosanase from *Streptomyces* sp. N174 reveals two residues essential for catalysis. J Biol Chem 270:31077–31082

Broekaert WF, Mariën W, Terras FR, De Bolle MF, Proost P, Van Damme J, Dillen L, Claeys M, Rees SB, Vanderleyden J et al (1992) Antimicrobial peptides from Amaranthus caudatus seeds with sequence homology to the cysteine/glycine-rich domain of chitin-binding proteins. Biochemistry 31:4308–4314

Brogue K, Chet I, Holliday M, Cressman R, Biddle P, Knowlton S, Mauvais CJ, Broglie R (1991) Transgenic plants with enhanced resistance to the fungal pathogen *Rhizoctonia solani*. Science 254:1194–1197

Davies GJ, Wilson KS, Henrissat B (1997) Nomenclature for sugar-binding subsites in glycosyl hydrolases. Biochem J 321:557–559

Davis LL, Bartnicki-Garcia S (1984) Chitosan synthesis by the tandem action of chitin synthetase and chitin deacetylase from *Mucor rouxii*. Biochemistry 23:1065–1073

Desaki Y, Miyata K, Suzuki M, Shibuya N, Kaku H (2018) Plant immunity and symbiosis signaling mediated by LysM receptors. Innate Immun 24:92–100

El Hadrami A, Adam LR, El Hadrami I, Daayf F (2010) Chitosan in plant protection. Mar Drugs 8:968–987

Fukamizo T, Honda Y, Goto S, Boucher I, Brzezinski R (1995) Reaction mechanism of chitosanase from *Streptomyces* sp. N174. Biochem J 311:377–383

Fukamizo T, Juffer AH, Vogel HJ, Honda Y, Tremblay H, Boucher I, Neugebauer WA, Brzezinski R (2000) Theoretical calculation of pKa reveals an important role of Arg205 in the activity and stability of *Streptomyces* sp. N174 chitosanase. J Biol Chem 275:25633–25640

Goormachtig S, Lievens S, Van de Velde W, Van Montagu M, Holsters M (1998) Srchi13, a novel early nodulin from *Sesbania rostrata*, is related to acidic class III chitinases. Plant Cell. 10:905–915

Hart PJ, Monzingo AF, Ready MP, Ernst SR, Robertus JD (1993) Crystal structure of an endochitinase from Hordeum vulgare L. seeds. J Mol Biol 229:189–193

Inamine S, Onaga S, Ohnuma T, Fukamizo T, Taira T (2015) Purification, cDNA cloning, and characterization of LysM-containing plant chitinase from horsetail (Equisetum arvense). Biosci Biotechnol Biochem 79:1296–1304

Jiménez-Barbero J, Javier Cañada F, Asensio JL, Aboitiz N, Vidal P, Canales A, Groves P, Gabius HJ, Siebert HC (2006) Hevein domains: an attractive model to study carbohydrate-protein interactions at atomic resolution. Adv Carbohydr Chem Biochem 60:303–354

Kasprzewska A (2003) Plant chitinases–regulation and function. Cell Mol Biol Lett 8:809–824

Katsumi T, Lacombe-Harvey ME, Tremblay H, Brzezinski R, Fukamizo T (2005) Role of acidic amino acid residues in chitooligosaccharide-binding to Streptomyces sp. N174 chitosanase. Biochem Biophys Res Commun 338:1839–1844

Kezuka Y, Kojima M, Mizuno R, Suzuki K, Watanabe T, Nonaka T (2010) Structure of full-length class I chitinase from rice revealed by X-ray crystallography and small-angle X-ray scattering. Proteins 78:2295–2305

Kitaoku Y, Fukamizo T, Numata T, Ohnuma T (2017) Chitin oligosaccharide binding to the lysin motif of a novel type of chitinase from the multicellular green alga, Volvox carteri. Plant Mol Biol 93:97–108

Kitaoku Y, Nishimura S, Hirono T, Suginta W, Ohnuma T, Fukamizo T. (2019) Structures and chitin binding properties of two N-terminal lysin motifs (LysMs) found in a chitinase from *Volvox carteri*. Glycobiology, in press

Kovrigin EL (2012) NMR line shapes and multi-state binding equilibria. J Biomol NMR 53:257–270

Lacombe-Harvey ME, Fukamizo T, Gagnon J, Ghinet MG, Dennhart N, Letzel T, Brzezinski R (2009) Accessory active site residues of Streptomyces sp. N174 chitosanase: variations on a common theme in the lysozyme superfamily. FEBS J 276:857–869

Lacombe-Harvey MÈ, Fortin M, Ohnuma T, Fukamizo T, Letzel T, Brzezinski R (2013) A highly conserved arginine residue of the chitosanase from Streptomyces sp. N174 is involved both in catalysis and substrate binding. BMC Biochem 14:23

Leah R, Tommerup H, Svendsen I, Mundy J (1991) Biochemical and molecular characterization of three barley seed proteins with antifungal properties. J Biol Chem 266:1564–1573

Lyu Q, Wang S, Xu W, Han B, Liu W, Jones DN, Liu W (2014) Structural insights into the substrate-binding mechanism for a novel chitosanase. Biochem J 461:335–345

Lyu Q, Shi Y, Wang S, Yang Y, Han B, Liu W, Jones DN, Liu W (2015) Structural and biochemical insights into the degradation mechanism of chitosan by chitosanase OU01. Biochim Biophys Acta 1850:1953–1961

Marcotte EM, Monzingo AF, Ernst SR, Brzezinski R, Robertus JD (1996) X-ray structure of an anti-fungal chitosanase from *Streptomyces* N174. Nat Struct Biol 3:155–162

Mauch F, Hadwiger LA, Boller T (1988a) Antifungal hydrolases in pea tissue: i. purification and characterization of two chitinases and two beta-1,3-glucanases differentially regulated during development and in response to fungal infection. Plant Physiol 87:325–333

Mauch F, Mauch-Mani B, Boller T (1988b) Antifungal hydrolases in pea tissue: ii. inhibition of fungal growth by combinations of chitinase and beta-1,3-Glucanase. Plant Physiol 88:936–942

Melchers LS, Apotheker-de Groot M, van der Knaap JA, Ponstein AS, Sela-Buurlage MB, Bol JF, Cornelissen BJ, van den Elzen PJ, Linthorst HJ (1994) A new class of tobacco chitinases homologous to bacterial exo-chitinases displays antifungal activity. Plant J 5:469–480

Monzingo AF, Marcotte EM, Hart PJ, Robertus JD (1996) Chitinases, chitosanases, and lysozymes can be divided into procaryotic and eucaryotic families sharing a conserved core. Nat Struct Biol 3:133–140

Neuhaus JM, Fritig B, Linthorst HJM, Meins F, Mikkelsen JD, Ryals J (1996) A revised nomenclature for chitinase genes. Plant Mol Biol Rep 14:102–104

Norberg AL, Karlsen V, Hoell IA, Bakke I, Eijsink VG, Sørlie M (2010) Determination of substrate binding energies in individual subsites of a family 18 chitinase. FEBS Lett 584:4581–4585

Ohnuma T, Taira T, Yamagami T, Aso Y, Ishiguro M (2004) Molecular cloning, functional expression, and mutagenesis of cDNA encoding class I chitinase from rye (*Secale cereale*) seeds. Biosci Biotechnol Biochem 68:324–332

Ohnuma T, Onaga S, Murata K, Taira T, Katoh E (2008) LysM domains from *Pteris ryukyuensis* chitinase-A: a stability study and characterization of the chitin-binding site. J Biol Chem 283:5178–5187

Ohnuma T, Numata T, Osawa T, Mizuhara M, Vårum KM, Fukamizo T (2011a) Crystal structure and mode of action of a class V chitinase from *Nicotiana tabacum*. Plant Mol Biol 75:291–304

Ohnuma T, Numata T, Osawa T, Mizuhara M, Lampela O, Juffer AH, Skriver K, Fukamizo T (2011b) A class V chitinase from *Arabidopsis thaliana*: gene responses, enzymatic properties, and crystallographic analysis. Planta 234:123–137

Ohnuma T, Ohnuma T, Sørlie M, Fukuda T, Kawamoto N, Taira T, Fukamizo T (2011c) Chitin oligosaccharide binding to a family GH19 chitinase from the moss *Bryum coronatum*. FEBS J 278:3991–4001

Ohnuma T, Numata T, Osawa T, Inanaga H, Okazaki Y, Shinya S, Kondo K, Fukuda T, Fukamizo T (2012) Crystal structure and chitin oligosaccharide-binding mode of a 'loopful' family GH19 chitinase from rye, *Secale cereale*, seeds. FEBS J 279:3639–3651

Ohnuma T, Umemoto N, Kondo K, Numata T, Fukamizo T (2013) Complete subsite mapping of a "loopful" GH19 chitinase from rye seeds based on its crystal structure. FEBS Lett 587:2691–2697

Ohnuma T, Umemoto N, Nagata T, Shinya S, Numata T, Taira T, Fukamizo T (2014) Crystal structure of a "loopless" GH19 chitinase in complex with chitin tetrasaccharide spanning the catalytic center. Biochim Biophys Acta 1844:793–802

Ohnuma T, Taira T, Umemoto N, Kitaoku Y, Sørlie M, Numata T, Fukamizo T (2017) Crystal structure and thermodynamic dissection of chitin oligosaccharide binding to the LysM module of chitinase-A from *Pteris ryukyuensis*. Biochem Biophys Res Commun 494:736–741

Onaga S, Taira T (2008) A new type of plant chitinase containing LysM domains from a fern (*Pteris ryukyuensis*): roles of LysM domains in chitin binding and antifungal activity. Glycobiology 18:414–423

Roberts WK, Selitrennikoff CP (1986) Isolation and partial characterization of two antifungal proteins from barley. Biochim Biophys Acta 880:161–170

Saito A, Ooya T, Miyatsuchi D, Fuchigami H, Terakado K, Nakayama SY, Watanabe T, Nagata Y, Ando A (2009) Molecular characterization and antifungal activity of a family 46 chitosanase from *Amycolatopsis* sp. CsO-2. FEMS Microbiol Lett 293:79–84

Sasaki C, Vårum KM, Itoh Y, Tamoi M, Fukamizo T (2006) Rice chitinases: sugar recognition specificities of the individual subsites. Glycobiology 16:1242–1250

Schlumbaum A, Mauch F, Vogeli U, Boller T (1986) Plant chitinases are potent inhibitors of fungal growth. Nature 324:365–367

Shinya S, Ohnuma T, Yamashiro R, Kimoto H, Kusaoke H, Anbazhagan P, Juffer AH, Fukamizo T (2013) The first identification of carbohydrate binding modules specific to chitosan. J Biol Chem 288:30042–30053

Shinya S, Nishimura S, Kitaoku Y, Numata T, Kimoto H, Kusaoke H, Ohnuma T, Fukamizo T (2016) Mechanism of chitosan recognition by CBM32 carbohydrate-binding modules from a Paenibacillus sp. IK-5 chitosanase/glucanase. Biochem J 473:1085–1095

Shinya S, Ghinet MG, Brzezinski R, Furuita K, Kojima C, Shah S, Kovrigin EL, Fukamizo T (2017) NMR line shape analysis of a multi-state ligand binding mechanism in chitosanase. J Biomol NMR 67:309–319

Suzukawa K, Yamagami T, Ohnuma T, Hirakawa H, Kuhara S, Aso Y, Ishiguro M (2003) Mutational analysis of amino acid residues involved in catalytic activity of a family 18 chitinase from tulip bulbs. Biosci Biotechnol Biochem 67:341–346

Taira T, Ohnuma T, Yamagami T, Aso Y, Ishiguro M, Ishihara M (2002) Antifungal activity of rye (*Secale cereale*) seed chitinases: the different binding manner of class I and class II chitinases to the fungal cell walls. Biosci Biotechnol Biochem 66:970–977

Taira T, Toma N, Ishihara M (2005a) Purification, characterization, and antifungal activity of chitinases from pineapple (*Ananas comosus*) leaf. Biosci Biotechnol Biochem 69:189–196

Taira T, Ohdomari A, Nakama N, Shimoji M, Ishihara M (2005b) Characterization and antifungal activity of gazyumaru (*Ficus microcarpa*) latex chitinases: both the chitin-binding and the antifungal activities of class I chitinase are reinforced with increasing ionic strength. Biosci Biotechnol Biochem 69:811–818

Taira T, Hayashi H, Tajiri Y, Onaga S, Uechi G, Iwasaki H, Ohnuma T, Fukamizo T (2009) A plant class V chitinase from a cycad (*Cycas revoluta*): biochemical characterization, cDNA isolation, and posttranslational modification. Glycobiology 19:1452–1461

Taira T, Fujiwara M, Dennhart N, Hayashi H, Onaga S, Ohnuma T, Letzel T, Sakuda S, Fukamizo T (2010) Transglycosylation reaction catalyzed by a class V chitinase from cycad, *Cycas revoluta*: a study involving site-directed mutagenesis, HPLC, and real-time ESI-MS. Biochim Biophys Acta 1804:668–675

Takenaka Y, Nakano S, Tamoi M, Sakuda S, Fukamizo T (2009) Chitinase gene expression in response to environmental stresses in Arabidopsis thaliana: chitinase inhibitor allosamidin enhances stress tolerance. Biosci Biotechnol Biochem 73:1066–1071

Terwisscha van Scheltinga AC, Kalk KH, Beintema JJ, Dijkstra BW (1994) Crystal structures of hevamine, a plant defence protein with chitinase and lysozyme activity, and its complex with an inhibitor. Structure. 2:1181–1189

Terwisscha van Scheltinga AC, Armand S, Kalk KH, Isogai A, Henrissat B, Dijkstra BW (1995) Stereochemistry of chitin hydrolysis by a plant chitinase/lysozyme and X-ray structure of a complex with allosamidin: evidence for substrate assisted catalysis. Biochemistry 34:15619–15623

Tremblay H, Yamaguchi T, Fukamizo T (2001) Brzezinski R. Mechanism of chitosanase-oligosaccharide interaction: subsite structure of *Streptomyces* sp. N174 chitosanase and the role of Asp57 carboxylate. J Biochem 130:679–686

Umemoto N, Kanda Y, Ohnuma T, Osawa T, Numata T, Sakuda S, Taira T, Fukamizo T (2015a) Crystal structures and inhibitor binding properties of plant class V chitinases: the cycad enzyme exhibits unique structural and functional features. Plant J 82:54–66

Umemoto N, Ohnuma T, Osawa T, Numata T, Fukamizo T (2015b) Modulation of the transglycosylation activity of plant family GH18 chitinase by removing or introducing a tryptophan side chain. FEBS Lett 589:2327–2333

van Aalten DM, Komander D, Synstad B, Gåseidnes S, Peter MG, Eijsink VG (2001) Structural insights into the catalytic mechanism of a family 18 exo-chitinase. Proc Natl Acad Sci U S A 98:8979–8984

Verburg JG, Huynh QK (1991) Purification and characterization of an antifungal chitinase from *Arabidopsis thaliana*. Plant Physiol 95:450–455

Yang H, Zhang T, Masuda T, Lv C, Sun L, Qu G, Zhao G (2011) Chitinase III in pomegranate seeds (*Punica granatum* Linn.): a high-capacity calcium-binding protein in amyloplasts. Plant J 68:765–776

Chapter 13
Chitinous Structures as Potential Targets for Insect Pest Control

Guillaume Tetreau and Ping Wang

Abstract Chitinous structures are physiologically fundamental in insects. They form the insect exoskeleton, play important roles in physiological systems and provide physical, chemical and biological protections in insects. As critically important structures in insects, chitinous structures are attractive target sites for the development of new insect-pest-control strategies. Chitinous structures in insects are complex and their formation and maintenance are dynamically regulated with the growth and development of insects. In the past few decades, studies on insect chitinous structures have shed lights on the physiological functions, compositions, structural formation, and regulation of the chitinous structures. Current understanding of the chitinous structures has indicated opportunities for exploring new target sites for insect control. Mechanisms to disrupt chitinous structures in insects have been studied and strategies for the potential development of new means of insect control by targeting chitinous structures have been proposed and are practically to be explored.

Keywords Insect chitinous structure · Cuticle · Peritrophic membrane · Chitin-binding proteins · Chitin synthase · Chitin deacetylase · Chitinase · N-acetylglucosaminidases

13.1 Introduction

Insects are major pests of agriculture and human health. Insect-pest-control heavily relies on the use of insecticides and control of insect pests becomes increasingly challenging, due to the rapid development of insecticide resistance in insect

G. Tetreau
University of Grenoble Alpes, CNRS, CEA, IBS, 38000 Grenoble, France
e-mail: guillaume.tetreau@gmail.com

P. Wang (✉)
Department of Entomology, Cornell University, Geneva, NY 14456, USA
e-mail: pingwang@cornell.edu

Q. Yang and T. Fukamizo (eds.), *Targeting Chitin-containing Organisms*, Advances in Experimental Medicine and Biology 1142,
https://doi.org/10.1007/978-981-13-7318-3_13

populations (Lacey et al. 2015; Popp et al. 2012). The development of insecticide resistance urges the search for new insect target sites for the development of novel insecticides to continue the success of insecticides for insect control (Casida and Durkin 2017).

Evolutionary success of insects is associated with their unique chitinous structures. In insects, chitin is the primary component of the cuticle to form the exoskeleton and internal chitinous structures, and is also the essential component of the midgut peritrophic membrane (PM) lining the midgut epithelium (Kelkenberg et al. 2015; Muthukrishnan et al. 2012; Wang and Granados 2001; Zhu et al. 2016). These chitinous structures play crucial roles in protecting the insects from physical damage, chemical toxicity and microbial infections from the environment (Kelkenberg et al. 2015; Kuraishi et al. 2013; Tellam et al. 1999; Terra 2001). The chitinous structures in insects are composed of chitin and other structural components including lipids, minerals, and, more importantly, various proteins (Terra and Ferreira 2005; Willis et al. 2005). The synthesis, modification, and recycling of the chitinous structures are dynamically regulated at each developmental stage to cope with the growth and development of insects (Zhu et al. 2016). Formation and dynamic regulation of chitinous structures in insects involve chitin and two major categories of proteins: chitin metabolism enzymes and structural chitin-binding proteins (Merzendorfer and Zimoch 2003; Willis et al. 2012; Zhu et al. 2016). As chitinous structures are essential in insects, identification of target sites to disrupt chitinous structures in insects is a logical approach for the development of new insect control technologies.

13.2 Chitin Metabolism Enzymes as Potential Target Sites

Chitin synthesis in insects is spatially and temporally regulated throughout different developmental stages. The biosynthesis, degradation, and recycling of chitin involve four major categories of enzymes—the chitin synthases (CHSs), chitin deacetylases (CDAs), chitinases (CHTs), and N-acetylglucosaminidases (NAGs) (Merzendorfer and Zimoch 2003; Muthukrishnan et al. 2012). These chitin metabolism enzymes are potential target sites to explore for development of insect control technologies.

13.2.1 Chitin Synthases (CHSs)

Chitin is synthesized from UDP-N-acetylglucosamine by a chitin synthase (CHS) (Merzendorfer 2006). Chitin synthesis in insects relies on two CHSs, CHS-1 and CHS-2 (also named CHS-A and CHS-B, respectively) (Muthukrishnan et al. 2012; Tetreau et al. 2015a). CHS-1 and CHS-2 differ from each other in their tissue-specific expression. CHS-1 is involved in chitin synthesis in the epidermis for the formation of cuticles. In contrast, CHS-2 is involved in chitin synthesis in the midgut only, for the formation of the PM (Merzendorfer 2006; Zhu et al. 2016; Zimoch et al. 2005).

The potential of CHS-1 and CHS-2 as target sites to suppress insect growth and development has been shown by inhibition of the expression of CHS-1 or CHS-2 in insects. Abnormal formation of cuticle and tracheal structures, slower development, and increased mortality were observed in insects when the expression of CHS-1 was inhibited (Chen et al. 2008; Hansen et al. 2009; Shi et al. 2016). Similarly, inhibition of CHS-2 gene expression in insects resulted in increased permeability of the PM and consequently in slower development and increased mortality, especially during the pupal-to-adult molting (Kato et al. 2006; Kelkenberg et al. 2015; Lee et al. 2017; Macedo et al. 2017; Shi et al. 2016).

13.2.2 Chitin Deacetylases (CDAs)

In insects, chitin can be partially deacetylated by CDAs (Zhu et al. 2016). Numerous CDAs have been identified and classified into five phylogenetic groups (Group I to V) (Dixit et al. 2008; Tetreau et al. 2015a), since the first insect CDA was identified from the cabbage looper, *Trichoplusia ni* (Guo et al. 2005).

The physiological roles of CDAs vary among the CDAs and may also differ in different insects. However, inhibition of some of the CDAs in insects has been reported to result in deleterious effects on growth and development. Inhibition of group I CDAs by RNAi may lead to disruption of molting and deformed cuticles in insects (Arakane et al. 2009; Dong et al. 2014; Luschnig et al. 2006; Quan et al. 2013; Wu et al. 2018; Xi et al. 2014; Yang et al. 2018). Inhibition of a group IV CDA by RNAi in the hemipteran *Nilaparvata lugens* has been shown to cause molting defects (Xi et al. 2014), and inhibition of a group IV CDA in the lepidopteran *Helicoverpa armigera* was also shown to lead to abnormal larval growth and delayed pupation (Chikate et al. 2016). The physiological roles of CDAs require to be further understood, but it is apparent that at least some of the CDAs can be potential target sites to insect control.

13.2.3 Chitinases (CHTs)

Insect chitinases (CHTs) belong to the GH18 family and are classified into 11 CHT groups (Groups I to X, plus Group h) (Tetreau et al. 2015a). The potential of CHTs as target sites for insect control has been indicated in various insects by the inhibition of specific CHT gene expression. Inhibition of a CHT from group I can disrupt pupal-to-adult molting in *Tribolium castaneum* and *N. lugens*, and larval-to-pupae and pupal-to-adult molting in *S. exigua* and *Locusta migratoria* (Li et al. 2015; Xi et al. 2015; Zhang et al. 2012; Zhu et al. 2008). Slower larval growth and increased mortality have also been observed in *Mythimna separata* by inhibition of the expression of a group I CHT (MsChi1) (Ganbaatar et al. 2017). Inhibition of the group II chitinase in *T. castaneum* (CHT10) by RNAi can prevent egg hatching and molting at every developmental stage (Zhu et al. 2008). It has also been observed that RNAi of the group III CHT in *Sogatella furcifera* (CHT7) could cause molting defects and impair wing development (Chen et al. 2017a), and inhibition of a group III CHT in *T. castaneum* (CHT7) disrupted the organization of high ordered chitin fibers for the formation of elytra (Noh et al. 2018; Zhu et al. 2008). RNAi of a group IV CHT in *Drosophila* led to deformed cuticles and molting defects in larvae and adults, and RNAi of a group IV CHT in *Ostrinia nubilalis* led to increased PM chitin content and decreased larval body weight (Khajuria et al. 2010; Pesch et al. 2016; Zhu et al. 2008). Inhibition of group V CHTs in *T. castaneum* and *Drosophila* could results in deformed cuticles, larval and adult molting defects, and mortality (Pesch et al. 2016; Zhu et al. 2008). Moreover, inhibition of a group h CHT has been reported to result in reduced body weight in *M. separata* (the CHT was named as MsChi2). The reported results from inhibition of CHTs in insects have provided evidence that CHTs can be target sites to further explore for insect control.

13.2.4 N-Acetylglucosaminidases (NAGs)

The N-acetylglucosaminidases (NAGs) in insects catalyze the removal of terminal sugars from chitooligosaccharides in chitin recycling (Zhu et al. 2016) and are classified in four phylogenetic groups (Groups I to IV). The NAGs from group I are the most abundantly expressed and appear to play a major role in chitin metabolism. RNAi of group 1 NAGs can result in severe molting defects (Hogenkamp et al. 2008).

13.3 Chitin-Binding Proteins (CBPs) as Potential Target Sites

Chitin is always associated with structural chitin-binding proteins (CBPs) in insects, and the function of chitinous structures relies on the proteins in the structure. Therefore, CBPs can be potentially important target sites to be explored for insect control. The chitin-binding affinity of the CBPs relies on two major consensus chitin-binding domains (CBDs)—chitin-binding domain 2 (ChtBD2) (pfam01607) (also known as peritrophin-A type CBD) and the Rebers and Riddiford (R&R) chitin-binding domain (ChtBD4). In addition to ChtBD2 and ChtBD4, another family of proteins, Tweedle proteins, has been identified to be CBPs in *B. mori* (Tang et al. 2010).

13.3.1 PM Proteins (PMPs)

PMPs, also known as peritrophins, contain 1–14 CBDs and are primarily expressed in the midgut, especially during the feeding stages (Jasrapuria et al. 2010; Tetreau et al. 2015b; Wang et al. 2004). PMPs are essential PM proteins cross-linking the chitin fibrils with the multiple CBDs to form the PM structure (Tellam et al. 1999; Wang and Granados 1997b; Wang et al. 2004). Therefore, PMPs are indispensable for the PM structure and function, and can be promising target sites for novel insecticides.

An insect may have multiple and redundant PMPs in the formation of the PM (Agrawal et al. 2014; Wang et al. 2004). Effective targeting of PMPs to functionally disrupt the PM structure may require targeting multiple PMPs in insects. It has been shown that inhibition of two of PMPs, TcPMP3 and TcPMP5-B, in *T. castaneum* by RNAi indeed led to abnormal larval growth and molting, and eventually to mortality (Agrawal et al. 2014).

13.3.2 Cuticular Proteins Analogous to Peritrophins (CPAPs)

CPAPs are divided into two types—CPAP1s and CPAP3s containing 1 or 3 CBDs, respectively (Jasrapuria et al. 2010). An insect generally contains 10 to 20 CPAP1s and 5 to 12 CPAP3s (Ioannidou et al. 2014; Tetreau et al. 2015b). CPAPs are found in different tissues and play important physiological functions (Dittmer et al. 2015; Jasrapuria et al. 2012; Rohrbough et al. 2007; Rushton et al. 2012).

Inhibition of CPAP genes can lead to significant deleterious effects in insects. In *T. castaneum*, inhibition of the expression of the CPAP genes TcCPAP1-C, TcCPAP1-H, TcCPAP1-J, and TcCPAP3-C by RNAi can result in mortality of the insect during the pupal-to-adult molting (Jasrapuria et al. 2012). Inhibition of the expression of TcCPAP3-A1, TcCPAP3-B, TcCPAP3-D1, or TcCPAP3-D2 can cause abnormal development of the insects, including weakened elytra, stiff joints, and increased mortality at different developmental stages (Jasrapuria et al. 2012). In *D. melanogaster*, loss of function of group A CPAP3s causes severe physiological defects (Petkau et al. 2012; Tiklova et al. 2013).

13.3.3 Cuticular Protein with R&R Motif (CPRs)

CPRs are the largest family of cuticle proteins (CPs) (Willis et al. 2012). The number of CPR genes ranges from as low as 32 in *Apis mellifera* up to 207 in *Manduca sexta*, and most insects generally have more than 100 (Cornman and Willis, 2008; Dittmer et al. 2015; Ioannidou et al. 2014). Inhibition of CPRs by

RNAi in *T. castaneum* can lead to malformation of adults with disrupted pore canal formation in the cuticle and weakened elytra (Noh et al. 2014, 2015). A dysfunctional mutation in a CPR gene in *B. mori* reduced the content of chitin in larval cuticle and affected its mechanical property (Qiao et al. 2014). Modification of expression of CPRs may potentially alter the penetration of insecticides in insects (Despres et al. 2014a, b; Koganemaru et al. 2013; Yahouédo et al. 2017).

13.3.4 Tweedle Proteins

Tweedle proteins have not been well studied. However, the potential of Tweedle proteins as targets to suppress insect growth and development has been indicated. It has been reported that RNAi of a Tweedle gene in *L. migratoria* (*LmTwdl1*) resulted in thinner epicuticle and disrupted chitin fiber arrangement in the procuticle, leading to high mortality during the larval–pupal molting (Song et al. 2016). Similarly, it has been observed in *Drosophila* that a Tweedle gene mutation led to abnormal cuticle organization and consequently to altered body morphology (Guan et al. 2006).

13.4 Targeting Chitin and Chitin-Associated Enzymes and Proteins for Pest Control

13.4.1 Chitinases as Insecticidal Proteins

Chitin is degraded by chitinases which are widely present in chitin-containing organisms, such as insects, crustaceans and fungi, and also in organisms without chitin, such as bacteria, plants, and humans (Adrangi and Faramarzi 2013). Chitinases from various sources have been explored as insecticidal enzymes by targeting the chitin in insects (da Silva et al. 2005; Jalil et al. 2015; Leger et al. 1986). Entomopathogenic fungi secrete chitinases to degrade the insect cuticle to gain entry into the insect host. Among the chitinases from microbial organisms, those from entomopathogenic fungi have been the most extensively studied for potential insect control (Hartl et al. 2012). Chitinases of plant origins, especially those over-expressed upon infection by fungi or attack by herbivores, are also attractive candidates for use in plants to confer resistance to pathogens and insect pests (Grover 2012). Insects are also a rich source of chitinases with the highest diversity characterized by different phylogenetic groups (de la Vega et al. 1998; Tetreau et al. 2015a). Chitinases vary biochemically in activity and their expression is complex. Chitinolytic activity is often synergized by multiple chitinases that are dynamically expressed (Oyeleye and Normi 2018). Continuing identification and understanding of chitinases from a diverse range of sources are required to provide a comprehensive resource of chitinases and necessary knowledge for the development of chitinase-based biotechnologies for pest control (Oyeleye and Normi 2018).

A number of plants have been engineered to express a chitinase to examine resistance to insect herbivory. Chitinase genes of various origins, including fungi, insects, plants, and virus, have been used as transgenes in tobacco, tomato, and maize plants to study the resistance to insect herbivory. The first reported transgenic plant expressing a chitinase was the tobacco, *Nicotiana tabacum* var. Xanthi, constitutively expressing a chitinase (GH18, group I) from the tobacco hornworm, *M. sexta*. The chitinase expressed in the tobacco plants appeared to be truncated, but chitinase activity was observed. Larval growth of the tobacco budworm, *Heliothis virescens*, and the merchant grain beetle, *Oryzaephilis mercator*, was strongly inhibited when they fed on the chitinase-expressing tobacco plants (Ding et al. 1998; Wang et al. 1996). Similarly, the expression of a chitinase (GH18; group I) from the cotton leafworm, *Spodoptera littoralis*, in maize led to a 50% mortality of the corn borer, *Sesamia cretica*, feeding on the plants (Osman et al. 2015). These reports have indicated that GH18 insect chitinases from group I can be insecticidal when expressed in plants. Expression and activity of other insect chitinase groups in transgenic plants have not been reported.

Chitinases of plant origins have also been expressed in transgenic plants to test resistance to insects. A chitinase (GH19; class I) from poplar has been expressed in tomato, *Lycopersicon esculentum*, to examine resistance to the Colorado potato beetle, *Leptinotarsa decemlineata* (Lawrence and Novak 2006). The transgenic tomato plants significantly slowed the growth and development of *L. decemlineata*. Tomato plants expressing a bean chitinase (GH19; class I) together with a snowdrop lectin reduced growth and lowered fecundity of the peach-potato aphid, *Myzus persicae* (Gatehouse et al. 1996), although whether the chitinase co-expressed with the lectin exhibited insecticidal activity is unknown (Gatehouse et al. 1997).

A GH18 chitinase from the baculovirus, *Autographa californica* nucleopolyhedrovirus (AcMNPV), has been expressed in tobacco, *N. tabacum* cv. Samsum NN, and the tobacco budworm, *H. virescens*, feeding on the transgenic plants showed increased mortality (Corrado et al. 2007). The baculovirus chitinase expressed in plants appears active against lepidopteran larvae but not against sucking insects, as the same AcMNPV chitinase-expressing tobacco plants had no effect on the peach-potato aphid *M. persicae* (Corrado et al. 2007).

Stacking of a chitinase gene with other insecticidal transgenes in plants have been reported. Expression of a chitinase gene from *M. sexta* together with the scorpion toxin gene from *Buthus martensii* in rapeseed, *Brassica napus*, has been shown to provide high resistance against the diamondback moth, *Plutella xylostella* (Wang et al. 2005). The gene stacking strategy has also been used to express a chitinase gene together with other transgenes in plants to confer resistance to both insects and plant diseases. Expression of a chitinase from the fungus *Paecilomyces javanicus* with two protease inhibitors in tobacco, *N. benthamiana*, conferred protection against the lepidopteran pests *Spodoptera exigua* and *S. litura* and against the leaf spot and soft rot diseases (Chen et al. 2014). Similarly, stacking of three transgenes coding for a chitinase, a bacterial toxin and the cell membrane receptor Xa21 in rice plants provided protection against the yellow stem borer, *Scirpophaga incertulas*, and against bacterial and fungal infections (Datta et al. 2002).

The insect midgut is an important portal of entry for microbial pathogens. The midgut is lined by the chitinous structure, the PM. Digestion of the chitin in the PM by chitinase may lead to disruption of the protective function of the PM. It has been reported that co-administration of the bacterial pathogen *Bacillus thuringiensis* (*Bt*) and the viral pathogen AcMNPV with a chitinase could increase the toxicity of *Bt* and the virulence of the virus in insects (Kramer and Muthukrishnan 1997). It has been observed that feeding chitinase to *S. littoralis* larvae synergized the toxicity of *Bt* in the larvae by perforating the PM (Regev et al. 1996; Tantimavanich et al. 1997). Therefore, chitinase genes have been introduced and expressed in *Bt* strains to improve the insecticidal activity of *Bt* (Ding et al. 2008; Lertcanawanichakul et al. 2004; Okay et al. 2007; Tantimavanich et al. 1997).

13.4.2 Chitinase Inhibitors and Their Insecticidal Activities

Chitinases in insects play crucial roles for the structure and function of the dynamically regulated chitinous structures. Chitinase inhibitors can interfere with the regulation of chitinous structures and consequently the normal physiology in insects. A series of chitinase inhibitors have been identified and their characteristics been studied (Andersen et al. 2005).

Chitinase inhibitors have been identified mainly from microorganisms and sponges. These chitinase inhibitors include allosamidin from *Streptomyces* sp. (Sakuda et al. 1987), styloguanidines from the sponge *Stylotella aurantium* (Kato et al. 1995), Cl-4 (cyclo-L-Arg-D-Pro) from *Pseudomonas sp.* (Izumida et al. 1996), psammaplin A from the sponge *Aplysinella rhax* (Tabudravu et al. 2002), and the cyclopentapeptides, argadin, and argifin, from the fungi *Clonostachys sp.* and *Gliocladium sp.* (Arai et al. 2000a, b; Omura et al. 2000). A chitinase inhibitor, phlegmacin B1, was more recently isolated from *Talaromyces sp.* (Chen et al. 2017b). Interestingly, chitinase-inhibiting activities have also been identified from compounds of plant origins. The plant-originated methylxanthine derivatives, theophylline, caffeine, and pentoxifylline, have been identified to exhibit chitinase inhibition (Rao et al. 2005) and antifungal activities (Tsirilakis et al. 2012). Berberine and derivatives from plants have also been found to inhibit insect chitinases (Duan et al. 2018).

Among the known chitinase inhibitors and their derivatives from biological resources, some have been tested and confirmed for their insecticidal activities (Saguez et al. 2008). Allosamidin and derivatives have a strong chitinase-inhibiting activity for insect chitinases, and show insecticidal activity in lepidopteran larvae by inhibiting ecdysis (Blattner et al. 1994; Sakuda et al. 1986, 1987). The argifin and argadin are confirmed to inhibit chitinase activity from the blowfly,

Lucilia cuprina, and have insecticidal activity in the American cockroach, *Periplaneta americana*, by inhibiting molting and subsequent mortality (Arai et al. 2000b; Shiomi et al. 2000). The recently identified microbial phlegmacin B1 was selected for inhibition of chitinase activity in the Asian corn borer, *Ostrinia furnacalis*, both by ingestion and by hemocoel injection (Chen et al. 2017b). The berberine and derivatives have also been shown to inhibit *O. furnacalis* larval growth and development, leading to mortality (Duan et al. 2018). In addition, psammaplin A and chitinase-inhibiting polysaccharides and mimetic peptides have been shown to inhibit larval growth and increase mortality in some insects (Saguez et al. 2008). Although chitinase inhibitors have not been utilized for insect pest control yet, the efficacy of some chitinase inhibitors to suppress insect growth and development, which eventually leads to insect mortality, has been well characterized. Continuing identification and further understanding of insecticidal chitinase inhibitors will provide the necessary foundation for development of chitinase inhibitor-based new insecticides for pest control.

13.4.3 Chitin Synthesis Inhibitors and Their Insecticidal Activities

Chitin synthesis inhibitors (CSIs) include eight classes of chemical compounds (Merzendorfer 2013). Among the CSIs, benzoylphenylureas (BPUs) have been the most commonly used CSIs as pesticides. Diflubenzuron was developed as the insecticide Dimilin in the 1970s (Marx 1977), and more derivatives of BPUs have been developed afterward, leading to commercialization of 15 BPUs worldwide for an estimated annual global market of half a million US dollars (Sun et al. 2015).

Intoxication of insects by CSIs is generally characterized by a decrease of chitin content and malformation of the PM and cuticles, leading to suppression of molting and egg hatching (Merzendorfer 2013). Different CSIs do not target the same sites in the chitin synthesis pathway. The CSIs pyrimidine-nucleoside peptides (PNPs) and thiophthalimides disrupt chitin formation by targeting the catalytic site of CHS, while BPUs, oxazolines, and thiazolidinones block the post-catalytic steps in the chitin synthesis (Cohen 2010; Merzendorfer et al. 2012; Nauen and Smagghe 2006). The target sites and mechanisms of inhibition of chitin synthesis by CSIs remain to be better understood (Merzendorfer 2013). As opposed to PNPs, the BPUs and etoxazole (oxazoline) are thought to interfere with the catalytic reaction of chitin synthesis indirectly, as these CSIs do not have the inhibiting activity in cell-free systems (Cohen 1982; Merzendorfer 2013). However, resistance to these insecticides in insects has been found to be linked to a mutation in the C-terminal transmembrane domain of CHS, which suggests presence of direct interaction of these CSIs with CHS to affect the activity of CHS (Douris et al. 2016; Van Leeuwen et al. 2012). Therefore, CSIs may differentially target the chitin synthesis in insects, but details of the mechanisms have yet to be elucidated.

CSIs have been used as insecticides, due to their efficacy in a broad range of insect pests and to their relatively low impact on higher animals and predatory insects (Sun et al. 2015). As CSIs disrupt the chitinous structures and their functions, CSIs can potentially synergize infections of bacterial and viral pathogens in insects (Arakawa et al. 2002; Kuraishi et al. 2011; Wang and Granados 2000). Practically, applications of CSIs with microbial insecticides may improve the efficacy of microbial pathogens for insect control.

13.4.4 Disruption of Chitinous Structures by Targeting CBPs

Chitinous structures in insects rely on the CBPs for their structure and functions. Therefore, disrupting the chitinous structures by targeting the CBPs or the binding of CBPs with chitin can be an approach for developing alternative insect control technologies (Wang and Granados 2001). Particularly, the chitinous structure in the PM depends on the non-covalent binding of CBPs with the chitin fibrils (Tellam et al. 1999; Wang and Granados 2000; Wang et al. 2004). Therefore, CBPs in the PM and the binding of PM CBPs to chitin can be prime targets to disrupt the important PM structure in insects (Wang and Granados 2001). Targeting the CBPs and the binding of CBPs in insect midgut to inhibit growth and disrupt the midgut defensive system has been experimentally confirmed to be effective (Wang and Granados 1997a, 2000).

The CBPs, including PMPs and CPAPs, have similar molecular structures with tandem CBD repeats across different species and different orders in insects (Hegedus et al. 2009; Jasrapuria et al. 2010; Wang et al. 2004). Targeting specific CBPs in an insect with antibodies in the sheep blowfly, *L. cuprina*, has been confirmed to be effective to suppress the growth of *L. cuprina* larvae and practically feasible for the control of this insect pest (Casu et al. 1997; East et al. 1993). Although effective inhibition of insect growth by antibody-based targeting of insect midgut CBPs has only been demonstrated in hematophagous insects, this approach is expected to have the same application potential in phytophagous insects. Target-specific antibodies can be sufficiently produced in plants and have been successfully used to generate plants resistant to pathogens (Cillo and Palukaitis 2014). Therefore, even though CBPs share similar domain and overall structures in insects, potentially CBPs from different insects can be specifically targeted to suppress specific target insects (Casu et al. 1997; East et al. 1993; Wang and Granados 1997a, Wang and Granados 1998). Alternatively, targeting CBP by RNAi may specifically interfere with the PM and cuticular organizations, so RNAi of CBP genes could be a potential strategy to further explore for development tactics for insect control.

One type of the PM CBPs is the insect intestinal mucins (IIMs). IIMs are PMPs containing highly O-glycosylated mucin domains (Wang and Granados 1997b).

IIMs are essential structural components for the PM structure and function (Wang and Granados 1997a, b, 2001). IIM-specific metalloproteases, the baculovirus enhancins, have been identified from baculoviruses (Hashimoto et al. 1991; Lepore et al. 1996; Roelvink et al. 1995; Wang and Granados, 1997a). The enhancin from the *T. ni* granulovirus (TnGV) can effectively degrade the IIM in *T. ni* larval midgut to cause disruption of the PM structure (Wang and Granados 1997a; Wang et al. 1994), and consequently increase the larval susceptibility to baculovirus infection in the insect (Derksen and Granados 1988; Wang and Granados 1997a, 1998).

The PM structure is basically formed by noncovalent binding of CBPs to chitin (Shen and Jacobs-Lorena 1998; Wang and Granados 2000, 2001; Wang et al. 2004). Therefore, blocking the CBP–chitin binding or disruption of the chitin-binding activity of CBPs to inhibit the PM structural formation has been proposed as a potential approach for insect control (Wang and Granados 2001). In the PM of *T. ni*, CBPs bind to the chitin fibrils with high affinity. The binding activity of the CBPs can be abolished by reduction of the intramolecular disulfate bonds in the CBDs of the CBPs (Wang and Granados 2000). It is expected that competitive binding of other chitin-binding molecules to the PM chitin in the midgut can disrupt the PM structure. As a matter of fact, it has been experimentally confirmed in *T. ni* that chitin-binding molecules could solubilize CBPs from PM structure and inhibit PM formation in insects by blocking the binding of CBPs to chitin (Wang and Granados 2000).

Insecticidal activity of the lectin WGA has been long known. The insecticidal activity of WGA in insects has been revealed to be targeting the PM by binding to the PM chitin (Harper et al. 1998; Hopkins and Harper 2001). In early 1990s, a group of fluorescent brighteners were found to significantly increase the susceptibility of lepidopteran larvae to baculovirus infections (Shapiro and Robertson 1992). Mechanistic studies have uncovered that fluorescent brighteners, such as Calcofluor, disrupt the binding of CBPs to chitin fibrils by competitively binding to chitin and consequently inhibiting the PM formation (Wang and Granados 2000). Therefore, both chitin-binding proteins and chitin-binding chemicals can be potentially used to disrupt the chitinous PM structure in insects and be developed as products for insect pest control.

13.4.5 Disruption of Chitinous Structures by RNAi

RNAi has provided a new revenue for development of insect control technologies by specifically targeting the chitinous structures in insects. Inhibition of CHS expression by RNAi in insects has been reported to strongly affect the chitinous structures, resulting in slow development and increased mortality of the insects (Chen et al. 2008; Hansen et al. 2009; Kato et al. 2006; Kelkenberg et al. 2015; Lee et al. 2017; Macedo et al. 2017; Shi et al. 2016). RNAi targeting the CHS-1 in *S. exigua* not only led to increased mortality but also to increased susceptibility to the fungal pathogen *Beauveria bassiana* (Lee et al. 2017). Similarly, the feeding of

Phthorimaea operculella larvae with dsRNAs targeting the CHS-1 gene led to varying levels of larval abnormality and increased mortality (Mohammed et al. 2017).

Inhibition of the expression of CDA genes, particularly those from group I and IV, by RNAi has been reported to cause altered cuticle organizations at different developmental stages, abnormal growth, deformed body shape, and disrupted molting in insects (Arakane et al. 2009; Dong et al. 2014; Luschnig et al. 2006; Quan et al. 2013; Wu et al. 2018; Xi et al. 2014; Yang et al. 2018). RNAi has been used to successfully downregulate the expression of various CHTs in insects. Phenotypes observed in the RNAi-treated insects include increased insect mortality (RNAi of CHTs in groups I, II, III), impaired molting (RNAi of CHTs in groups I, II, III), reduced hatching (RNAi of CHTs in group II), abnormal wing development (RNAi of CHTs in group III), and disrupted organization of chitin fibers (RNAi of CHT in group III) (Chen et al. 2017a; Ganbaatar et al. 2017; Li et al. 2015; Noh et al. 2018; Xi et al. 2015; Zhang et al. 2012; Zhu et al. 2008). However, inhibition of some other CHTs, especially those in the group IV and V, did not lead to developmental defects (Khajuria et al. 2010; Pesch et al. 2016; Zhu et al. 2008, 2016). It has also been reported that inhibition of NAGs by RNAi may lead to severe molting defects in insects (Hogenkamp et al. 2008; Zhu et al. 2016).

Physiological effects from inhibition of chitin-binding protein expression in insects by RNAi have been studied. It has been reported that inhibition of the expression of PMPs by RNAi in insects caused slower growth and impaired molting, and inhibition of the expression of most CPAPs led to disrupted molting, extended insect growth, impaired development, and wing formation, and ultimately increased mortality (Jasrapuria et al. 2012; Petkau et al. 2012; Tiklova et al. 2013). Inhibition of CPRs by RNAi also led to the destabilization of the cuticles in adults resulting from abnormal chitin content, cuticle organization and mechanical properties of the cuticles (Noh et al. 2014, 2015; Qiao et al. 2014). Similarly, RNAi of a Tweedle protein in insects resulted in higher mortality during molting due to disorganization of chitin fibers (Song et al. 2016).

The effects from RNAi of chitin-associated protein genes reported in the literature have indicated potential applications of RNAi to target chitin-associated proteins for insect control. Transgenic plants engineered to target chitin-associated enzyme genes by RNAi have been reported and their insect resistance been tested. *H. armigera* larvae feeding on transgenic tobacco plants expressing miRNA specific to a *H. armigera* CHT gene showed suppression of larva-to-larva molting and final larval mortality (Agrawal et al. 2015). Transgenic tobacco and tomato plants expressing dsRNA specific to an *H. armigera* CHT (group I) both induced malformation of *H. armigera* larvae on the plants, leading to suppressed growth and increased mortality (Mamta et al. 2015). *M. separata* CHT-specific siRNAs have been expressed in tobacco plants and *M. separata* larvae showed inhibited growth on the plants (Bao et al. 2016). Direct feeding of *M. separata* larvae with dsRNA targeting the CHT confirmed inhibition of larval growth by RNAi in the insect (Ganbaatar et al. 2017).

Production of dsRNAs of CHSs in plants to confer RNAi of CHSs in insects has also be studied. Expression dsRNA to the *H. armigera* CHS-2 gene in chloroplasts of tobacco plants significantly inhibited the growth of *H. armigera* on the plants (Jin et al. 2015).

13.5 Conclusion

Chitinous structures are unique and physiologically essential in insects, and are therefore potential target sites for insect pest control. Insect chitinous structures are all composed of chitin and proteins, but the specific structures and compositions are diverse and their synthesis and regulation are complex. The unique structures and structural components, and the biosynthesis and regulation pathways of the chitinous structures can be exploited to identify potential molecular targets for the development of technologies to disrupt these essential structures in insects. Strategies to disrupt chitinous structures in insects have been studied and compounds with properties to disrupt the chitinous structures have been developed as insecticides. However, molecular target sites in the chitinous structures for the purpose of insect pest control are largely not explored. Ample opportunities to disrupt chitinous structures in insects for insect pest control are to be explored.

Acknowledgements This work was supported in part the USDA AFRI Foundational Program competitive grant no. 2016-67013-24754 and USDA NIFA Hatch Project.

References

Adrangi S, Faramarzi MA (2013) From bacteria to human: a journey into the world of chitinases. Biotechnol Adv 31(8):1786–1795

Agrawal A, Rajamani V, Reddy VS, Mukherjee SK, Bhatnagar RK (2015) Transgenic plants over-expressing insect-specific microRNA acquire insecticidal activity against *Helicoverpa armigera*: an alternative to Bt-toxin technology. Transgenic Res 24(5):791–801

Agrawal S, Kelkenberg M, Begum K, Steinfeld L, Williams CE, Kramer KJ et al (2014) Two essential peritrophic matrix proteins mediate matrix barrier functions in the insect midgut. Insect Biochem Mol Biol 49:24–34

Andersen OA, Dixon MJ, Eggleston IM, van Aalten DM (2005) Natural product family 18 chitinase inhibitors. Nat Prod Rep 22(5):563–579

Arai N, Shiomi K, Iwai Y, Omura S (2000a). Argifin, a new chitinase inhibitor, produced by Gliocladium sp. FTD-0668. II. Isolation, physico-chemical properties, and structure elucidation. *J Antibiot (Tokyo)* 53(6): 609–614

Arai N, Shiomi K, Yamaguchi Y, Masuma R, Iwai Y, Turberg A et al (2000b) Argadin, a new chitinase inhibitor, produced by *Clonostachys* sp. FO-7314. Chem Pharm Bull (Tokyo) 48(10):1442–1446

Arakane Y, Dixit R, Begum K, Park Y, Specht CA, Merzendorfer H et al (2009) Analysis of functions of the chitin deacetylase gene family in *Tribolium castaneum*. Insect Biochem Mol Biol 39(5–6):355–365

Arakawa T, Furuta Y, Miyazawa M, Kato M (2002) Flufenoxuron, an insect growth regulator, promotes peroral infection by nucleopolyhedrovirus (BmNPV) budded particles in the silkworm, *Bombyx mori* L. J Virol Methods 100:141–147

Bao W, Cao B, Zhang Y, Wuriyanghan H (2016) Silencing of *Mythimna separata* chitinase genes via oral delivery of in planta-expressed RNAi effectors from a recombinant plant virus. Biotech Lett 38(11):1961–1966

Blattner R, Furneaux RH, Kemmitt T, Tyler PC, Ferrier RJ, Tidén A-K (1994) Syntheses of the fungicide/insecticide allosamidin and a structural isomer. J Chem Soc Perkin Trans 1(23): 3411–3421

Casida JE, Durkin KA (2017) Pesticide chemical research in toxicology: lessons from nature. Chem Res Toxicol 30(1):94–104

Casu R, Eisemann C, Pearson R, Riding G, East I, Donaldson A et al (1997) Antibody-mediated inhibition of the growth of larvae from an insect causing cutaneous myiasis in a mammalian host. Proc Natl Acad Sci USA 94(17):8939–8944

Chen C, Yang H, Tang B, Yang W-J, Jin D-C (2017a) Identification and functional analysis of chitinase 7 gene in white-backed planthopper, *Sogatella furcifera*. Comp Biochem Physiol B Biochem Mol Biol 208–209:19–28

Chen L, Liu T, Duan Y, Lu X, Yang Q (2017b) Microbial secondary metabolite, Phlegmacin B1, as a novel inhibitor of insect chitinolytic enzymes. J Agric Food Chem 65(19):3851–3857

Chen P-J, Senthilkumar R, Jane W-N, He Y, Tian Z, Yeh K-W (2014) Transplastomic *Nicotiana benthamiana* plants expressing multiple defence genes encoding protease inhibitors and chitinase display broad-spectrum resistance against insects, pathogens and abiotic stresses. Plant Biotechnol J 12(4):503–515

Chen X, Tian H, Zou L, Tang B, Hu J, Zhang W (2008) Disruption of *Spodoptera exigua* larval development by silencing chitin synthase gene A with RNA interference. Bull Entomol Res 98(6):613–619

Chikate YR, Dawkar VV, Barbole RS, Tilak PV, Gupta VS, Giri AP (2016) RNAi of selected candidate genes interrupts growth and development of *Helicoverpa armigera*. Pestic Biochem Physiol 133:44–51

Cillo F, Palukaitis P (2014) Transgenic resistance. Adv Virus Res 90:35–146

Cohen E (1982) Chitin synthetase activity and inhibition in different insect microsomal preparations. EXS 41:470–472

Cohen E (2010) Chitin biochemistry: synthesis, hydrolysis and inhibition. In: Casas J, Simpson SJ (eds) Advances in insect physiology: insect integument and colour, vol 38, pp 5–74

Cornman RS, Willis JH (2008) Extensive gene amplification and concerted evolution within the CPR family of cuticular proteins in mosquitoes. Insect Biochem Mol Biol 38(6):661–676

Corrado G, Arciello S, Fanti P, Fiandra L, Garonna A, Digilio MC et al (2007) The Chitinase A from the baculovirus AcMNPV enhances resistance to both fungi and herbivorous pests in tobacco. Transgenic Res 17(4):557–571

da Silva MV, Santi L, Staats CC, da Costa AM, Colodel EM, Driemeier D et al (2005) Cuticle-induced endo/exoacting chitinase CHIT30 from *Metarhizium anisopliae* is encoded by an ortholog of the chi3 gene. Res Microbiol 156(3):382–392

Datta K, Baisakh N, Maung Thet K, Tu J, Datta S (2002) Pyramiding transgenes for multiple resistance in rice against bacterial blight, yellow stem borer and sheath blight. Theor Appl Genet 106(1):1–8

de la Vega H, Specht CA, Liu Y, Robbins PW (1998) Chitinases are a multi-gene family in *Aedes*, *Anopheles* and *Drosophila*. Insect Mol Biol 7(3):233–239

Derksen AC, Granados RR (1988) Alteration of a lepidopteran peritrophic membrane by baculoviruses and enhancement of viral infectivity. Virology 167(1):242–250

Despres L, Stalinski R, Faucon F, Navratil V, Viari A, Paris M et al (2014a) Chemical and biological insecticides select distinct gene expression patterns in *Aedes aegypti* mosquito. Biol Lett 10(12):20140716

Despres L, Stalinski R, Tetreau G, Paris M, Bonin A, Navratil V et al (2014b). Gene expression patterns and sequence polymorphisms associated with mosquito resistance to *Bacillus thuringiensis israelensis* toxins. BMC Genom 15:926

Ding X, Gopalakrishnan B, Johnson LB, White FF, Wang X, Morgan TD et al (1998) Insect resistance of transgenic tobacco expressing an insect chitinase gene. Transgenic Res 7:77–84

Ding X, Luo Z, Xia L, Gao B, Sun Y, Zhang Y (2008) Improving the insecticidal activity by expression of a recombinant cry1Ac gene with chitinase-encoding gene in acrystalliferous *Bacillus thuringiensis*. Curr Microbiol 56(5):442–446

Dittmer NT, Tetreau G, Cao X, Jiang H, Wang P, Kanost MR (2015) Annotation and expression analysis of cuticular proteins from the tobacco hornworm, *Manduca sexta*. Insect Biochem Mol Biol 62:100–113

Dixit R, Arakane Y, Specht CA, Richard C, Kramer KJ, Beeman RW et al (2008) Domain organization and phylogenetic analysis of proteins from the chitin deacetylase gene family of *Tribolium castaneum* and three other species of insects. Insect Biochem Mol Biol 38(4): 440–451

Dong B, Miao G, Hayashi S (2014) A fat body-derived apical extracellular matrix enzyme is transported to the tracheal lumen and is required for tube morphogenesis in *Drosophila*. Development 141(21):4104–4109

Douris V, Steinbach D, Panteleri R, Livadaras I, Pickett JA, van Leeuwen T et al (2016) Resistance mutation conserved between insects and mites unravels the benzoylurea insecticide mode of action on chitin biosynthesis. Proc Natl Acad Sci 113(51):14692–14697

Duan Y, Liu T, Zhou Y, Dou T, Yang Q (2018) Glycoside hydrolase family 18 and 20 enzymes are novel targets of the traditional medicine berberine. J Biol Chem 293(40):15429–15438

East IJ, Fitzgerald CJ, Pearson RD, Donaldson RA, Vuocolo T, Cadogan LC et al (1993) *Lucilia cuprina*: inhibition of larval growth induced by immunization of host sheep with extracts of larval peritrophic membrane. Int J Parasitol 23(2):221–229

Ganbaatar O, Cao B, Zhang Y, Bao D, Bao W, Wuriyanghan H (2017) Knockdown of *Mythimna separata* chitinase genes via bacterial expression and oral delivery of RNAi effectors. BMC Biotechnol 17(1):9

Gatehouse AMR, Davison GM, Newell CA, Merryweather A, Hamilton WDO, Burgess EPJ et al (1997) Transgenic potato plants with enhanced resistance to the tomato moth, *Lacanobia oleracea*: growth room trials. Mol Breeding 3(1):49–63

Gatehouse AMR, Down RE, Powell KS, Sauvion N, Rahbé Y, Newell CA et al (1996) Transgenic potato plants with enhanced resistance to the peach-potato aphid *Myzus persicae*. Entomologia Experimentalis Et Applicata 79:295–307

Grover A (2012) Plant chitinases: genetic diversity and physiological roles. Crit Rev Plant Sci 31(1):57–73

Guan X, Middlebrooks BW, Alexander S, Wasserman SA (2006) Mutation of TweedleD, a member of an unconventional cuticle protein family, alters body shape in *Drosophila*. Proc Natl Acad Sci 103(45):16794–16799

Guo W, Li GX, Pang Y, Wang P (2005) A novel chitin-binding protein identified from the peritrophic membrane of the cabbage looper, *Trichoplusia ni*. Insect Biochem Mol Biol 35(11): 1224–1234

Hansen IA, Tian H, Peng H, Yao Q, Chen H, Xie Q et al (2009) Developmental control of a lepidopteran pest *Spodoptera exigua* by ingestion of bacteria expressing dsRNA of a non-midgut gene. PLoS ONE 4(7):e6225

Harper MS, Hopkins TL, Czapla TH (1998) Effect of wheat germ agglutinin on formation and structure of the peritrophic membrane in European corn borer (*Ostrinia nubilalis*) larvae. Tissue Cell 30(2):166–176

Hartl L, Zach S, Seidl-Seiboth V (2012) Fungal chitinases: diversity, mechanistic properties and biotechnological potential. Appl Microbiol Biotechnol 93(2):533–543

Hashimoto Y, Corsaro BG, Granados RR (1991) Location and nucleotide sequence of the gene encoding the viral enhancing factor of the *Trichoplusia ni* granulosis virus. J Gen Virol 72(Pt 11):2645–2651

Hegedus D, Erlandson M, Gillott C, Toprak U (2009) New insights into peritrophic matrix synthesis, architecture, and function. Annu Rev Entomol 54:285–302

Hogenkamp DG, Arakane Y, Kramer KJ, Muthukrishnan S, Beeman RW (2008) Characterization and expression of the beta-N-acetylhexosaminidase gene family of *Tribolium castaneum*. Insect Biochem Mol Biol 38(4):478–489

Hopkins TL, Harper MS (2001) Lepidopteran peritrophic membranes and effects of dietary wheat germ agglutinin on their formation and structure. Arch Insect Biochem Physiol 47(2):100–109

Ioannidou ZS, Theodoropoulou MC, Papandreou NC, Willis JH, Hamodrakas SJ (2014) CutProtFam-Pred: Detection and classification of putative structural cuticular proteins from sequence alone, based on profile Hidden Markov Models. Insect Biochem Mol Biol 52:51–59

Izumida H, Imamura N, Sano H (1996) A novel chitinase inhibitor from a marine bacterium, *Pseudomonas* sp. J Antibiot (Tokyo) 49(1):76–80

Jalil SU, Mishra M, Ansari MI (2015) Current view on chitinase for plant defence. Trends in Biosciences 8(24):6733–6743

Jasrapuria S, Arakane Y, Osman G, Kramer KJ, Beeman RW, Muthukrishnan S (2010) Genes encoding proteins with peritrophin A-type chitin-binding domains in *Tribolium castaneum* are grouped into three distinct families based on phylogeny, expression and function. Insect Biochem Mol Biol 40(3):214–227

Jasrapuria S, Specht CA, Kramer KJ, Beeman RW, Muthukrishnan S (2012) Gene families of cuticular proteins analogous to peritrophins (CPAPs) in *Tribolium castaneum* have diverse functions. PLoS One 7(11):e49844

Jin S, Singh ND, Li L, Zhang X, Daniell H (2015) Engineered chloroplast dsRNA silences cytochrome p450 monooxygenase, V-ATPase and chitin synthase genes in the insect gut and disrupts *Helicoverpa armigera* larval development and pupation. Plant Biotechnol J 13(3): 435–446

Kato N, Mueller CR, Fuchs JF, Wessely V, Lan Q, Christensen BM (2006) Regulatory mechanisms of chitin biosynthesis and roles of chitin in peritrophic matrix formation in the midgut of adult *Aedes aegypti*. Insect Biochem Mol Biol 36(1):1–9

Kato T, Shizuri Y, Izumida H, Yokoyama A, Endo M (1995) Styloguanidines, new chitinase inhibitors from the marine sponge *Stylotella aurantium*. Tetrahedron Lett 36:2133–2136

Kelkenberg M, Odman-Naresh J, Muthukrishnan S, Merzendorfer H (2015) Chitin is a necessary component to maintain the barrier function of the peritrophic matrix in the insect midgut. Insect Biochem Mol Biol 56:21–28

Khajuria C, Buschman LL, Chen MS, Muthukrishnan S, Zhu KY (2010) A gut-specific chitinase gene essential for regulation of chitin content of peritrophic matrix and growth of *Ostrinia nubilalis* larvae. Insect Biochem Mol Biol 40(8):621–629

Koganemaru R, Miller DM, Adelman ZN (2013) Robust cuticular penetration resistance in the common bed bug (*Cimex lectularius* L.) correlates with increased steady-state transcript levels of CPR-type cuticle protein genes. Pestic Biochem Physiol 106(3):190–197

Kramer KJ, Muthukrishnan S (1997) Insect chitinases: Molecular biology and potential use as biopesticides. Insect Biochem Mol Biol 27(11):887–900

Kuraishi T, Binggeli O, Opota O, Buchon N, Lemaitre B (2011) Genetic evidence for a protective role of the peritrophic matrix against intestinal bacterial infection in *Drosophila melanogaster*. Proc Natl Acad Sci 108(38):15966–15971

Kuraishi T, Hori A, Kurata S (2013) Host-microbe interactions in the gut of *Drosophila melanogaster*. Front Physiol 4:375

Lacey LA, Grzywacz D, Shapiro-Ilan DI, Frutos R, Brownbridge M, Goettel MS (2015) Insect pathogens as biological control agents: back to the future. J Invertebr Pathol 132:1–41

Lawrence SD, Novak NG (2006) Expression of poplar chitinase in tomato leads to inhibition of development in colorado potato beetle. Biotech Lett 28(8):593–599

Lee J-B, Kim HS, Park Y (2017) Down-regulation of a chitin synthase a gene by RNA interference enhances pathogenicity of *Beauveria bassiana* ANU1 against *Spodoptera exigua* (HÜBNER). Arch Insect Biochem Physiol 94(2):e21371

Lepore LS, Roelvink PR, Granados RR (1996) Enhancin, the granulosis virus protein that facilitates nucleopolyhedrovirus (NPV) infections, is a metalloprotease. J Invertebr Pathol 68(2):131–140

Lertcanawanichakul M, Wiwat C, Bhumiratana A, Dean DH (2004) Expression of chitinase-encoding genes in *Bacillus thuringiensis* and toxicity of engineered *B. thuringiensis* subsp. *aizawai* toward *Lymantria dispar* larvae. Curr Microbiol 48(3):175–181

Li D, Zhang J, Wang Y, Liu X, Ma E, Sun Y et al (2015) Two chitinase 5 genes from *Locusta migratoria*: molecular characteristics and functional differentiation. Insect Biochem Mol Biol 58:46–54

Luschnig S, Batz T, Armbruster K, Krasnow MA (2006) *Serpentine* and *vermiform* encode matrix proteins with chitin binding and deacetylation domains that limit tracheal tube length in *Drosophila*. Curr Biol 16(2):186–194

Macedo LLP, de Souza Antonino, Junior JD, Coelho RR, Fonseca FCA, Firmino AAP, Silva MCM et al (2017) Knocking down chitin synthase 2 by RNAi is lethal to the cotton boll weevil. Biotechnol Res Innovat 1(1):72–86

Mamta Reddy KRK, Rajam MV (2015) Targeting chitinase gene of *Helicoverpa armigera* by host-induced RNA interference confers insect resistance in tobacco and tomato. Plant Mol Biol 90(3):281–292

Marx JL (1977) Chitin synthesis inhibitors: new class of insecticides. Science 197(4309): 1170–1172

Merzendorfer H (2006) Insect chitin synthases: a review. J Comparat Physiol B-Biochem Syst Environ Physiol 176(1):1–15

Merzendorfer H (2013) Chitin synthesis inhibitors: old molecules and new developments. Insect Sci 20(2):121–138

Merzendorfer H, Kim HS, Chaudhari SS, Kumari M, Specht CA, Butcher S et al (2012) Genomic and proteomic studies on the effects of the insect growth regulator diflubenzuron in the model beetle species *Tribolium castaneum*. Insect Biochem Mol Biol 42(4):264–276

Merzendorfer H, Zimoch L (2003) Chitin metabolism in insects: structure, function and regulation of chitin synthases and chitinases. J Exp Biol 206(24):4393–4412

Mohammed AMA, Diab MR, Abdelsattar M, Khalil SMS (2017) Characterization and RNAi-mediated knockdown of chitin synthase A in the potato tuber moth, *Phthorimaea operculella*. Sci Rep 7(1):9502

Muthukrishnan S, Merzendorfer H, Arakane Y, Kramer KJ (2012) 7 - Chitin Metabolism in Insects. In: Gilbert LI (ed) Insect molecular biology and biochemistry. Academic Press, San Diego, pp 193–235

Nauen R, Smagghe G (2006) Mode of action of etoxazole. Pest Manag Sci 62(5):379–382

Noh MY, Kramer KJ, Muthukrishnan S, Kanost MR, Beeman RW, Arakane Y (2014) Two major cuticular proteins are required for assembly of horizontal laminae and vertical pore canals in rigid cuticle of *Tribolium castaneum*. Insect Biochem Mol Biol 53:22–29

Noh MY, Muthukrishnan S, Kramer KJ, Arakane Y (2015) *Tribolium castaneum* RR-1 cuticular protein TcCPR4 Is required for formation of pore canals in rigid cuticle. PLoS Genet 11(2): e1004963

Noh MY, Muthukrishnan S, Kramer KJ, Arakane Y (2018) A chitinase with two catalytic domains is required for organization of the cuticular extracellular matrix of a beetle. PLoS Genet 14(3): e1007307

Okay S, Tefon BE, Özkan M, Özcengiz G (2007). Expression of chitinase A (chiA) gene from a local isolate of *Serratia marcescens* in Coleoptera-specific *Bacillus thuringiensis*. J Appl Microbiol 104(1):161–170

Omura S, Arai N, Yamaguchi Y, Masuma R, Iwai Y, Namikoshi M et al (2000) Argifin, a new chitinase inhibitor, produced by Gliocladium sp. FTD-0668. I. Taxonomy, fermentation, and biological activities. J Antibiot (Tokyo) 53(6):603–608

Osman GH, Assem SK, Alreedy RM, El-Ghareeb DK, Basry MA, Rastogi A et al (2015). Development of insect resistant maize plants expressing a chitinase gene from the cotton leaf worm, *Spodoptera littoralis*. Sci Rep 5(1):18067

Oyeleye A, Normi YM (2018). Chitinase: Diversity, limitations and trends in engineering for suitable applications. Biosci Rep BSR20180323

Pesch Y-Y, Riedel D, Patil KR, Loch G, Behr M (2016) Chitinases and imaginal disc growth factors organize the extracellular matrix formation at barrier tissues in insects. Sci Rep 6(1)

Petkau G, Wingen C, Jussen LCA, Radtke T, Behr M (2012) Obstructor-A is required for epithelial extracellular matrix dynamics, exoskeleton function, and tubulogenesis. J Biol Chem 287(25):21396–21405

Popp J, Pető K, Nagy J (2012) Pesticide productivity and food security: a review. Agronom Sustain Develop 33(1):243–255

Qiao L, Xiong G, R-x Wang, S-z He, Chen J, X-l Tong et al (2014) Mutation of a cuticular protein, BmorCPR2, alters larval body shape and adaptability in silkworm, *Bombyx mori*. Genet 196 (4):1103–1115

Quan G, Ladd T, Duan J, Wen F, Doucet D, Cusson M et al (2013) Characterization of a spruce budworm chitin deacetylase gene: Stage- and tissue-specific expression, and inhibition using RNA interference. Insect Biochem Molecul Biol

Rao FV, Andersen OA, Vora KA, Demartino JA, van Aalten DM (2005) Methylxanthine drugs are chitinase inhibitors: investigation of inhibition and binding modes. Chem Biol 12(9):973–980

Regev A, Keller M, Strizhov N, Sneh B, Prudovsky E, Chet I et al (1996) Synergistic Activity of a *Bacillus thuringiensis* delta-endotoxin and a bacterial endochitinase against *Spodoptera littoralis* larvae. Appl Environ Microbiol 62(10):3581–3586

Roelvink PW, Corsaro BG, Granados RR (1995) Characterization of the *Helicoverpa armigera* and *Pseudaletia unipuncta* granulovirus enhancin genes. J Gen Virol 76(11):2693–2705

Rohrbough J, Rushton E, Woodruff E, Fergestad T, Vigneswaran K, Broadie K (2007) Presynaptic establishment of the synaptic cleft extracellular matrix is required for post-synaptic differentiation. Genes Dev 21(20):2607–2628

Rushton E, Rohrbough J, Deutsch K, Broadie K (2012) Structure-function analysis of endogenous lectin mind-the-gap in synaptogenesis. Develop Neurobiol 72(8):1161–1179

Saguez J, Vincent C, Giordanengo P (2008) Chitinase inhibitors and chitin mimetics for crop protection. Pest Technol 2(2):81–86

Sakuda S, Isogai A, Matsumoto S, Suzuki A (1987) Search for microbial insect growth regulators. II. Allosamidin, a novel insect chitinase inhibitor. J Antibiot (Tokyo) 40(3):296–300

Sakuda S, Isogai A, Matsumoto S, Suzuki A, Koseki K (1986) The structure of allosamidin, a novel insect chitinase inhibitor, produced by *Streptomyces Sp*. Tetrahedron Lett 27(22):2475–2478

Shapiro M, Robertson JL (1992) Enhancement of gypsy moth Lepidoptera Lymantriidae baculovirus activity by optical brighteners. J Econ Entomol 85:1120–1124

Shen Z, Jacobs-Lorena M (1998) A type I peritrophic matrix protein from the malaria vector *Anopheles gambiae* binds to chitin. Cloning, expression, and characterization. J Biol Chem 273(28):17665–17670

Shi J-F, Mu L-L, Chen X, Guo W-C, Li G-Q (2016) RNA interference of chitin synthase genes inhibits chitin biosynthesis and affects larval performance in *Leptinotarsa decemlineata* (Say). Int J Biol Sci 12(11):1319–1331

Shiomi K, Arai N, Iwai Y, Turberg A, Kolbl H, Omura S (2000) The structure of argifin, a new chitinase inhibitor, produced by *Gliocladium* sp. Tetrahedron Lett 41:2141–2143

Song T-Q, Yang M-L, Wang Y-L, Liu Q, Wang H-M, Zhang J et al (2016) Cuticular protein LmTwdl1 is involved in molt development of the migratory locust. Insect Sci 23(4):520–530

St. Leger RJ, Cooper RM, Charnley AK (1986) Cuticle-degrading enzymes of entomopathogenic fungi: Cuticle degradation in vitro by enzymes from entomopathogens. J Inverteb Pathol 47(2):167–177

Sun R, Liu C, Zhang H, Wang Q (2015) Benzoylurea chitin synthesis inhibitors. J Agric Food Chem 63(31):6847–6865

Tabudravu JN, Eijsink VG, Gooday GW, Jaspars M, Komander D, Legg M et al (2002) Psammaplin A, a chitinase inhibitor isolated from the Fijian marine sponge *Aplysinella rhax*. Bioorg Med Chem 10(4):1123–1128

Tang L, Liang J, Zhan Z, Xiang Z, He N (2010) Identification of the chitin-binding proteins from the larval proteins of silkworm, *Bombyx mori*. Insect Biochem Mol Biol 40(3):228–234

Tantimavanich S, Pantuwatana S, Bhumiratana A, Panbangred W (1997) Cloning of a chitinase gene into *Bacillus thuringiensis* subsp. *aizawai* for enhanced insecticidal activity. J Gen Appl Microbiol 43:341–347

Tellam RL, Wijffels G, Willadsen P (1999) Peritrophic matrix proteins. Insect Biochem Mol Biol 29(2):87–101

Terra WR (2001) The origin and functions of the insect peritrophic membrane and peritrophic gel. Arch Insect Biochem Physiol 47(2):47–61

Terra WR, Ferreira C (2005) Biochemistry of digestion. In: Gilbert LI, Iatrou K and Gill SS (eds) Comprehensive molecular insect science. Elseviers B.V., Amsterdam: New York. vol. 4, pp 171–224

Tetreau G, Cao X, Chen Y-R, Muthukrishnan S, Jiang H, Blissard GW et al (2015a). Overview of chitin metabolism enzymes in *Manduca sexta*: identification, domain organization, phylogenetic analysis and gene expression. Insect Biochem Molecul Biol 62:114–126

Tetreau G, Dittmer NT, Cao X, Jasrapuria S, Chen Y-R, Muthukrishnan S et al (2015b). Analysis of chitin-binding proteins from *Manduca sexta* provides new insights into evolution of peritrophin A-type chitin-binding domains in insects. Insect Biochem Molecul Biol 62: 127–141

Tiklova K, Tsarouhas V, Samakovlis C (2013) Control of airway tube diameter and integrity by secreted chitin-binding proteins in *Drosophila*. PLoS One 8(6):e67415

Tsirilakis K, Kim C, Vicencio AG, Andrade C, Casadevall A, Goldman DL (2012) Methylxanthine inhibit fungal chitinases and exhibit antifungal activity. Mycopathologia 173(2–3):83–91

Van Leeuwen T, Demaeght P, Osborne EJ, Dermauw W, Gohlke S, Nauen R et al (2012) Population bulk segregant mapping uncovers resistance mutations and the mode of action of a chitin synthesis inhibitor in arthropods. Proc Natl Acad Sci 109(12):4407–4412

Wang J, Chen Z, Du J, Sun Y, Liang A (2005) Novel insect resistance in *Brassica napus* developed by transformation of chitinase and scorpion toxin genes. Plant Cell Rep 24:549–555

Wang P, Granados RR (1997a) An intestinal mucin is the target substrate for a baculovirus enhancin. Proc Natl Acad Sci USA 94(13):6977–6982

Wang P, Granados RR (1997b) Molecular cloning and sequencing of a novel invertebrate intestinal mucin cDNA. J Biol Chem 272(26):16663–16669

Wang P, Granados RR (1998) Observations on the presence of the peritrophic membrane in larval *Trichoplusia ni* and its role in limiting baculovirus infection. J Invertebr Pathol 72(1):57–62

Wang P, Granados RR (2000) Calcofluor disrupts the midgut defense system in insects. Insect Biochem Mol Biol 30:135–143

Wang P, Granados RR (2001) Molecular structure of the peritrophic membrane (PM): identification of potential PM target sites for insect control. Arch Insect Biochem Physiol 47(2):110–118

Wang P, Hammer DA, Granados RR (1994) Interaction of *Trichoplusia ni* granulosis virus-encoded enhancin with the midgut epithelium and peritrophic membrane of four lepidopteran insects. J Gen Virol 75(Pt 8):1961–1967

Wang P, Li GX, Granados RR (2004) Identification of two new peritrophic membrane proteins from larval *Trichoplusia ni*: structural characteristics and their functions in the protease rich insect gut. Insect Biochem Mol Biol 34(3):215–227

Wang X, Ding X, Gopalakrishnan B, Morgan TD, Johnson L, White FF et al (1996) Characterization of a 46 kDa insect chitinase from transgenic tobacco. Insect Biochem Mol Biol 26(10):1055–1064

Willis JH, Iconomidou VA, Smith RF, Hamodrakas SJ (2005) Cuticular proteins. In: Gilbert LI, Iatrou K and Gill SS (eds) Comprehensive molecular insect science. Elseviers B.V., Amsterdam: New York. vol 4, pp 79–109

Willis JH, Papandreou NC, Iconomidou VA, Hamodrakas SJ (2012) Cuticular Proteins. In: Gilbert LI (ed) Insect molecular biology and biochemistry. Academic Press, San Diego, pp 134–166

Wu J-J, Chen Z-C, Wang Y-W, Fu K-Y, Guo W-C, Li G-Q (2018). Silencing chitin deacetylase 2 impairs larval-pupal and pupal-adult molts in *Leptinotarsa decemlineata*. Insect Molecul Biol 28(1):52–64

Xi Y, Pan P-L, Ye Y-X, Yu B, Zhang C-X (2014) Chitin deacetylase family genes in the brown planthopper, *Nilaparvata lugens* (Hemiptera: Delphacidae). Insect Molecul Biol 23(6): 695–705

Xi Y, Pan PL, Ye YX, Yu B, Xu HJ, Zhang CX (2015) Chitinase-like gene family in the brown planthopper, *Nilaparvata lugens*. Insect Molecular Biology 24(1):29–40

Yahouédo GA, Chandre F, Rossignol M, Ginibre C, Balabanidou V, Mendez NGA et al (2017). Contributions of cuticle permeability and enzyme detoxification to pyrethroid resistance in the major malaria vector *Anopheles gambiae*. Scientif Rep 7(1):11091

Yang W-J, Xu K-K, Yan X, Chen C-X, Cao Y, Meng Y-L et al (2018) Functional characterization of chitin deacetylase 1 gene disrupting larval–pupal transition in the drugstore beetle using RNA interference. Comp Biochem Physiol B: Biochem Mol Biol 219–220:10–16

Zhang D, Chen J, Yao Q, Pan Z, Chen J, Zhang W (2012) Functional analysis of two chitinase genes during the pupation and eclosion stages of the beet armyworm *Spodoptera exigua* by RNA interference. Arch Insect Biochem Physiol 79(4–5):220–234

Zhu KY, Merzendorfer H, Zhang W, Zhang J, Muthukrishnan S (2016) Biosynthesis, turnover, and functions of chitin in insects. Annu Rev Entomol 61:177–196

Zhu Q, Arakane Y, Beeman RW, Kramer KJ, Muthukrishnan S (2008) Functional specialization among insect chitinase family genes revealed by RNA interference. Proc Natl Acad Sci USA 105(18):6650–6655

Zimoch L, Hogenkamp DG, Kramer KJ, Muthukrishnan S, Merzendorfer H (2005) Regulation of chitin synthesis in the larval midgut of *Manduca sexta*. Insect Biochem Mol Biol 35(6): 515–527

FSC
www.fsc.org
MIX
Papier | Fördert
gute Waldnutzung
FSC® C083411

Kolibri 360 GmbH
Ferdinand-Jühlke-Straße 7
99095 Erfurt, Deutschland
produktsicherheit@kolibri360.de